THE WORLD
IN OUTLINE

A TEXT-BOOK OF GEOGRAPHY

by

E. D. LABORDE, PH.D.

Author of
The Cambridge School Geographies
etc.

CAMBRIDGE
AT THE UNIVERSITY PRESS
1946

CAMBRIDGE UNIVERSITY PRESS
Cambridge, New York, Melbourne, Madrid, Cape Town,
Singapore, São Paulo, Delhi, Mexico City

Cambridge University Press
The Edinburgh Building, Cambridge CB2 8RU, UK

Published in the United States of America by Cambridge University Press, New York

www.cambridge.org
Information on this title: www.cambridge.org/9781107697584

First edition 1935
Second edition 1946
First published 1946
First paperback edition 2013

A catalogue record for this publication is available from the British Library

ISBN 978-1-107-69758-4 Paperback

PREFACE TO FIRST EDITION

This book has been written to meet the needs of candidates preparing in Secondary Schools for the School Certificate Examination. Teachers —and examiners—are not infrequently puzzled by the cryptic phrase 'the world in outline' which occurs in the syllabuses of some examining bodies. Obviously, an outline of world geography must differ at every stage of educational progress. The solution which is offered here for the School Certificate standard will, it is hoped, find approval from teachers of experience.

The continents have been arranged in the conventional order, and there is no gradation of difficulty in treatment. Hence, the teacher is free to take the continents as he pleases. Special emphasis has been laid on the characteristic features, whether physical or human, in each land mass. A skeleton of physical geography has been added in an appendix for revisional purposes.

Several of the illustrations are from photographs taken by the author. Acknowledgements of the source of the others are made beneath them. The author takes this opportunity of expressing a particular obligation to the Information Department of the office of the High Commissioner for Australia, to the Canadian Pacific Railway, to the South African Railways, to the Orient Line, and to the Czecho-Slovak Legation.

E. D. L.

HARROW-ON-THE-HILL

December 1934

PREFACE TO SECOND EDITION

The lapse of a decade and the upheaval of a world war have in the main left the original text of this book unaffected. Revision might seem desirable in three directions: statistics, airways, and frontiers. No advantage appears to be gained, however, by substituting one set of pre-war figures for another, and post-war statistics, even where available, would merely serve to demonstrate the abnormalities due to the disturbance of war. As the statistics given in the book are dated, they cannot mislead. Airways had not reached full development in 1939, but the main trends are bound to persist. It has seemed best to leave the pre-war routes unaltered, except in the few instances where modifications are known to have occurred. Frontiers are still in a state of flux. He would be a bold man who would venture to predict what the international boundaries eventually will be. Consequently, it has seemed best to make a few cautious modifications and to leave the rest for revision when world affairs have become more stable.

E. D. L.

HARROW-ON-THE-HILL

March 1946

CONTENTS

PART III. AFRICA

PART IV. NORTH AMERICA

PART V. SOUTH AMERICA

PART VI. AUSTRALIA, NEW ZEALAND, AND THE PACIFIC ISLANDS

PLATES

PART I

EUROPE

THE principal feature in the study of the geography of Europe is the progress which man has achieved in the control of nature and his increasing ability to make full use of the gifts that she offers him. Not that the influence of geography has been or ever will be removed. The distribution of fertile lowland will ever determine the centres of agriculture, deposits of minerals will continue to decide the position of industrial areas; routes will always follow the lines of most unbroken relief, and climate will place limits on the extension of the various kinds of crops. Yet in no other continent has man's use of the various natural advantages and his removal of geographical disadvantages reached so high a degree of achievement. Great forest tracts may remain almost untouched by man in Brazil or Siberia; but in Europe forestry, with its selection of species, its systematic felling and replacement of trees, has become a full-fledged science. The largest coalfields in the world lie almost waste in Siberia, China, and Alaska; but every mineral deposit in Europe is being exploited to the full economic capacity. Scientific farming and stock-rearing have replaced the old haphazard methods, and even fishing and fur production are being controlled on lines similar to those of stock-rearing. Industrial progress is every day wresting further secrets from nature. The sands of Saxony are made into glass, the clay of the London Basin forced to yield aluminium, the smaller trees of Norway and Sweden which were formerly used merely as firewood are now being turned into plywood, match sticks, and pulp from which an infinite number of articles are manufactured. Transport, too, has been made so easy along the railways, great motor roads, and airways that it is now difficult to understand why in former times a ridge of hills of moderate height and gradient, like the Pennines, formed so great an obstacle to the passage of men. In fact, man has almost completely triumphed over nature. This is the chief interest of the following pages.

Position and Size. Europe is only a western peninsula of Asia, being separated from that continent by no natural feature, since the Ural Mountains are no real barrier. But the practice of treating Europe separately is established by ancient usage and is convenient both on account of the importance of that 'continent' and because of the marked difference between its civilisation and that of Asia proper.

Regarded as a continent, Europe is small. It extends from Long. 10° W. to Long. 60° E., and from Lat. 36° N. to Lat. 71° N. Like Australia, therefore, it has its greatest length from east to west, and hence does not contain such a wide range of climate as Asia or North America. In fact, its mainland lies wholly within the temperate belt. Its area is only 3,750,000 square miles as compared with Asia's 17,000,000 and North America's 8,000,000. But, on the other hand, it has an enormously long coastline, the effect of which is to carry sea influence far into the land and to provide calm inland seas as training grounds for maritime enterprise. Only a small area of peninsular Europe (i.e. Europe outside Russia) is more than 250 miles from the coast, while parts of Asia are no less than 1500 miles from the sea.

Build and Relief. Europe is built of four main east-and-west structural belts. These are (1) a tangle of fold mountains in the south, (2) a series of block formations and rift valleys which in some cases lie among the folds, but for the most part are to the north of them, (3) the Great European Plain, and (4) a mass of old mountains and uplands which fill the northwest corner.

The fold ranges are a continuation of those of Asia and enter Europe at two points. The Taurus Mountains of Asia Minor, after partial submergence in the Ægean, rise again to form the Pindus range and the Dinaric Alps. Farther north, the Caucasus Mountains reappear in the Crimea and in the Balkan range, which continues north as the Carpathians. The two lines of fold unite in the Alps. From the western end of this culminating range issue two folds, one of which forms the Pyrenees, while the other after passing through the Balearic Isles and the Sierra Nevada reappears in Africa as the Atlas Mountains. This fold can be traced through Sicily and the Apennines back to the Alps.

The earth movement which caused these folds exerted great strain on the older ridge of uplands, known as the Hercynian Mountains, which stretched from southern Ireland to the Carpathians. Extensive faulting occurred, followed by the uplift

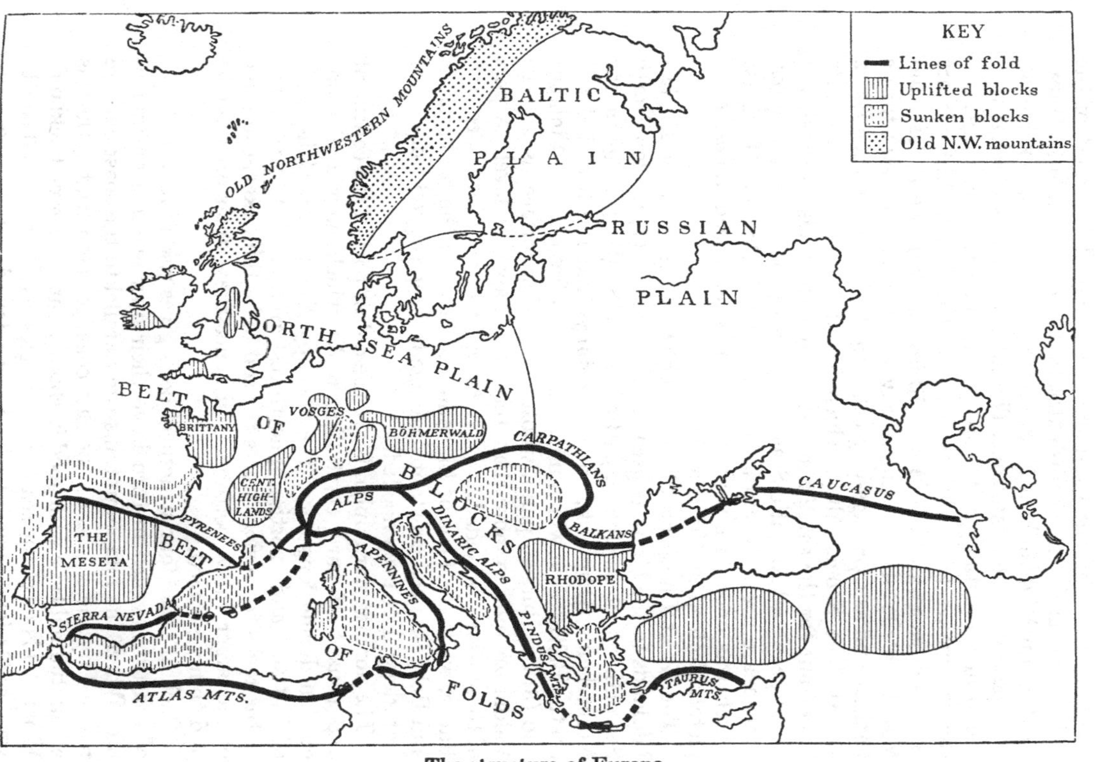

The structure of Europe

of some areas and the subsidence of others. The largest of the uplifted areas are the Spanish Meseta, the peninsula of Brittany, the Central Highlands of France, the Vosges, the Black Forest, and the Bohemian Mountains. The surface of the old rock is very rough in the details of its topography, and the shallowness of the soil makes for infertility. Hence, these block uplands are poor and backward. The Middle Rhine valley between Basle and Mainz is a rift whose sediment-covered floor is important for its fertility and as a passage-way between the Vosges and the Black Forest. The Saône valley, which is only partly a rift, is also important for the same reasons. Within the folds the basins of the Ægean and Adriatic Seas and most of the western Mediterranean are areas of subsidence. So is the plain of Hungary, which was once an inland sea, but has now been almost completely drained by the Danube.

The Great European Plain, which stretches from the English Midlands to the Urals, consists of three parts: (1) the low North Sea plain which occupies a part of eastern England, Belgium, the Netherlands, Denmark, north Germany, and much of Poland; (2) the Baltic plain which includes most of Sweden and Finland; and (3) the Russian plain, whose undulating surface rises in the Valdai uplands to more than 1000 feet above sea-level. A part of these wide lowlands has been drowned by the North Sea and Baltic, while the Fens of England and much of the Netherlands and Belgium need constant drainage to keep the soil from being waterlogged or even flooded.

The northwestern uplands are the oldest part of the continent and consist of hard crystalline rocks which have undergone erosion during long ages. They occupy Norway, the Highlands of Scotland, and a portion of northwestern Ireland. The resistance of the rocks to erosion causes the soil to be thin, except in some of the larger valley bottoms, and hence these areas are infertile and thinly peopled.

Each of these four structural belts gives rise to its own types of coastline, which in turn have their own reaction on human life. Where the fold mountains run parallel to the coast and are reached by the sea, as on the east coast of the Adriatic, there is a longitudinal type with good harbours, but little or no backland. Where folds meet the sea at right angles, as in Greece and southwestern Ireland, good harbours are found with a moderate amount of fertile backland. The most important type occurs when the block uplands touch the sea, as in the southwest of England, Brittany, and northern Spain. The *rias*, or drowned estuaries, which characterise this type, make excellent harbours with access to productive and otherwise important backlands. The encouragement given by this type to maritime enterprise is proved by the popularity of fishing as an occupation along these coasts and historically by the number of famous seamen and adventurers who have had their origin in them. The Great European Plain is marked by low, even coasts fringed with sand-banks. Except where river mouths occur, few harbours exist and the population is not seafaring. The coasts of Norway and the Highlands of Scotland are deeply indented by fjords, which owe their origin to the erosive action of ice. They make good harbours and may breed a race of seamen like the Norwegians, but they suffer from a lack of productive backland.

The topography of much of northern Europe has been influenced by the ice-cap which overlay the area during a part of the Tertiary age. While the cap was at its greatest extent (see map on page 6), it deposited boulder clay on the lowlands of eastern England, the Netherlands, and north Germany. When the ice began to melt, the streams issuing from it laid down a thick film of rich black earth, known as *tchernoziom*, in southern Russia, while in north Germany it piled up moraines across the drainage and rearranged the former river system. The moraines are barren, while the low ground behind them tends to be swampy. In Finland the damming of the natural drainage gave rise to a vast number of shallow lakes, large and small, the most important of which are Onega and Ladoga within the borders of Russia. In the British Isles the final stages of the ice-cap gave rounded outlines to the hills and caused fjords, glacier valleys, and lakes, and various other traces of ice topography.

The earth movements which have affected the continent have resulted at various times in volcanic activity. Basaltic lava flows from the surface of Antrim and Mull, while the Cuillin Hills in Skye and the Renfrew, Campsie, Ochill, and Sidlaw Hills in Scotland are of volcanic origin. Throughout the Midland plain of Scotland and the Central Highlands of France ancient craters and necks attest to former activity. The crust is even now unstable in the Mediterranean, on whose shores still exist the active

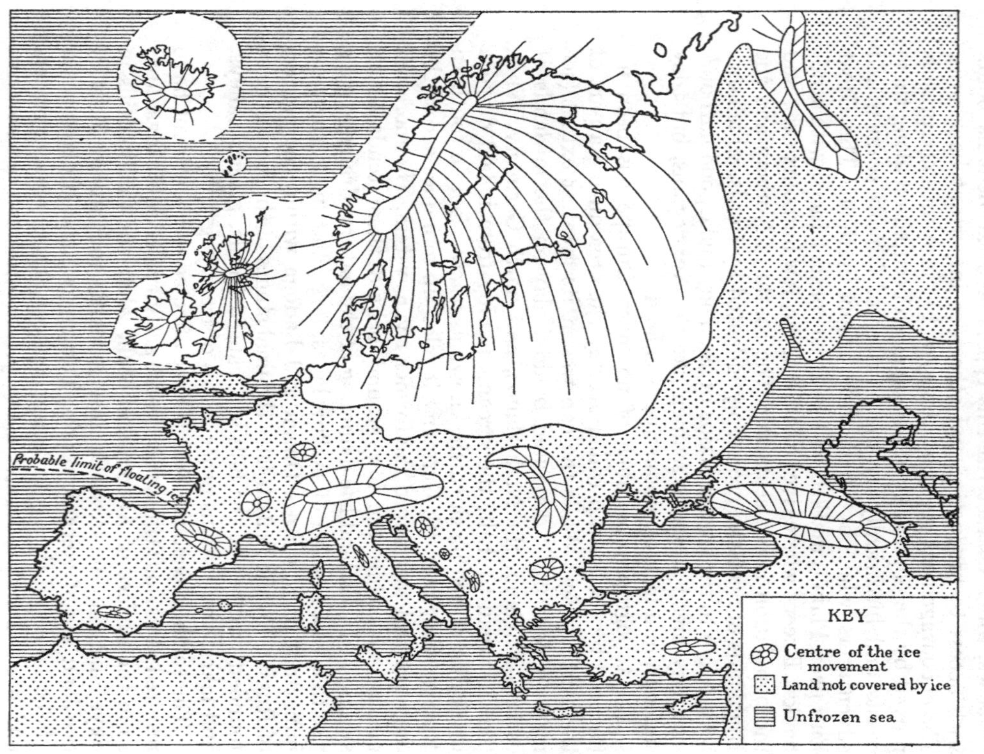

The Tertiary Ice Cap at its maximum

cones of Vesuvius, Stromboli, Etna, and Santorin. Earthquakes are also of frequent occurrence.

The rivers of Europe are of five types: (1) the western, of which the Thames and Seine are examples, is fed by a good rainfall throughout the year and keeps a fairly even volume, though with a tendency to flood in winter; (2) the northern, which is frozen over in winter and in spring suffers from floods owing to the thawing of the ice in the upper course first; (3) the eastern, which, like the Don and Volga, has a maximum in summer, this being the season of rains, and is low in winter when frost binds the moisture in the ground; (4) the mountain stream, which rushes down broken gradients and has spring floods owing to the melting snows in that season; and (5) the wadi type, which is found in the Mediterranean coast lands and which is characterised by winter spates and low summer volume. Most of the big rivers belong to more than one type; for instance, the Rhone is a mountain river as far as Lyon, but is thenceforward of the western type, while its right-bank feeders from the Cévennes are wadis.

Climate. The position of Europe in the north temperate belt is the main influence on the climate of the continent. But the wind system is an important moderating factor. Normally, a belt of high pressure runs across the Atlantic in the latitude of Spain, and this belt has a centre of intensity around the Azores. On the other hand, a centre of low pressure usually exists near Iceland and stretches an arm eastwards towards the north of Europe. The winds which pass from the high- to the low-pressure centres, being deflected by the rotation of the Earth, become the southwesterly and westerly breezes which prevail in nearly every corner of the continent.

But these normal conditions are frequently interrupted by four subordinate factors: (1) The Atlantic high- and low-pressure areas often change their shape and size; for instance, the Azores high-pressure area may extend northwards so as to cover the British Isles and give a period of calm, fine, but perhaps foggy weather. (2) In winter the great high-pressure centre which gives Asia its winter monsoon extends into Russia and may spread westwards as far as the British Isles. When this happens, the weather becomes bright and frosty, with an east wind. (3) Low-pressure disturbances, of which more will be said later, often pass over the continent in different directions, bringing with them the familiar 'unsettled' weather of the British Isles or causing the foehn winds of Switzerland. (4) The Mediterranean, being

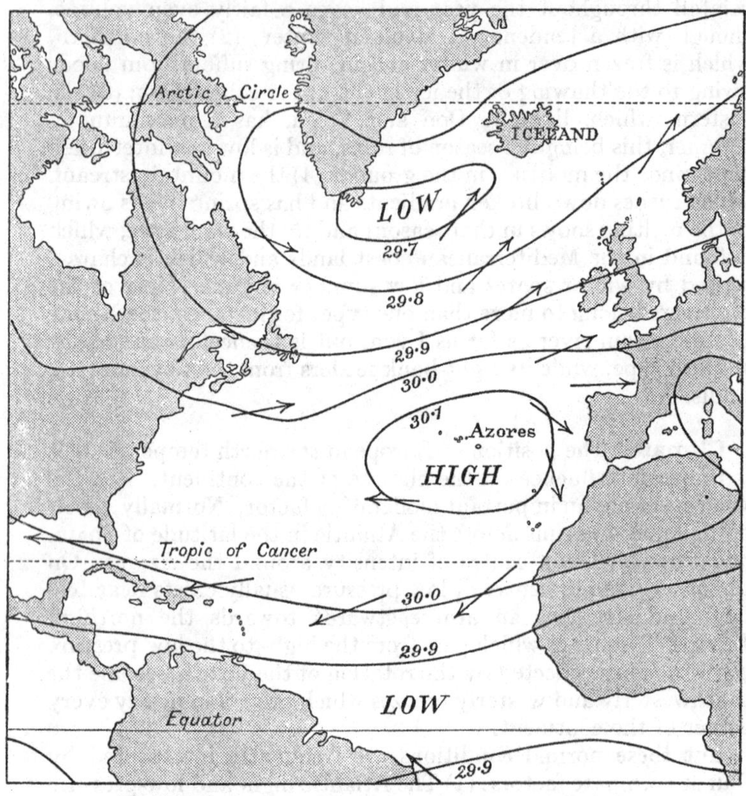

The North Atlantic. Mean annual pressure and winds

shut off from the cold bottom waters of the ocean, forms a storehouse of warmth in winter and sometimes causes the formation of a low-pressure area over its surface. The cold winds from the north are thus drawn towards the south and rush down the valley of the Rhone as the fierce wind known as the *mistral*, or through the passes leading to the Adriatic. The latter winds are known as the *bora*.

Besides these causes of anomalous conditions, the annual swing of the planetary wind belts moves the whole system southwards in winter and northwards in summer. Consequently, the Mediterranean and its coasts are removed from the influence of the Westerlies in summer and form part of the tropical high-pressure belt at that season, thus becoming a temporary extension of the Sahara. Furthermore, the difference in pressure between the Azores high-pressure area and the Icelandic low-pressure centre is less in summer than in winter, and as a result the whole system of winds is weaker, the breezes are not so strong, and the direction from which they blow is far more variable.

The sea influence exerted by the wind system on the climate of Europe is reinforced by the presence of a vast flow of relatively warm surface water from the tropical latitudes of America towards the shores of western Europe. The Gulf Stream Drift, as this ocean current is called, keeps the sea free from ice as far as the North Cape, a point well within the Arctic Circle, while the Gulf of St Lawrence and the coastal waters of Labrador are icebound. It is impossible to calculate how much of the relative warmth of the Westerlies in winter comes from the Drift and how much is due to the southern origin of the winds themselves, but there can be no reasonable doubt that both factors play an important part.

The influence of the sea, acting through the winds, is greatest on the west and lessens gradually towards the east. As a result, the range of temperature increases from 14° F. at Valentia Island to 33° F. in Berlin and 54° F. in Moscow. While on the extreme west the winters are mild and frost is comparatively rare, in the east the ground is covered with snow for several months in the year, the inland waters are icebound, and all out-of-door work must cease. Between these extremes there is a gradual transition. In summer, the temperature is more even along the parallels of latitude, since, as we have seen, the weakened system of winds does not transmit the moderating influence of the sea as readily at that season as it does in winter. Nevertheless, the west is slightly cooler than the east.

If the Atlantic provides the chief sea influence, the Mediter-

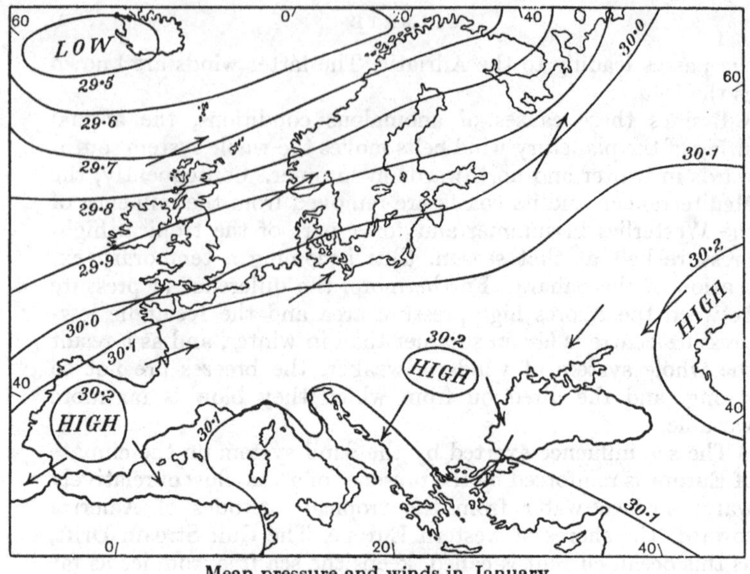

Mean pressure and winds in January

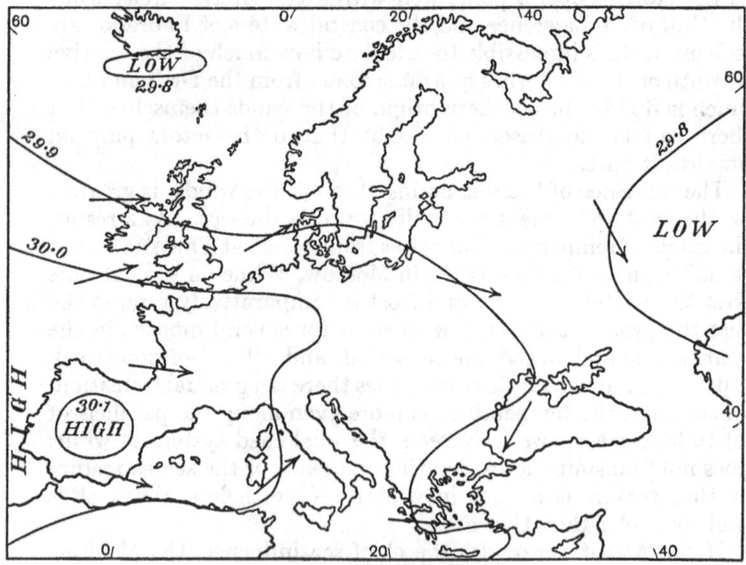

Mean pressure and winds in July

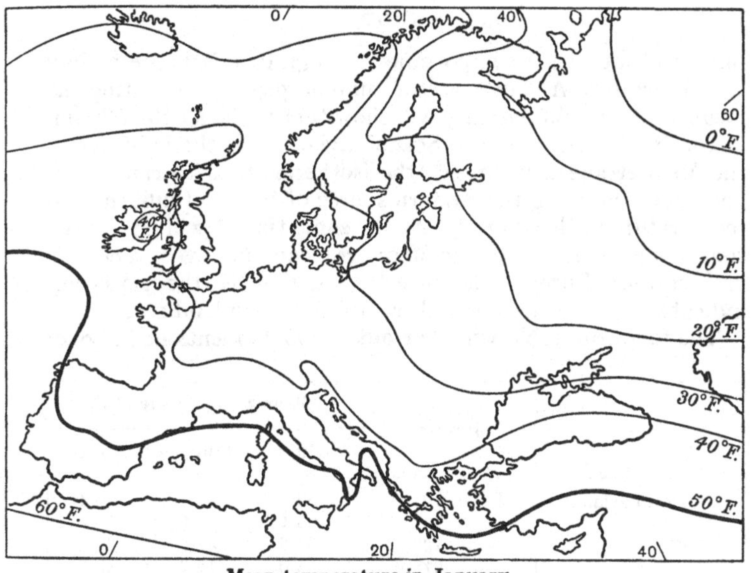

Mean temperature in January

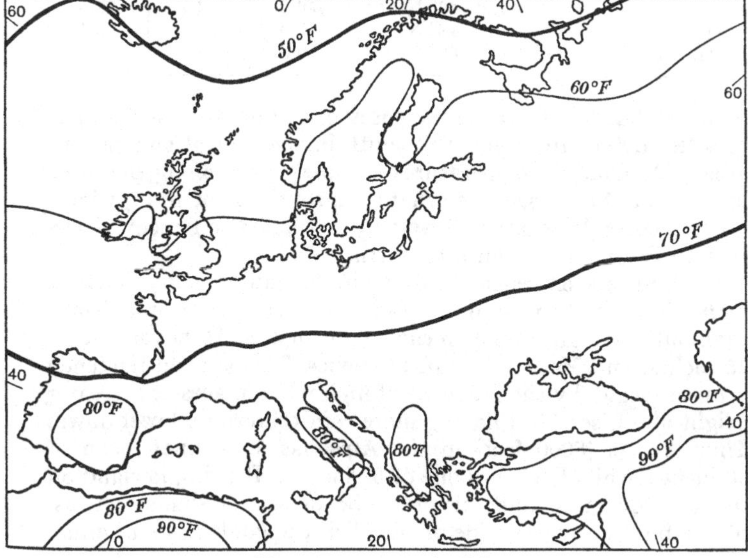

Mean temperature in July

ranean is not without importance, though its effects are mainly seen in winter. A glance at the map on page 11 indicating the mean isotherms for January will show that in the north of Spain, in the British Isles, and in Scandinavia, where the influence of the Mediterranean is absent, the isotherms tend to run north-and-south, marking the western source of the chief influence on temperature in that month. At the same time it will be noticed that along the shores of the Mediterranean, and even in central and eastern Europe, the direction taken by the isotherms indicates an influence derived from the great inland sea.

The following table will bear out the facts mentioned above:

Place	Position	Mean temperature (° F.)		
		July	January	Range
Valentia Island	Long. 10° W.	58·8	44·4	14·4
Hanover	10° E.	63·1	32·7	30·4
Berlin	13° E.	64·6	31·3	33·3
Moscow	37° E.	66·0	12·2	53·8
Rome	Lat. 42° N.	76·6	44·1	32·5
Vienna	48° N.	67·3	28·9	38·4
Haparanda	66° N.	59·0	10·6	48·4

namely, that (1) the range increases from west to east and also to a less extent from north to south in the central and eastern areas; (2) summer temperatures are fairly even along the same latitudes, but decrease somewhat from south to north; (3) winter temperatures decrease from west to east and also in central and eastern Europe from south to north.

Sea influence as transmitted by the prevailing winds causes a high rainfall in the west and a gradual decrease eastwards. Thus, Plymouth has an annual mean of 36 inches, Hanover one of 25 inches, and Moscow one of 21 inches. This general tendency is upset locally by the influence of high relief, places at a greater height above sea-level having more rain than those lower down. Thus, Säntis (8000 feet) in the Alps has an annual mean of 96 inches, while Basle has one of 32 inches. Besides, lowland on the lee side of uplands tends to be in a rain shadow; thus, Shrewsbury on the lee side of the Welsh uplands has an annual

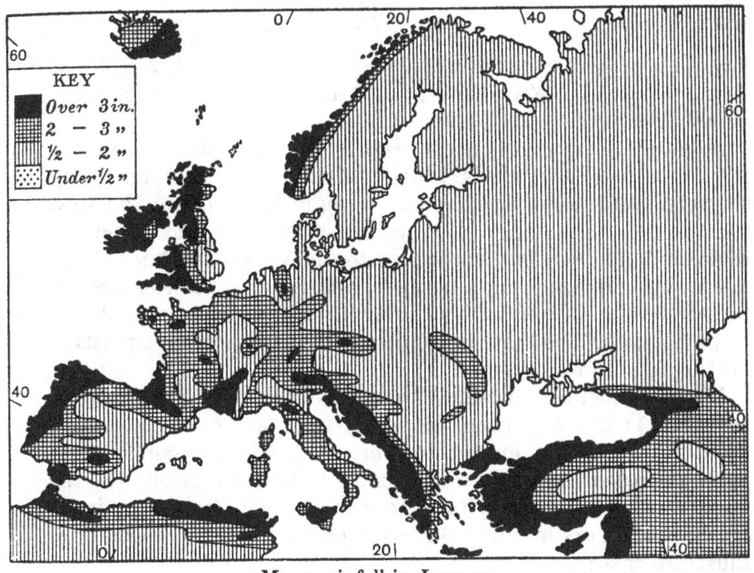

Mean rainfall in January

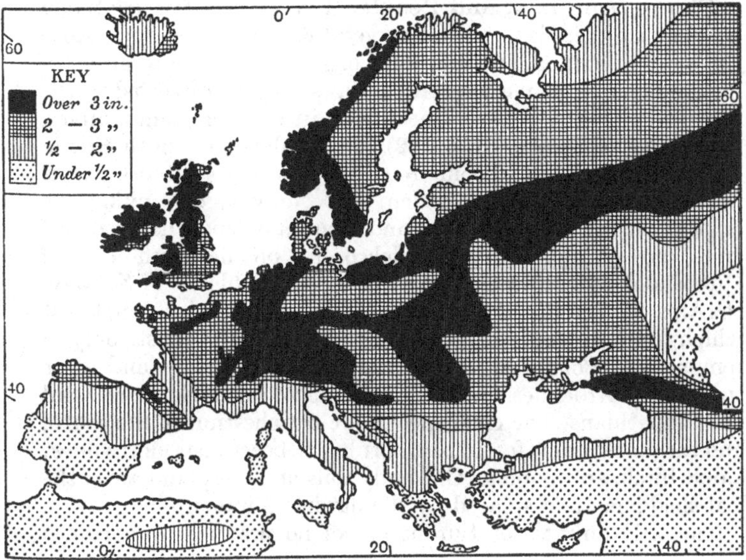

Mean rainfall in July

mean of 25 inches, and Frankfurt protected by the uplands through which the Main flows gets 22 inches, while its unsheltered neighbours average 27 inches.

Much of the rainfall in the west is due to relief and to the low-pressure disturbances which are usually accompanied by showers. As these disturbances are more frequent on the west coast and also since they lose their intensity when they penetrate over the land, they give more rain to the Atlantic seaboard than elsewhere. On the plains farther east the disturbances are repelled by the winter high-pressure centre which then extends from Asia, and these areas are dependent for their precipitation on the action of convection.

These influences combine to cause three types of rainfall régime: (1) the maritime, in which rain falls throughout the year, but which has an autumn maximum; (2) the continental, with a marked summer maximum due to convection; (3) the Mediterranean, in which the summers are dry and the rainfall maximum occurs in winter. The diagrams on the opposite page illustrate these types.

Plants, Animals, and People. Except in a few relatively small areas, there is no primitive vegetation left in Europe, since the hand of man has modified the original plant life to an enormous extent. Yet five vegetation types may be observed in the continent: (1) the Mediterranean with its southern pines, olives, vines, and evergreen shrubs; (2) the grasslands of south Russia and Hungary; (3) the coniferous forest which stretches across north Russia, Finland, Sweden, and Norway, continuing the *taiga* of Siberia; (4) the *tundra* and mountain type with its dwarf trees, mosses, and annuals, which occurs on the north coast of Russia and just below the snow line in the Highlands of Norway, the Alps, Pyrenees, and Caucasus; (5) the vast leaf-shedding forest which formerly covered most of Europe, but now exists only in remnants in Poland and western Russia and in highland areas such as the Ardennes, the block uplands of south Germany, and the Carpathians. The chief species are the chestnut of the south, and the more widely found oak, hornbeam, beech, ash, elm, willow, alder, birch, and poplar. On sandy soils and in upland areas the leaf-shedders are displaced by the hardier conifers.

The wild animals of Europe are of no practical importance

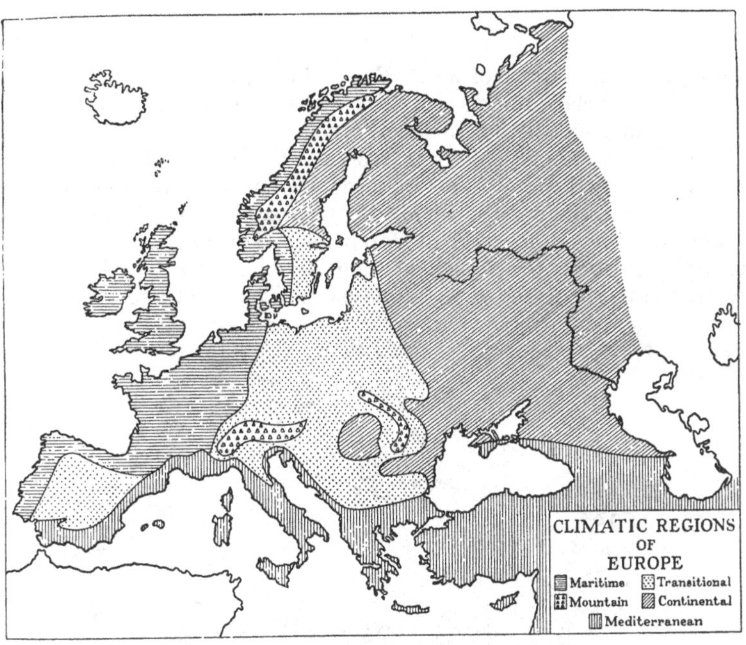

Main climatic regions of Europe

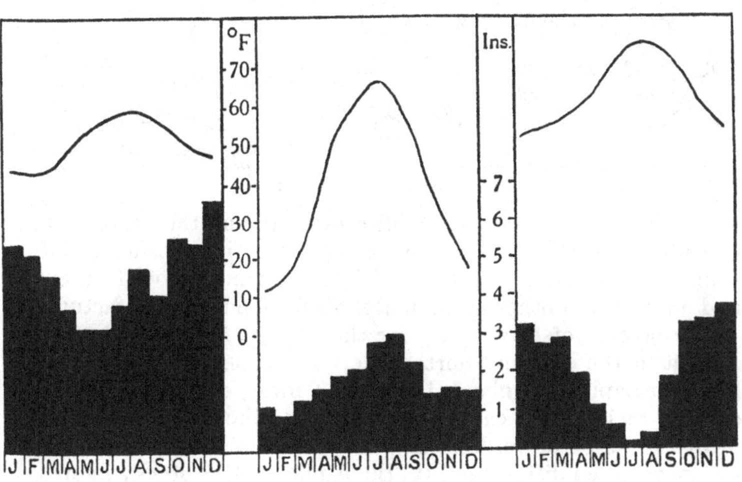

Temperature and rainfall types

to-day. In the forests of north Russia bears, martens, ermines, and other fur-clad animals still exist, but are restricted in number by the value set on their pelts. Wolves also survive, though in diminished numbers, chiefly in Russia. Elsewhere, wild boars, foxes, deer, and many kinds of birds are 'preserved' either by law or public opinion, usually for the purposes of sport. Domestic animals, however, are of great importance. The chief are the ox, sheep, pig, horse, goat, ass, cat, dog, and reindeer. The last is restricted to the north of Scandinavia, and the goat and the ass

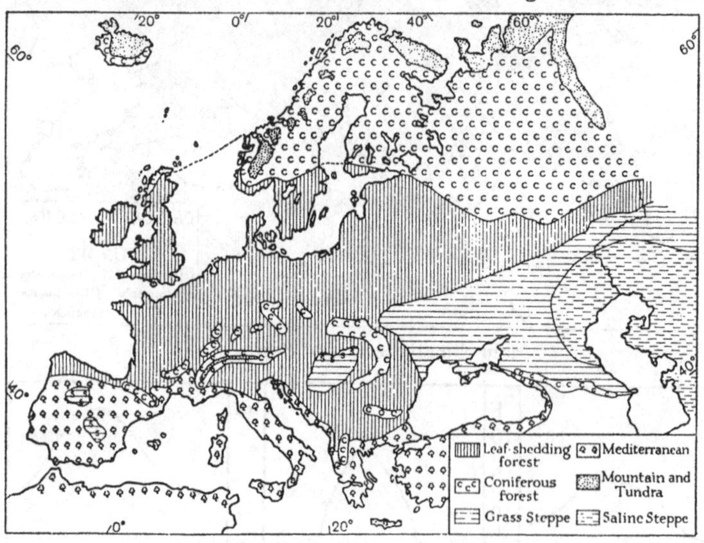

Distribution of vegetation

are found chiefly near the Mediterranean; but the others are spread widely over the continent. The most important are, of course, the ox, sheep, and pig, which provide meat, wool, leather, and a number of other useful materials for food or manufacture.

The peoples of Europe are for the most part of mixed race. Except in the east and north, where intrusions of Asiatics are found in Lapland, Finland, Bulgaria, Turkey, and Russia, three distinct racial types can be observed: (1) the Nordic, characterised by tall stature, fair complexion and hair, blue eyes, and a calm reserved disposition; (2) the Alpine, with a stocky figure,

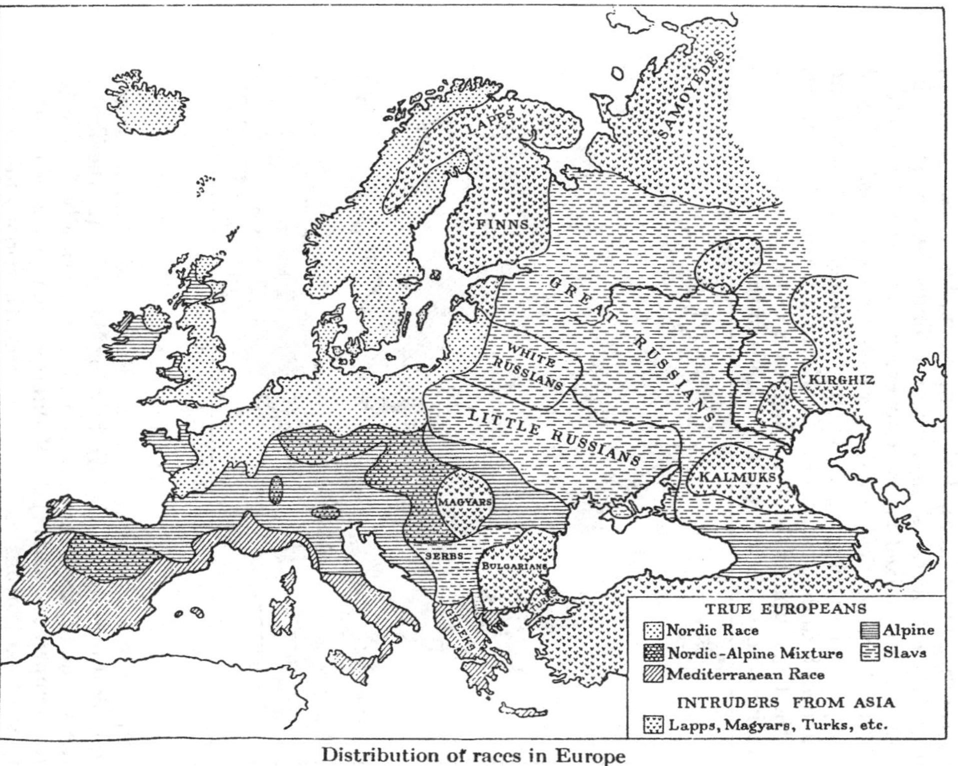

Distribution of races in Europe

rather coarse features, a round head, a moderately fair complexion, and brown or red hair; (3) the Mediterranean, marked by shortness of stature, dark complexion, black hair and eyes, delicately cut features, finely shaped limbs, and a quick disposition. But mixed types occur everywhere, and race is often obscured by political divisions which cut across ethnological groups, and by language. Broadly speaking, the Nordics are found in Scandinavia and on the North Sea plain, the Alpines in the block uplands and the highlands of the south, while the Mediterraneans occupy the coast lands of that sea.

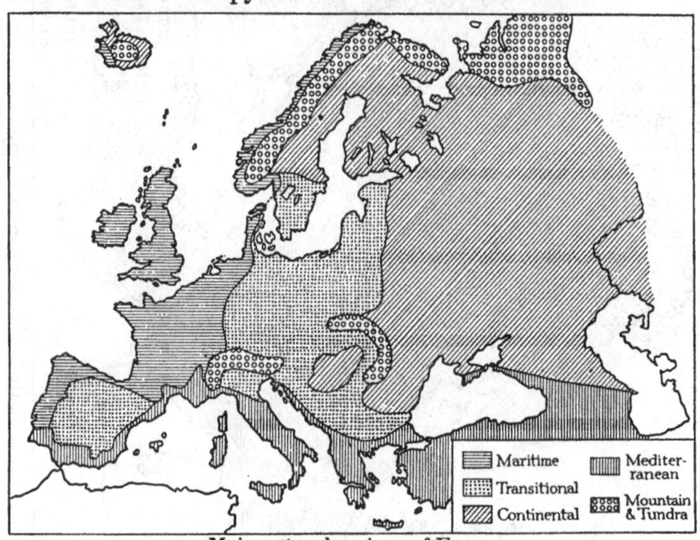

Main natural regions of Europe

THE REGIONS

Differences of climate mark off Europe into four main geographical regions: (1) the Western, or Maritime, (2) the Transitional, and (3) the Continental—which are all subdivisions of the cool temperate—and (4) the Mediterranean, which is of a distinct type. Relief breaks each of these climatic regions into a considerable number of natural subdivisions whose physical and human differences are often great. In the following pages the main subdivisions will be indicated and described.

The Western Region

This consists of the western seaboard of the continent from the south of Portugal to the North Cape. Its breadth varies, being least in Portugal and Norway where highlands lie across the path of the prevailing winds, and greatest in northern France where plains stretch eastwards. It includes, besides practically all of

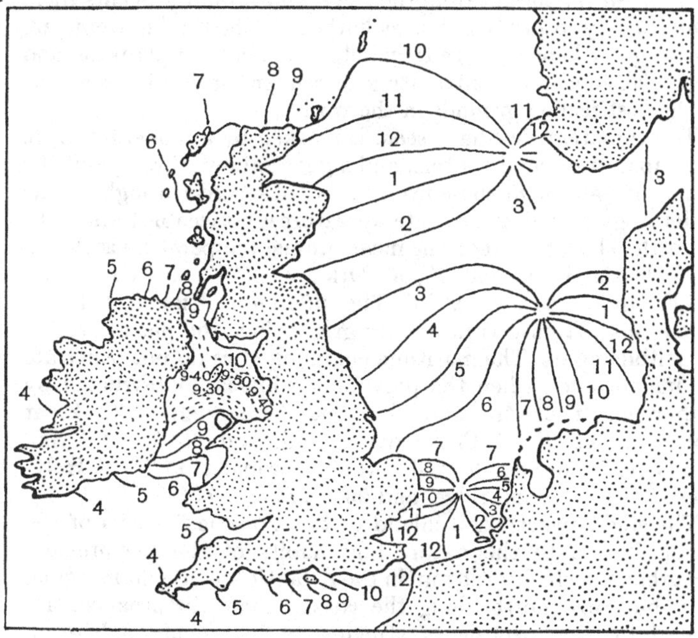

Tides of the North Sea

The lines connect places with high tide at the same time. Note that the high tide in the Straits of Dover is the same as that in the Moray Firth, while those of the North Sea belong to the previous tidal wave.

Portugal, France, and Denmark, a strip of northern Spain and the whole of the British Isles, Belgium, the Netherlands, and Norway.

Off the coast of Norway and the Iberian Peninsula the land falls steeply away to considerable depths below the sea surface, but between these extremities there is a broad continental shelf on which stand the British Isles. The tides which approach from the southwest are heightened partly by the shallowing of the sea and partly by the constriction of the water in narrow channels, and a forward movement is thus given to the water, which results in tidal currents. Estuaries, like those of the Severn, Thames, and Seine, have high tides which enable large ships to penetrate farther into the land than would otherwise be possible. Besides, the ebb sweeps the silt from the river mouths and prevents them from quickly losing their use as harbours. The tidal currents are also of great benefit, since they help to spread the plankton and other fish foods, so contributing to render the British seas one of the best fishing grounds in the world.

The climate, as we have seen, is marked by a moderate range of temperature, mild winters, and a good rainfall throughout the year. But although these features are observed throughout the region, a gradual decrease in temperature is noticeable from south to north. Thus, at Lisbon the mean annual temperature is 59° F., at Valentia Island 50° F., at Orkney 45° F., at Trondheim 40° F., and at Tromsö 36° F. The difference of 23° F. between Lisbon and Tromsö is not really great, since the places are 32° of latitude apart. The maritime characteristics do not penetrate far inland, even when the prevailing wind is unobstructed by high relief. Thus, the annual temperature range is 27° F. at Bordeaux, 30° F. at Clermont-Ferrand, and 33° F. at Lyon. Similarly, farther north we find that the range is 17° F. at Holyhead, 24° F. at Cambridge, and 29° F. at Utrecht.

The slight penetration inland of the climatic features of the region is due to the nature and habits of the low-pressure disturbances which spring up in the Atlantic, invade the marginal seas, but turn north along the coast. These disturbances are caused by the meeting of streams of cold polar and warm equatorial air. When two such currents flow side by side in opposite directions, any slight bend in the cold stream tends to grow larger and to form a kind of gulf of warm air within the cold air. Since warm air is lighter than cold air, the warm air in the gulf rises over the cold air in front of it. At the same time the cold air behind the gulf pushes under the warm air. In the end, the cold air meets from all directions under the gulf of warm air, so that the latter floats like a solid saucer in the former.

Wherever the warm, moist equatorial air touches the cold polar air, condensation follows. The high outer edges of the

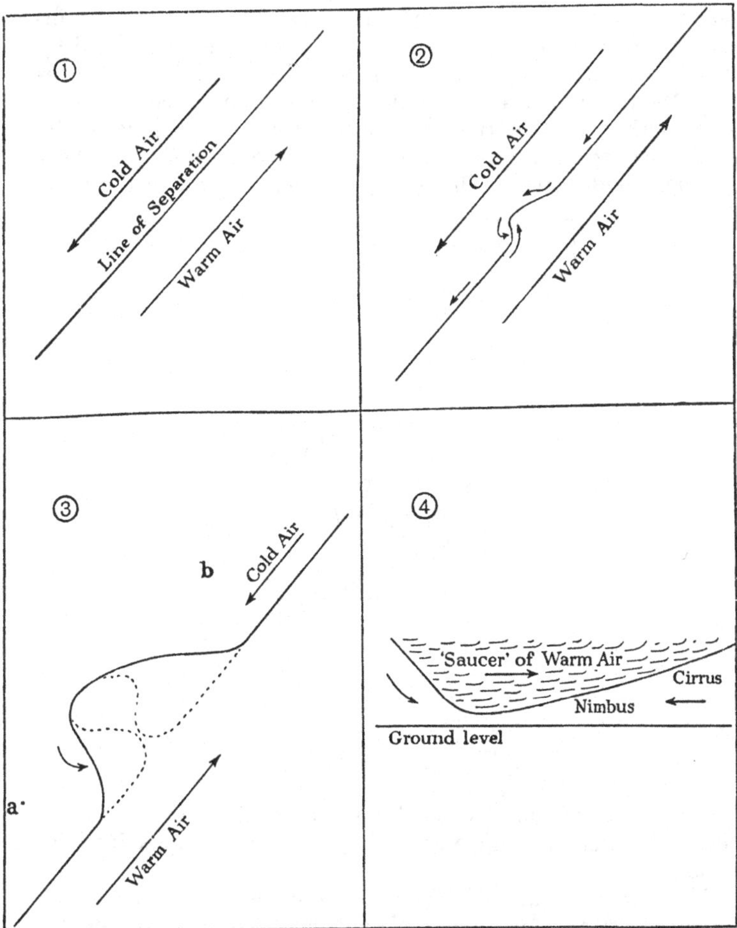

Phases of a low-pressure disturbance. (1) Preparatory phase. (2) Birth of a disturbance: a kink forms in the line of separation. (3) The kink grows. The continuous line of separation shows a disturbance such as is found over the Atlantic. The dotted line marks the state in which it usually reaches the British Isles, after the pressure of the cold air from behind has cut off a 'saucer' of warm air. (4) Cross section of the 'saucer' along a line joining *ab*. Note that (1), (2), and (3) show the disturbance in plan and (4) in section.

'saucer' are marked by cirro-stratus clouds, and cloudiness increases towards the centre. This increase is soon accompanied by drizzle and finally by a sharp shower. The weather which goes with the passage of a disturbance is shown on the chart below. During their passage the temperature is lowered partly by the presence of the warm air and partly by the release of latent heat through condensation. They may occur at any time of the year,

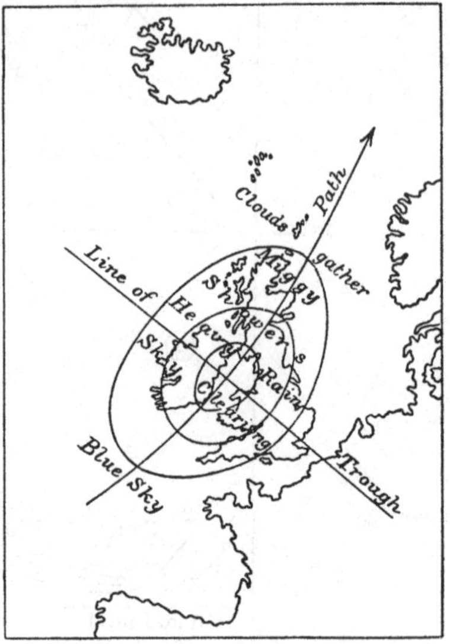

Plan of a low-pressure disturbance

but the winter months usually see a constant succession of them. In October and March they are accompanied by high winds.

The maritime climate is also marked by loss of sunshine owing to cloudiness and fogs. These phenomena occur most frequently on the coast, but the neighbourhood of industrial towns is also afflicted with thick, dirty-looking fogs due to the smoke from their factory chimneys. Fogs are depressing and hinder traffic on land and sea.

The region may be subdivided into fourteen subregions, seven of which are in France and five in the British Isles.

Portugal and Northern Spain from the Cantabrian Mountains to the sea form the first. The subregion is formed by the edges of the Spanish Meseta, which is bordered on the north by the fold ranges of the Cantabrian Mountains, but whose faulted side falls away steeply to the west. The ground is everywhere broken, rocky, and thinly peopled, except around the mouth of

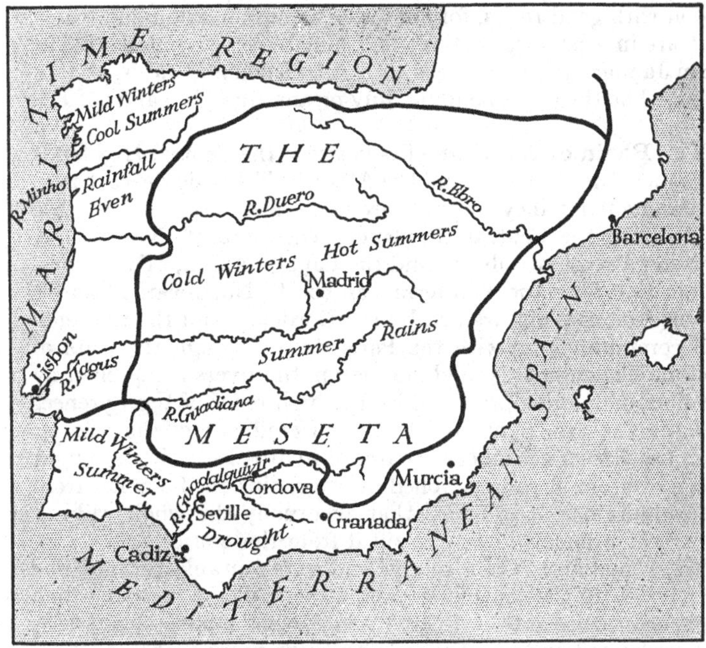

Regional divisions of the Iberian peninsula

the Tagus, where a lowland area forms the nucleus of Portugal. In the south of this republic the shores are low and sandy, but farther north they become straight and rocky. The northwestern corner of the peninsula, which forms the Spanish province of Galicia, has a ria coast and is therefore a fishing district with well-known centres at Vigo and Corunna. The Cantabrian coast is massive and rocky, with few harbours. Its backland is too hilly and broken to be of much use for agriculture, but contains mineral deposits. Bilbao, with its famous iron mines and its manufactures of iron goods, chemicals, and glass, is the largest town in the Spanish portion of the subregion. The valleys of northern Portugal are famous for their wine, whose name (port) is derived from the centre of the industry at Oporto on the Duero. Lisbon at the mouth of the Tagus is the capital city and chief port. Its immediate backland is the only part of the subregion with good roads, for elsewhere communication and transport are in a backward state. The climate is wet, but mild. The mean January temperature is 49° F. at Lisbon and 45° F. at Santiago, while the corresponding July means are 70° F. and 65° F.

The Basin of Aquitaine lies between the Central Highlands of France, the Pyrenees, and the Atlantic. Towards the east runs the triangular valley of the Garonne, whose feeders penetrate deeply into the Central Highlands. The Gate of Carcassonne, carrying a road, a railway, and the Canal du Midi, affords communication with the Mediterranean coast. Northwards there is a monotonous plain drained by the Dordogne and the Charente and communicating with the Paris Basin through the Gate of Poitou. The climate is mild and damp, the rivers being filled by the abundant rains produced by the high relief of the Pyrenees and the Central Highlands. The rapid gradients of the streams from the former are an easy source of hydro-electricity and light and power are already supplied over a radius of fifty miles from the mountains. It is hoped that eventually the whole railway system of France will be electrified from this source.

The coastline is even, sandy, and unfavourable to maritime enterprise. In the southwest stretch the *landes*, a district which was once turned into a desert by inblowing shore sands, but is now reclaimed and planted with southern pines that yield resin and timber for pit-props. Elsewhere the land is used for agriculture. Maize and wheat are the chief cereals and, with the vine, form the chief crops. Wine (claret) is the main economic product, but some of the liquor is made into brandy (cognac). The population lives in scattered homesteads known locally as *bordes*, though a few large towns exist. The chief of these is Bordeaux, the focus and outlet, which stands at the head of the Gironde estuary and, with its outport at Pauillac, is one of the largest ports in France. Its population was 263,000 in 1931. Its trade is mainly with America. Toulouse (pop. 195,000 in 1931), near the Gate of Carcassonne, is a gap town as well as a centre for agricultural produce.

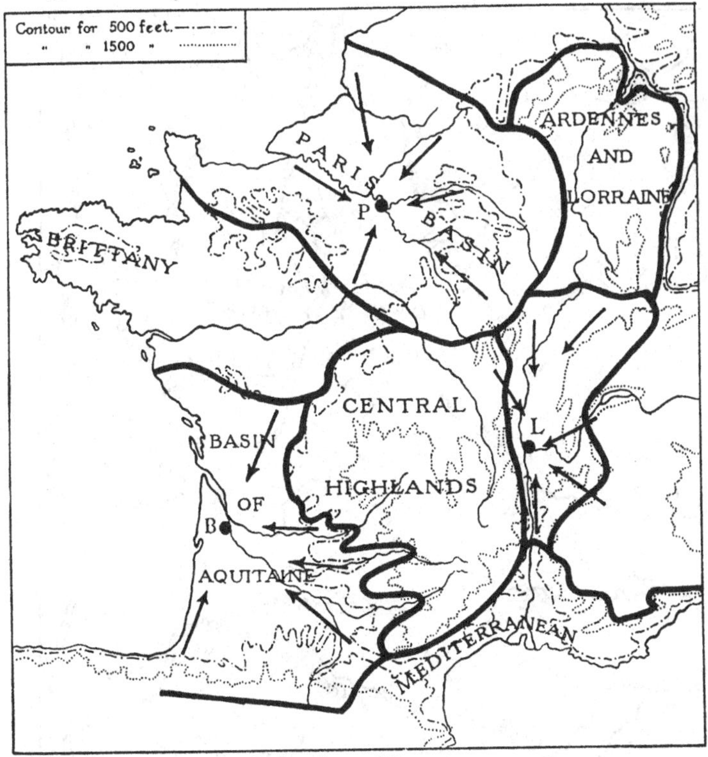

Subregions of France

The Central Highlands of France are a mass of crystalline rock whose general surface has been reduced to a peneplain. The whole block was raised by the earth movement which elevated the Alps, and the rejuvenation of the cycle of erosion which resulted has caused the rivers to carve deep, narrow valleys in its edges. The steep face in the east and southeast is due to faulting, while the surface of the tableland itself is scarred with rifts and pitted with old volcanoes. A downward tilt towards the northwest makes approach easier in that direction and leads most of the drainage into the Loire and its feeder, the Allier. The climate is less maritime than that of the previous subregion, for

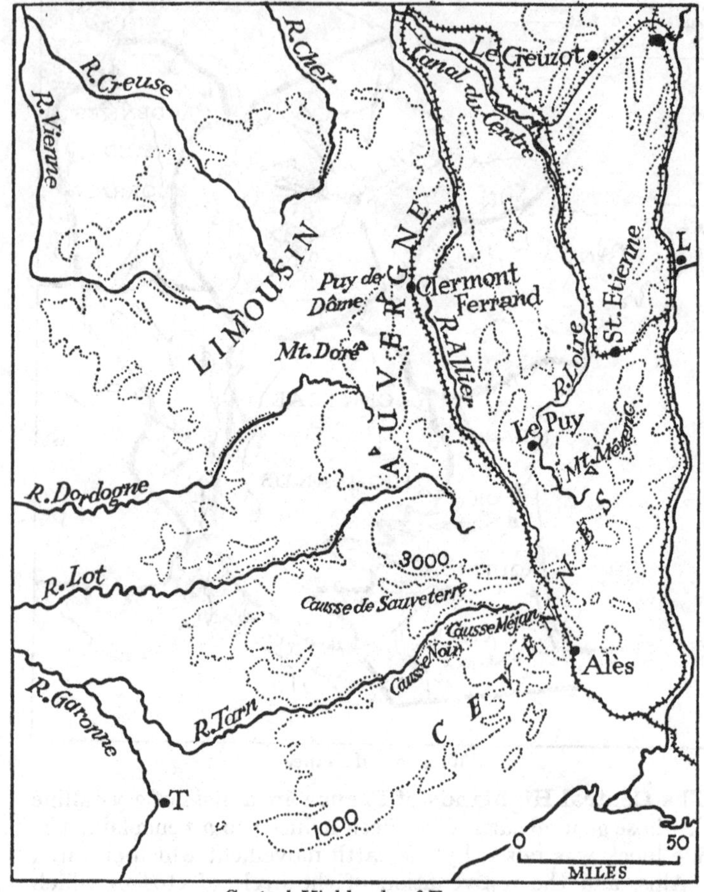

Central Highlands of France

the summers are apt to be hot and the winters cold and snowy; but precipitation is heavy, and the clayey valley bottoms are often sodden. Four subdivisions are to be distinguished: (1) the fairly regular tableland of Limousin, where cattle-breeding is the chief occupation; (2) Auvergne, the volcanic district, where the chain of ancient cones (*puys*) and necks is a characteristic feature and where Clermont-Ferrand (pop. 103,000 in 1931), the chief town of the subregion, carries on its growing industries of cutlery, leather, and macaroni; (3) the Causses, two arid limestone districts in the south, where sheep-rearing is the chief occupation; and (4) the Cévennes, or southeastern edge of the tableland, where a pocket of coal of poor quality has caused the growth of the little industrial town of Alès. Much of the country is wooded with chestnut trees up to the 1500-foot contour and with beeches above that line. A railway plods its way up the Loire to descend into the Rhone valley at St Étienne, while a canal leads from the Loire to the Saône through the town of le Creuzot, famous for its manufacture of ordnance.

Brittany consists largely of a Hercynian block of similar structure to Limousin, but lower in height. Its surface is that of a peneplain. Much of the subregion is forest-clad, though the higher ground is moorland. The population is scattered in homesteads, inland towns being all very small. Along the Loire valley is the château country of France centring about Tours, while the port of Nantes stands at the mouth of the river. The coastal area of the Breton peninsula, marked by deeply indented rias, rocky islets, and other ragged features, is too bleak to encourage agriculture and too exposed to the wind to carry forest. Its people live by fishing and supply most of the personnel of the French naval and mercantile marine. In this area are the naval port of Brest and the fishing towns of St Malo and St Brieuc. So distinct is this coastal belt that it is known as *Armor*, or sea country, as opposed to the inland *Arcoët*, or woodland. The people of this subregion look seawards rather than towards France, and they have to a large extent preserved their Celtic language, their local costumes, and peculiar ways. Their keynote is still that conservatism which led to the *chouan* revolt against the French Revolution, for they regularly send to the Chamber of Deputies members who are opposed to change. Roads are bad throughout the subregion, and the meshes of the railway system are less closely set than elsewhere in France, except in the Central Highlands.

The Saône Valley and the Jura consist of a wide river-plain and a crescent of fold ridges. The western boundary is formed by the fault line of the Central Highlands and its continuation in the Côte d'Or and Plateau de Langres. The slopes of these uplands are some of the best wine lands in France, the Côte d'Or being especially famous for its burgundy. Dijon, the old capital of the Duchy of Burgundy, is associated with Beaune and Mâcon as the centres of the local wine trade. The valley floor itself is often too wet for agriculture and is used for pasturing cattle. Around Chalon-sur-Saône the silted bed of an old lake forms a fertile, if damp, area. On the east bank of the river the plain is overlaid by boulder clay, as a result of the action of the Tertiary glaciers which descended from the Alps. The covering is thickest in the *pays de Dombes*, between the Rhone and the Saône, which was formerly a swampy area, but is now the market garden of Lyon. This city, the third in France, with a population of 580,000 in 1931, occupies what was once a strong defensive position between the rivers and is the meeting point of important international routes. It is famous for its silk manufactures which to-day are carried out partly on the factory and partly on the domestic system of industry, electric power being drawn for the looms from the Rhone and other Alpine streams. Just below the town the Rhone enters its gorge from which it issues near Valence. A pocket of coal on the west bank gives rise to important industries of silk and lace at St Étienne. The valley has an even less maritime climate than the Central Highlands and lies in a rain shadow. Lyon has a mean temperature of 35° F. in January and of 68° F. in July, with a rainfall of 31 inches a year, October being the wettest month.

The Jura consists of a number of fold ridges and valleys, the former being well wooded and the latter rich in agriculture. The range is an outlying portion of the Alpine system, but its peaks do not reach 6000 feet. The valleys are thickly peopled and produce, besides the yield of the farmyard and field, a number of objects, like watches and carefully made knick-knacks, whose manufacture helps the inhabitants to while away the long winter days. Besançon is the headquarters of the trade in such articles.

The Saône valley is part of the important corridor leading from the Mediterranean to northwestern Europe. Road, railway, and river afford a passage along this route, whose main line crosses the Côte d'Or into the Yonne valley, while branches go eastwards to the Rhine through the important Belfort Gap and westwards past le Creuzot to the Loire. The Upper Rhone valley takes the main international rail and road route over the Mont Cénis Pass to Italy.

The Lorraine and Ardennes subregion is here taken, for want of a better name, to include the upland area immediately west of the Rhine. It is for the most part a forested country, in which agriculture is carried on in the valleys and clearings. Its drainage is northwards into the Meuse and Moselle, the valley slopes of which latter are noted for their wine. The climate is even less of the extreme maritime type than that of the Saône valley, snow lying for weeks at a time in winter, while the summers are hot. Relief rains give heavy precipitation. Two facts give this subregion its importance: (1) its deposits of iron, coal, and salt, and (2) the disputed international frontiers which pass through it.

Among the series of rocks which outcrop in the neighbourhood of Metz are layers which bear various minerals. Salt is found around Nancy and, besides providing two-thirds of the salt supply of France, is used in chemical industries. Iron deposits are more extensive, stretching from the Côtes de Moselle between Nancy and Thionville as far as the Woëvre plateau. The output of ore is about 40,000,000 tons a year. Coal is found in the basin of the Saar a few miles to the east. The economic convenience of working coal and iron in close proximity has been upset by the jealousy of the French and German nations, each of whom covets the whole area. After the Franco-Prussian War of 1870–1 the frontier was run along the crest of the Vosges and then northwestwards in a great sweep to Longwy, so as to give to Germany both the coal and the iron fields (see map on page 112). The defeat of Germany in 1918 has led to a rearrangement of the frontiers, by which France recovered the iron fields, while the fate of the Saar basin was left to be decided by popular vote in 1935. In the meantime France was to get the output of the coal mines in compensation for the destruction of her own mines during the Great War. The popular vote, as was fully expected, gave the Saar basin back to Germany. North of the disputed area is the little Duchy of Luxembourg which is now attached by customs union to Belgium. The rest of the Ardennes is divided between the latter country and Germany.

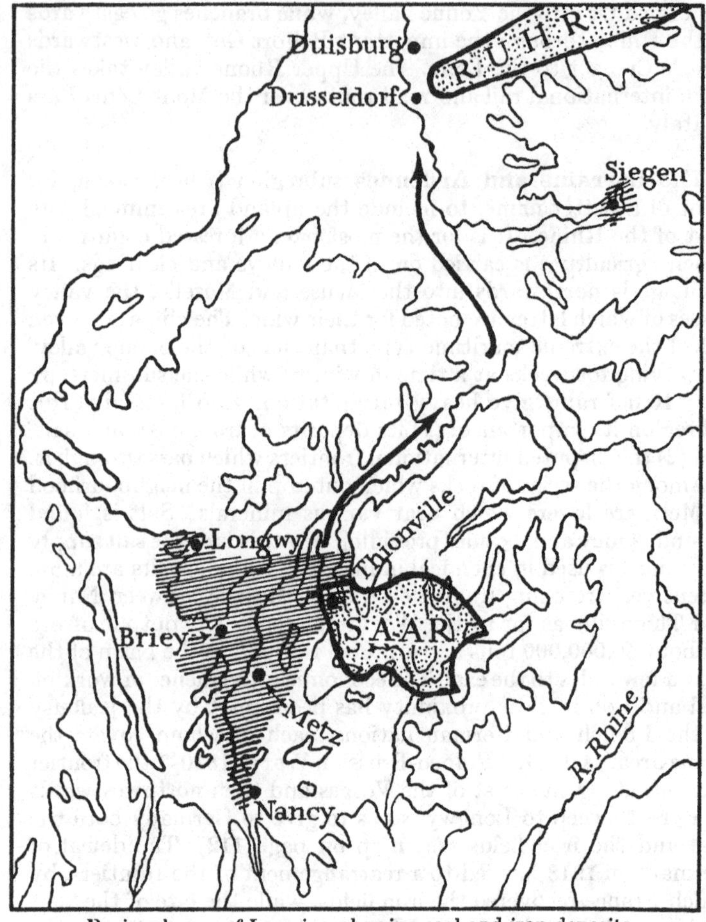

Regional map of Lorraine, showing coal and iron deposits

The Paris Basin is a horseshoe-shaped subregion enclosed by the Sill of Artois, the limestone uplands of Lorraine, the Central Highlands, and the hills of Normandy. Its structure is related to that of the southeast of England and consists of a series of layers whose upturned outer edges come to the surface to form a number of semicircular ridges one within the other (see diagram). The drainage is therefore inwards, with outlets to the west through the Seine and the Somme. The central area is of the youngest rocks: clay and sandstone; the next encircling ridges are chalk; while the outer lines are of Jurassic formation. In the north and in the Orleannais the surrounding ridges scarcely exceed 600 feet above sea-level. Gradients in the central part of the basin are so gentle that the Seine moves sluggishly through huge meanders, especially near Paris and Rouen, and the river regularly overflows its banks every winter. Its chief feeders are the Marne, Oise, and Yonne.

Only the coast strip has the extreme maritime climate, the inland portions showing a slight tendency to extremes in winter and summer. Thus, at Paris the mean January temperature is 35° F. and the mean July temperature 65° F. The protection of

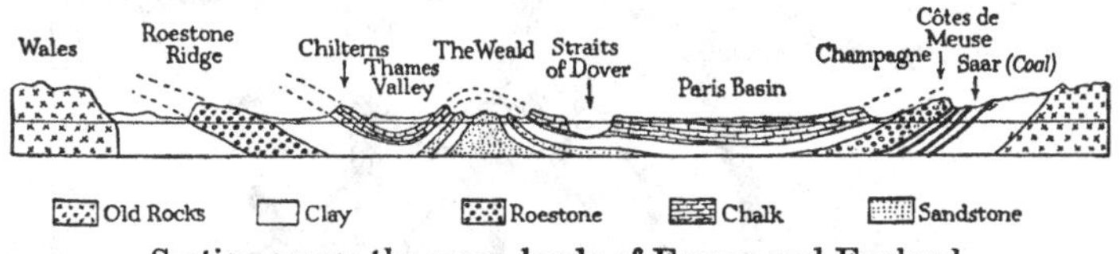

Section across the scarp lands of France and England

the hills of Normandy reduces the rainfall of the Paris area to 22 inches, but the chalk hills of Champagne get rather more. The slopes of these hills form one of the most famous wine-producing districts in the world, but Normandy at the other end of the basin is outside the area in which the vine flourishes and produces cider instead. The rest of the basin is devoted to cereals, wheat being the chief crop, followed by rye and barley.

The great city of Paris is the natural focus of the basin. Centring on the little Île de la Cité, this town has gradually outgrown its successive belts of walls and to-day has a population of over two million persons. In the past it has been, owing to ease of communication, the administrative centre of France and also the focus of commerce, science, and art; but to-day factories are springing up on its outskirts and it is being turned into an industrial town. Small vessels can ascend the Seine, thus making Paris into a seaport, though large craft are forced to use the outports of Rouen, Havre, and Cherbourg. Rouen is now well on its way to becoming an industrial town for the manufacture of raw material from America. The Channel coast is lined with seaside resorts, of which the best known are Deauvillc and Trouville. Calais, Boulogne, and Dieppe are ferry ports with English connexions.

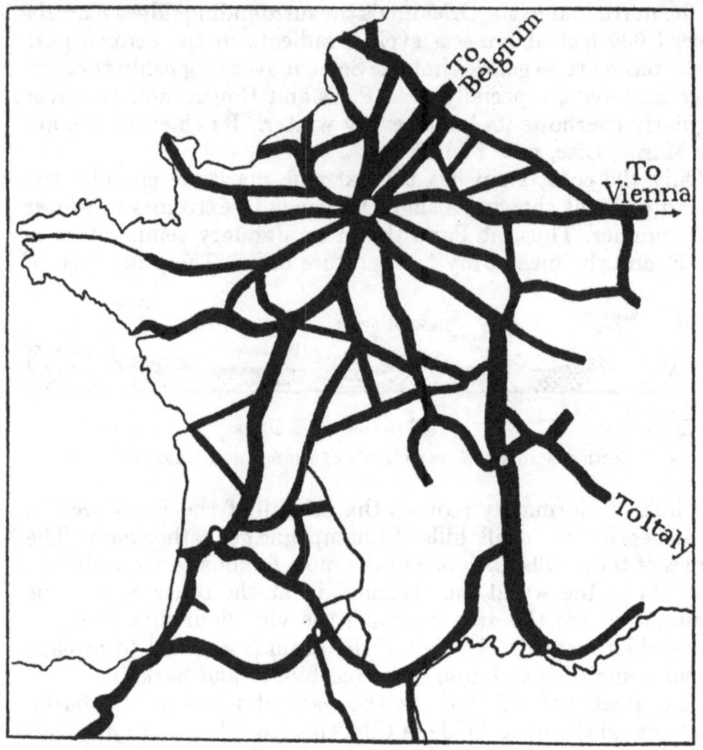

Main railways of France

The whole road and railway system of France focuses on Paris. Besides the line which runs through Lyon to Marseille, with branches to Switzerland and Italy, there are others going east through Strasbourg and Cologne to Germany, southwards through Orleans to Bordeaux and Spain, westwards to Brest, Cherbourg, and Havre, and northwards to Calais and Brussels. The network

Plate I

(*Canadian Pacific Railway*)

GRAVOSA (RAGUSA), JUGOSLAVIA

A typical Mediterranean port. Note the breakwater and the fort

Plate II

PRAG
The Charles Bridge and Hradçany

is close-meshed, but so bound up with Paris is the system that it is often quicker to take an express to Paris and out again than to undertake a far shorter cross-country journey. The roads are devised on a similar plan and are laid out with unusual directness.

Paris is also the centre of the waterways of France. Low

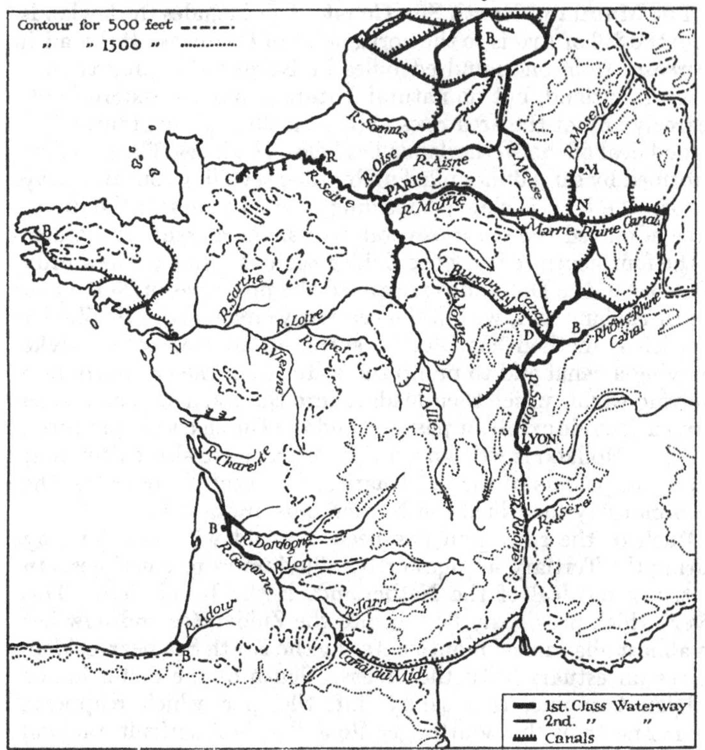

Waterways of France and Belgium

watersheds allow canals to pass north, south, and east to join the Seine with all the neighbouring rivers. The Orleannais offers no difficulty to the connexion of the Yonne with the Loire, and the Burgundy Canal threads the Côte d'Or to the Saône and thence onwards through the Doubs to the Rhine valley. Eastwards the Marne-Rhine Canal goes through Toul, Nancy, and

Strasbourg, while in the northeast the Aisne is joined to the Meuse. Readers of R. L. Stevenson will remember his account of the canal running from the Oise northwards into Belgium. So complete is the system that it is possible to tour France by boat.

The Mainland North Sea Coast. This includes the lowlands from the Sill of Artois to the north point of Denmark. Its breadth decreases from one hundred miles in Belgium to some twenty miles in Jutland, but no natural feature marks its eastern limit. Scarcely any of the area rises above the 300-foot contour, while a good deal of the Netherlands lies below sea-level. The coastline is formed by sand dunes piled up by the prevailing southwesterly winds. The land is slowly subsiding, and breaches in the dunes have admitted the sea in several places. As a result of a storm in the fourteenth century the Zuider Zee was formed. The sturdy diligence of the Dutch has reclaimed the provinces of North and South Holland and, within the last few years, most of the Zuider Zee. Their method has been to surround an area with a dyke carrying a canal and to pump the water from the enclosure into the ring canal, which then leads it into the sea. The 'polders' so formed provide excellent pasture, and the Dutch cheese produced in North Holland is well known. Root crops are also cultivated; but in the polders of the old Haarlemmer Meer is carried on the bulb industry for which the Netherlands are famous.

Much of the subregion has been overlaid with boulder clay during the Tertiary ice ages or by alluvium in recent times. In fact, a good deal of the Netherlands is the Rhine delta. This river, which previously flowed into the Zuider Zee and now has an almost abandoned mouth entering the North Sea near Leiden, shares an estuary with the Meuse. South of the delta, in the Belgian Campine, is a sandy, infertile area which reappears farther north as the Bourtanger Moor. The best agricultural land is in Belgium, where the maritime plain grows cereals, flax, and beet, while Flanders adds tobacco and hops to these products. The province of Brabant is largely occupied by the intensive cultivation of fruit and market vegetables. Farther north the damp soil favours pasturage, horses being reared in Gelderland and cattle in the rest of the Netherlands and Jutland. In the latter country the Danes have established a system of co-operative dairying with centres for the production of butter, cheese, and bacon.

Along the foot of the uplands forming the southern boundary of the subregion, outcrops of mineral-bearing rocks have given rise to a rich industrial area. A line of coalfields, known as the Franco-Belgian, extends from near Béthune to Aachen and re-appears across the Rhine in the Ruhr valley. In the Middle Ages

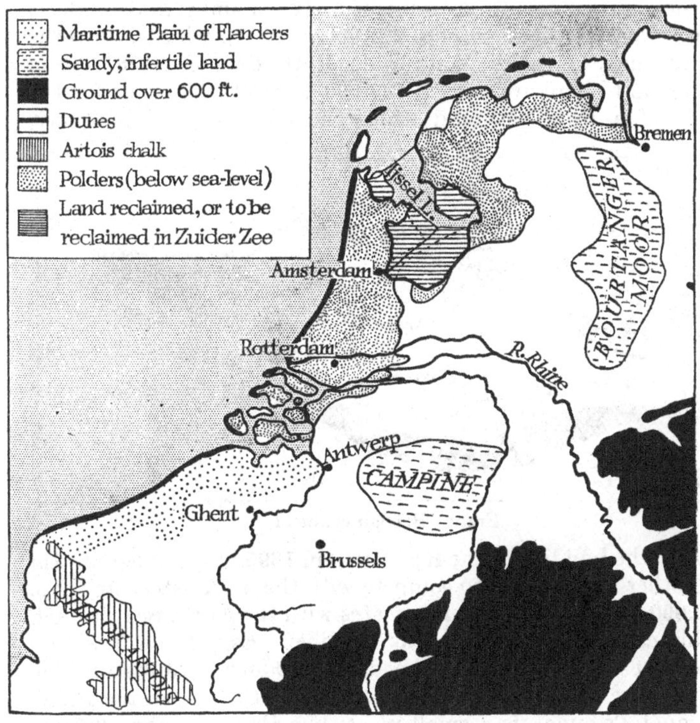

Belgium and the Netherlands. Some features of the land surface

the wool produced in the Ardennes had led to the establishment of the weaving industry, and to-day Liége, Namur, and Charleroi are weaving towns, though they have also metallurgical industries. Nearness to the sea at the port of Dunkirk enables Lille, Roubaix, and Tourcoing to import raw cotton, and Valenciennes is still famous for its lace. Other well-known industrial towns here are Arras, Cambrai, and Mons. Outside the coal area brick-making is important, and Delft makes fine quality pottery. Amsterdam and Rotterdam import and refine or manufacture sugar, gin, tobacco, quinine, rubber, cocoa, and other tropical products from the Dutch East Indies.

The subregion contains some of the most important seaports in the world. Antwerp (pop. 284,000 in 1931), which is also the world centre of the diamond trade, Rotterdam (pop. 587,000 in 1932), and Hamburg (pop. 1,080,000 in 1932) are rivals for the trade of a wide backland which includes most of Germany,

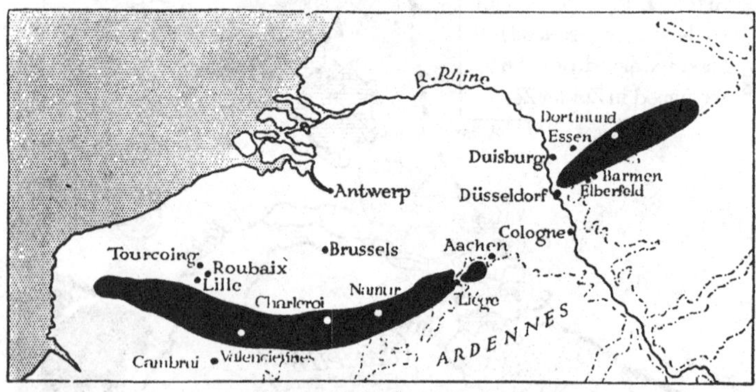

Franco-Belgian coalfield

though the building of the Kiel Canal in 1895 has enabled Stettin and other Baltic ports to compete with them. Amsterdam (pop. 766,000 in 1932) now communicates with the North Sea through a ship canal whose mouth is at Ymuiden. Esbjerg has a purely local backland, exporting the dairy produce of Jutland. The Hague, the capital of the Netherlands and the seat of the International Tribunal, is a small place, but Brussels (pop. 835,000 in 1932) is a large city and the cultural and commercial focus of Belgium.

The English Plain includes the southeast of Britain and is bounded on the north and west by the so-called Tees-Exe line, an irregular curve joining the mouths of the two rivers (see map on page 37). The surface is by no means flat, but is lined by scarped hills which run northeast, east, and south from a complicated knot of high ground in Wiltshire and Somerset. Between the

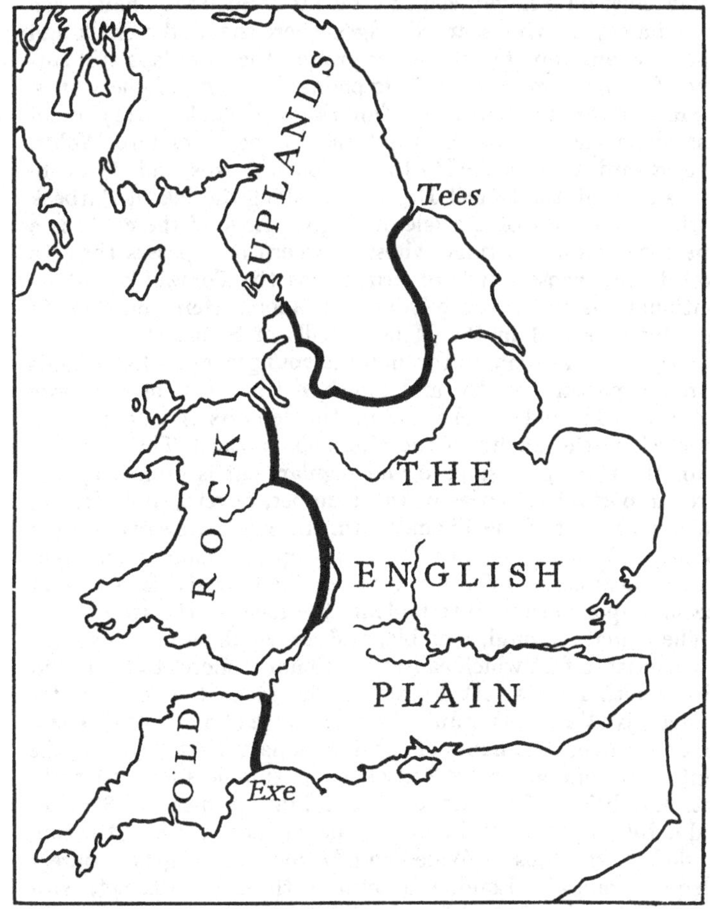

Tees-Exe line

ridges lie vales of clay or marl. Layers of clay, chalk, and limestone formerly extended in a huge downfold from the Pennines to the Thames, rising again southwards in an upfold whose remains now form the Weald. A second and smaller downfold makes the Hampshire Basin. The highest layers of this formation, which must once have reached a height of 3000 feet, have been eroded away, leaving scarped ridges where the harder chalk and limestone outcrop. On the lower ground the soft clays still survive. The diagram on page 31 explains the origin of the scarps.

From the knot in Salisbury Plain ridges of chalk run (1) north-east along the line of the Chilterns to the Yorkshire Wolds, (2) eastwards to form the North and South Downs, and (3) southwards through the Dorset Downs and along the Isle of Purbeck to the south coast of the Isle of Wight. Beyond the chalk is a limestone (roestone) ridge whose southern arm passes through Dorset and whose northern arm forms the Cotswolds and its continuations as far as the Yorkshire Moors. Here and there in the Midlands and in the Mendip Hills of Somerset the older underlying rocks crop up through the younger layers like islands of rough pasture, heath, and forest in the midst of the more fertile soil. The older rocks contain the deposits of coal and iron upon which the wealth of the Midlands is based. The coastline is for the most part smooth and regular, but is broken by the three important estuaries of the Humber, Severn, and Mersey, which with that of the Thames form the chief gateways into the country. A canal system which joins up the main rivers gives cheap water communication between the four rivers, a fact of special importance to Birmingham, the focus of the system.

The climate is mild, equable, and above all sunny along the south coast, a fact which causes the Channel shores to be dotted with health resorts. At Plymouth the mean temperature for January is 44° F. and for July 60° F., a range of only 16° F. There are on the average 5 hours of sunshine a day. Farther north, the winters become much colder, while the summers are distinctly hot. Cambridge, for instance, has a January mean of 37·6° F. and a July mean of 62° F. A good deal of the area is in the rain shadow of the hills of Wales and Devon and so gets a barely adequate rainfall. London is actually the driest district, with a mean annual fall of 24·5 inches.

We shall now proceed to consider the various parts into which the subregion may be divided, and we naturally begin with the Weald, since this district is the approach through which early man and later culture reached Britain. Its structure is best shown by

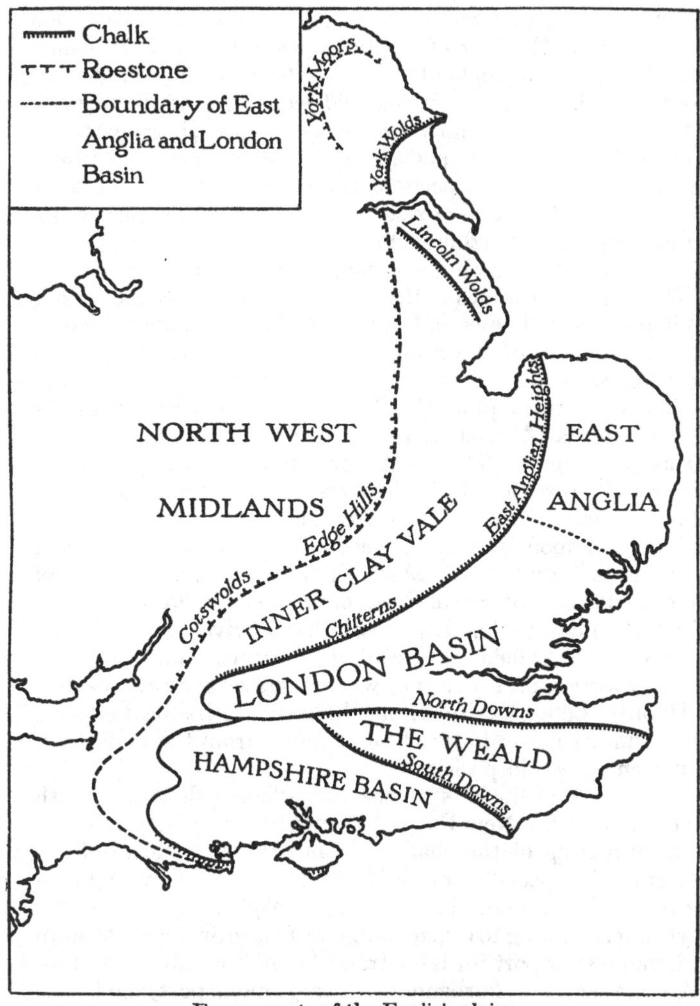

Escarpments of the English plain

a diagram (see page 31). The clay and sand areas were formerly an expanse of forest, but now the clays are used for dairy pasture and the sandy areas together with the chalk rims are being rapidly built over with residences for the more well-to-do business folk of London. The Vale of Holmesdale and the Medway valley form 'the garden of England' owing to their fertile yield of fruit, flowers, and hops. A small coalfield leads to manufactures at Ashford. The Downs, which are chiefly used as sheep pastures, are crossed by several important gaps which carry routes from London to (1) the ferry ports of Dover, Folkestone, and Newhaven, and (2) the seaside resorts of Margate, Hastings, Eastbourne, Brighton, Worthing, etc.

The London Basin, whose triangular clay plain is drained by the Thames, consists of (1) the Vale of Kennet, (2) the broad, silt-floored lower Thames valley, and (3) the low, estuary-pierced shores of Essex. In the lower Thames valley patches of the sand layer which once covered the whole area still survive and are clad in heath or pine. The largest patch is devoted to the military camp at Aldershot, while some of the smaller ones are used as golf courses. The great city dominates the life of the basin, for the dwellings of its workers now extend as far as the chalk hills and much of the surrounding country is used for supplying its food and other needs. Thus, a belt of market gardens runs round the outskirts of the city, and the Vale of Kennet and the cornfields of Essex are engaged in growing wheat to feed its inhabitants. High rents have driven many of the factories as far afield as Reading, Watford, and Croydon. Originally a river port on a spot where a gravel spur approached the Thames, London's growth has been due to its focal relation to its basin, its natural attraction of routes from the Continent, and its central world position.

The Hampshire Basin is another clay-floored downfold, with a sandy area in the New Forest. Formerly, its life depended on the sheep-rearing of the chalk hills and the agriculture of the lower ground, especially the early vegetables and flowers grown for the London market in the Isle of Wight. But nowadays everything is tending to centre round the fast-growing Southampton, London's outport for large trans-Atlantic mailboats and for the African services. Portsmouth is a large naval port, and Cowes is the yachting centre on the Solent.

Between the Chilterns and the roestone ridge of the Cotswolds and its continuations is the Inner Clay Vale, which consists of (1) the Avon Basin, (2) the Oxford Basin, and (3) the Fen lands.

To these for convenience sake may be added East Anglia. The Avon Basin is a small area enclosed by chalk and limestone hills.

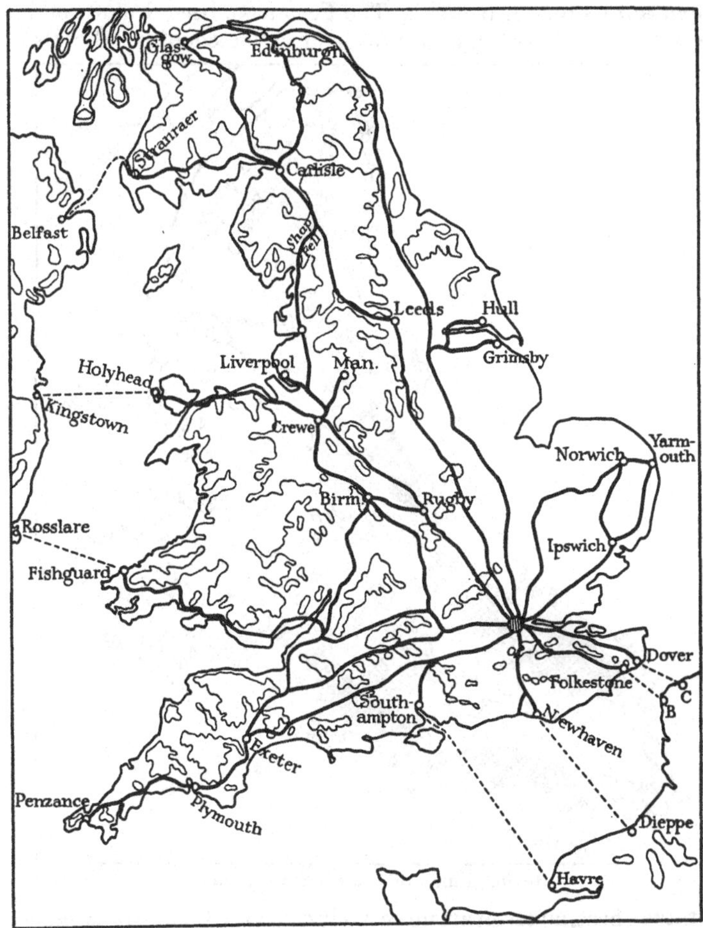

Routes focusing on London

Its focus is Bristol, a port whose old connexion with the import trade of tobacco and cocoa has been revived by the working of a small coalfield. Larger ships, which cannot ascend the river, unload at the outport of Avonmouth. Bath has medicinal springs. The Oxford Basin is mainly agricultural, but its focal town has now important motor-car factories. The Fens have been drained and are some of the best agricultural land in the country. Cereals,

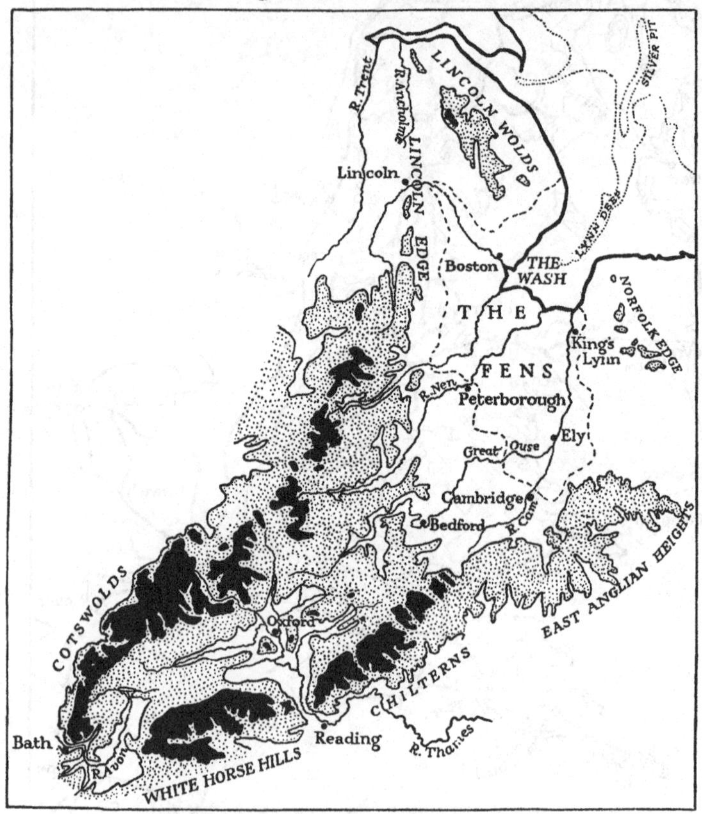

Regional map of the Inner Clay Vale

potatoes, beet, and fruit are intensively cultivated. The towns, none of them large, are placed on former gravel islands (e.g. Ely) or at the edge of the Fens (e.g. Cambridge, Peterborough). The cattle reared on the drier ground give rise to dairying at Aylesbury and Bedford and to the manufacture of boots and shoes at

Northampton and Kettering. East Anglia varies in the quality and therefore in the utilisation of its soil. The sandy Breckland is occupied by heaths, while the chalk hills are sheepwalks or

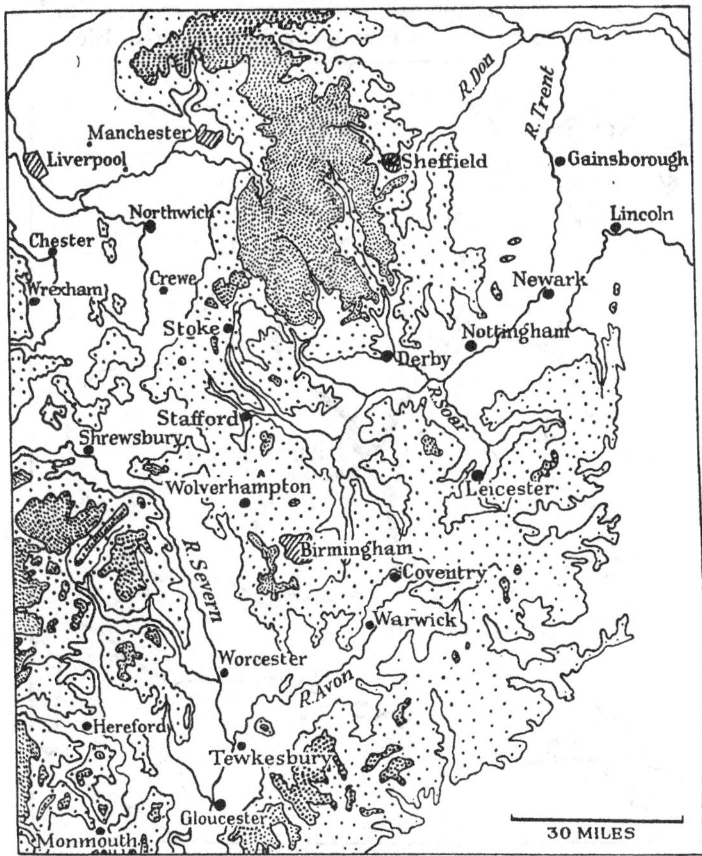

The northwest Midlands

cornlands. In the northeast lies the Broadland district, whose streams and 'broads' make a popular holiday resort and whose drained flats are pasture grounds for cattle. Elsewhere, East Anglia is fertile and is one of the chief productive areas of cereals and root crops in the British Isles. Norwich and Ipswich, together with Lincoln on the northern edge of the Fens, manufacture farm implements and other requisites, and Yarmouth and Lowestoft are fishing ports and seaside resorts. Grimsby on the Humber is the largest fishing port in the British Isles and

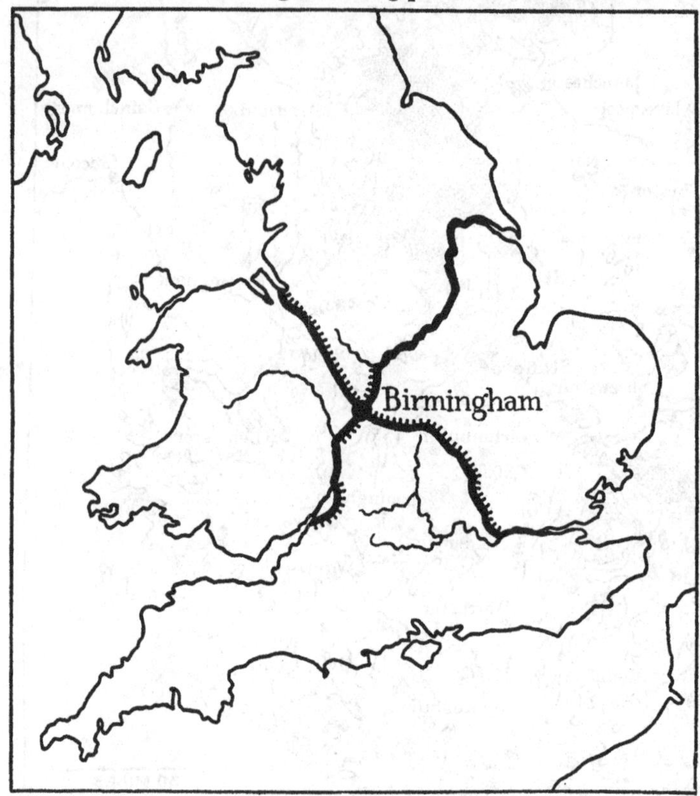

Position of Birmingham

exports goods to the north of Europe from the Midlands; but it is surpassed as a commercial seaport by Hull on the north shore of the estuary.

The northwest Midlands include the valleys of the Warwick Avon, the Upper Trent, Lower Severn, the plains of Shrewsbury and Cheshire, and a triangular mass of irregular broken uplands. The Upper Trent valley occupies the coalfields that lie to the south of the Pennines and is therefore an industrial district. The clays of the Staffordshire coalfield have given rise to the manufacture of china and earthenware, and the neighbourhood of the triple town of Burslem-Stoke-Hanley is known as the Potteries. Nottingham specialises in lace-making. Farther down the valley, agriculture takes first place and has its centres at Newark and Gainsborough. The plains of Cheshire and Shrewsbury are pasture lands on which dairying is carried on and 'store' cattle fattened for the market. The salt deposits near Nantwich and Middlewich are the basis of chemical industries at Warrington and St Helens. The valleys of the Warwick Avon and Lower Severn are rich agricultural areas abundantly productive of wheat, barley, potatoes and other root crops, and fruit. The Wye valley is a pastoral district famous for its breed of cattle. The central uplands, once devoted merely to forest, are now productive of coal and iron, and the district of the Clent Hills is known as the Black Country from its connexion with mining and manufacture. Birmingham (pop. 1,000,000 in 1931) owes its enormous recent growth not only to the presence nearby of coal and iron, but also to its position relative to the four great estuaries. It is a sprawling town surrounded by satellites, of which Wolverhampton is the chief. They are all engaged in the manufacture of metal goods. So is Coventry in the Avon valley; but Leicester and Stafford make boots and shoes, while Worcester is famous for its high-grade porcelain.

The Southwest, Wales, and the North. These three separate areas lie to the north and west of the Tees-Exe line and have the common trait of being uplands of old rock. Exposed to the full influence of the prevailing westerly winds, they have a milder and more equable climate than the rest of England. Thus:

	Mean January temperature	Mean July temperature	Range	Mean annual rainfall
Scilly Is.	46° F.	61° F.	15° F.	32 in.
Holyhead	42° F.	58° F.	16° F.	35 in.
Ft. William	39° F.	57° F.	18° F.	80 in.

These figures should be compared with those given for Cambridge and London on page 38. Cornwall enjoys such a mild winter that palms will grow in the open. Some of the higher parts of the subregion have a mean annual rainfall of between one and two hundred inches. The coastline is rocky in most parts and contains many good harbours, like those of Plymouth and Milford Haven.

The southwest of England, often called the West Country, is formed mainly of an old Hercynian block whose structure has almost been effaced by erosion. Dartmoor, Bodmin Moor, the

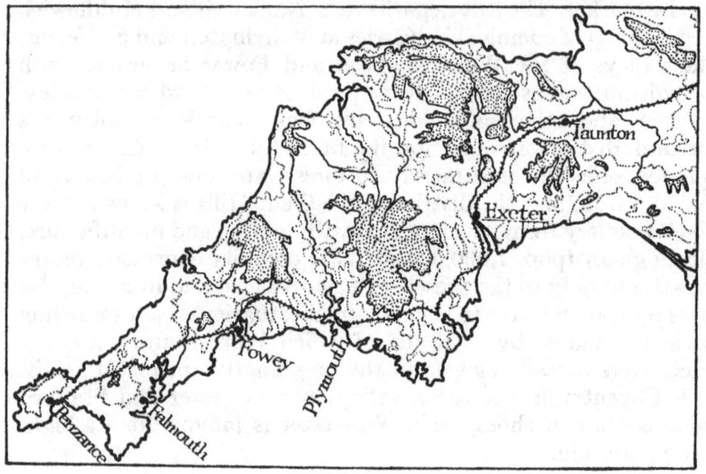

The West Country

little peninsula of Land's End, and the Scilly Isles are granite humps protruding through the newer rocks. The decay of the granite has caused the kaolin deposits which are mined at St Austell and shipped from Fowey to Worcester and the Potteries. At one time tin was mined near Bodmin, but most of the veins are now exhausted. A great deal of fishing is carried on from the little ports of Cornwall and Devon, but the shore population tends to concentrate on the cultivation of early vegetables for the London market and on attending to the needs of the holiday-makers and invalids staying at places like Torquay and Weymouth. In Devonshire cattle are reared and apples grown in the deeply cut, clay-bottomed valleys. The plain of Somerset, the one extensive area of lowland, lies between Exmoor and the Mendips and is an agricultural area, with wheat as the chief crop; but the Taunton enclave is a dairying district

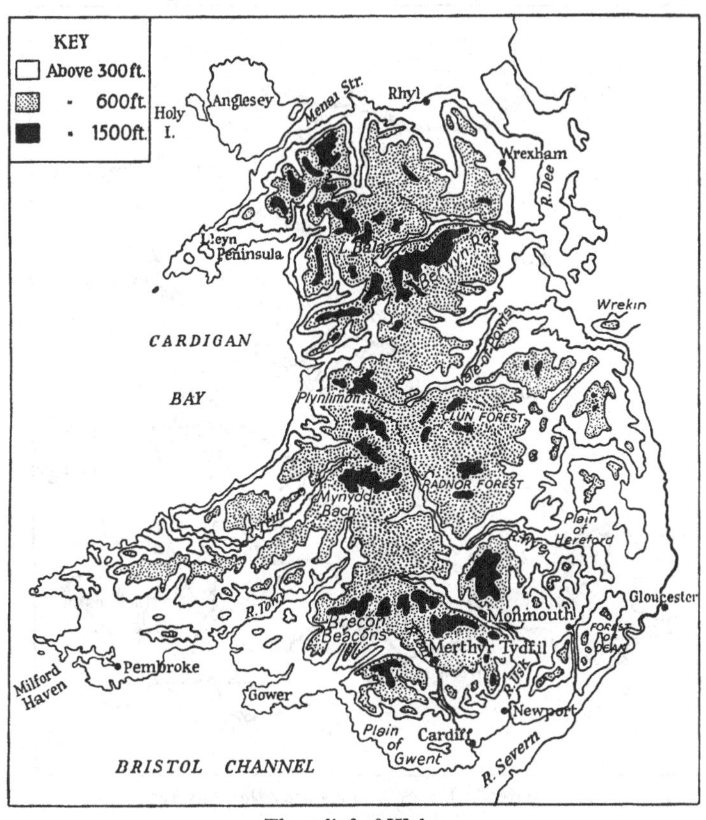

The relief of Wales

famous for its West Country cream. The main route from London uses this enclave to reach Exeter, the principal town of the peninsula.

Wales is divided by its structure into (1) the north and centre, where the rocks are oldest and the grain of the relief runs from northeast to southwest, and (2) the south, where the rocks are coal-bearing and the grain runs east and west. The country consists of five upland masses separated by deep, narrow valleys and edged on the north and south by strips of coastal plain. The

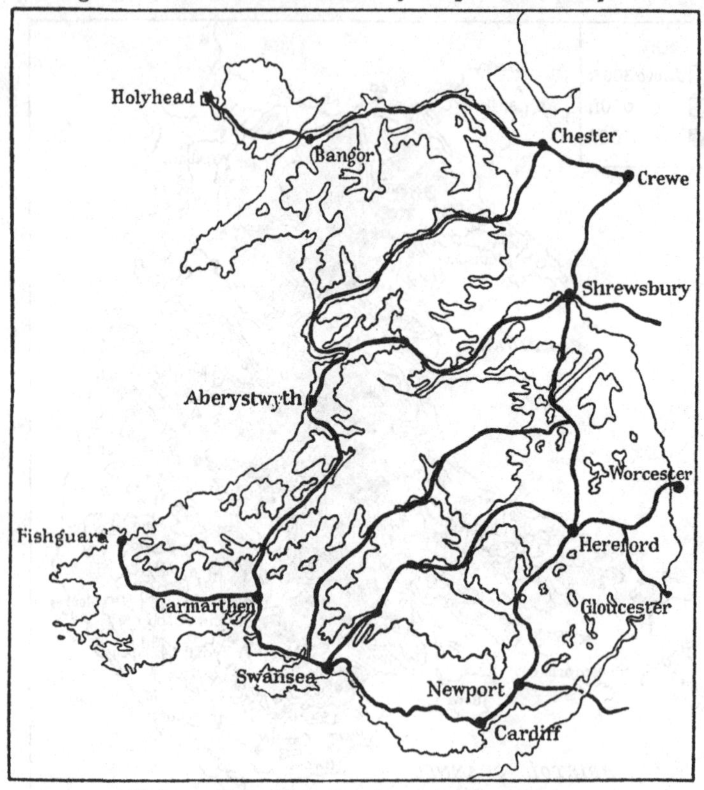

Main routes in Wales, as shown by the railways

uplands, and especially the Snowdon district, show many signs of recent glaciation. They are wind-swept, infertile, and boggy, but provide excellent pasture for sheep. The valleys form strips of cultivation, but none of them, except the Vale of Powis, are greatly productive, and they are more important as routes. The high valleys are isolated. Owing to the path of the main watershed being at no great distance from the coast, the larger streams flow towards England, while those running north, west, and south are short and unimportant. This fact, together with the division of the area into upland masses, deprives Wales of a national focus and has made its permanent conquest an easy matter.

A large coalfield stretching along the south coast from Pontypool to Kidwelly makes Glamorgan the chief part of Wales, no less than one-third of the population of the country inhabiting that county. The mines occur in long narrow valleys, like that of the Rhondda, whose bottoms have become continuous urban strips. Where they converge on the coast, there have sprung up towns of which Cardiff and Newport are the largest. These two places export steam coal and the tin plates manufactured at Merthyr Tydfil. Farther west, the coal is of the anthracite variety, the main outlet being at Swansea. A main railway from London enters the plain of Gwent by the Severn tunnel and after touching at the larger ports reaches Fishguard, a ferry port whence boats run to Rosslare in Ireland. The northern coast strip is similarly used by another railway which makes for Holyhead and Dublin. Here the coast towns, Rhyl and Llandudno, are seaside resorts.

The north of England and the south of Scotland consist largely of four upland areas: the Pennines, Cheviots, hills of Cumberland and Isle of Man, and the Southern Uplands. The Pennines are a fold range whose upper layers, though removed by erosion from the anticline, appear on the surface (see diagram) at the

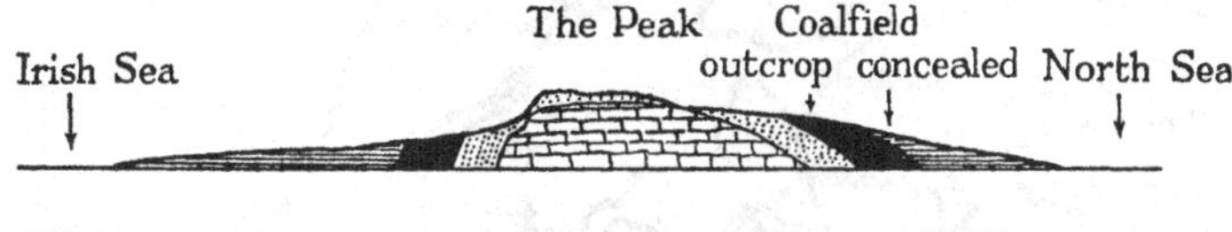

Section across the Pennines

foot of the slopes. Hence, the layer of carboniferous limestone which formerly overlay the whole range now merely outcrops at its four corners, forming the five coalfields of Cumberland, South Lancashire, Staffordshire, York-Notts-Derby, and Northumberland-Durham. The highest ground in the north and south

Pennines is formed by horizontal beds of sandstone, while in the centre the older mountain limestone is exposed. The range is

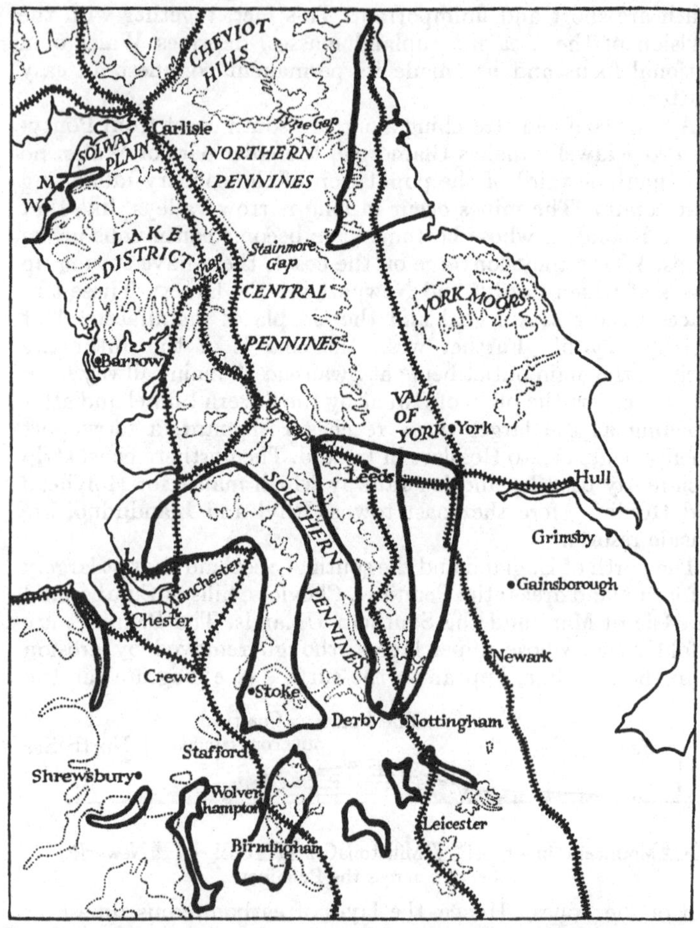

Pennine subdivisions, gaps and coalfields. The dotted line is the 500 foot contour. The coalfields are ringed about with thick lines

broken by faults which have made the Tyne and Stainmore Gaps and the steep western slopes which at Crossfell Edge form a precipice a thousand feet high. The Cheviots are of older rock and are marked by outflows of lava. The Southern Uplands have a northeast to southwest grain and are separated from the Cheviots by the valley of the Tweed. The hills of Cumberland are a dome of old rock whose ancient radial valleys have been deepened by glacial action and contain in their bottoms the beautiful lakes for which the district is famous. The ridge of Shap Fell connects the hills with the Pennines. The Isle of Man is an outlying portion of the hills of Cumberland and is divided into two upland masses separated by a belt of lowland.

Between the uplands and the sea are several fairly large and fertile lowland areas. The two most important are the Vale of York and the plain of Lancashire. A rim of lowland runs round the Cumberland peninsula and expands in the north into the Solway plain. It is continued on the other side of the Solway Firth as Galloway. Between the Southern Uplands and the Cheviots lies the Tweed valley with many secondary arms penetrating into the hills. Such dales are also a feature of the eastern slopes of the Pennines. The chief human development is the industrial region which has sprung up on either side of the Pennines. Textiles were early produced here, owing to the sheep-rearing carried out on the upland moors, and Leeds is the centre of the world's greatest woollen cloth manufacturing area. In Lancashire cotton cloth is produced, the centre of the industry being at Manchester. The two districts communicate with each other by means of railways which tunnel through the topmost Pennine ridge or else follow the gap cut by the River Aire. The latter route also carries a road and canal. At Sheffield the presence of sandstone which makes good grindstones has fostered the manufacture of cutlery.

On the Northumberland and Durham coalfield ports have grown up at the mouths of the Tyne, Wear, and Tees for the export of the mineral, and the ease with which timber and iron could be imported from Scandinavia has developed shipbuilding and the manufacture of heavy iron goods, an industry due primarily to the deposits of iron in the Cleveland Hills. Durham, on the peninsula formed by a bend in the Wear, arose as a stronghold. Iron manufactures are also carried on at Barrow-in-Furness, owing to the presence of high quality ore, and at Workington on the Cumberland coalfield. Maryport exports coal to Ireland.

In the Tweed valley the wool obtained from the sheep on the uplands has given rise to woollen manufactures which, here as in Yorkshire, depend now on imported raw material. The chief centres are Hawick and Galashiels. Berwick, near the mouth of the Tweed, is an old Border fortress.

Generally speaking, sheep-rearing is the main occupation of the uplands, and agriculture goes on everywhere in the lowlands, though Galloway is a dairy country, producing bacon, eggs, and milk. The Vale of York continues the cereal and root crops of the lower Trent valley, whilst the plain of Lancashire is largely given up to potatoes or else to dairy and market gardening for the supply of the needs of the industrial towns. The fishing centre at Fleetwood also has this aim, while Blackpool is a large seaside holiday resort for the workers. The Merse, or lower Tweed valley, produces cereals, its climate being dry and surprisingly mild.

The great routes between England and Scotland pass on either side of the Pennines. One follows the east coast from York to Edinburgh, while the other crosses Shap Fell and the Solway Moss to send a branch, known as the Waverley Route, through the Scott country to Edinburgh and another over the Beattock Pass to Glasgow or Edinburgh. A third branch follows the Galloway coast to the ferry port of Stranraer, which is connected with Larne, an outport for Belfast. The paths used by these routes emphasise the influence of relief on overland lines of communication, while the presence of towns like Carlisle, Berwick, Alnwick, Newcastle, and Durham is a reminder of the former need to hold strategic military points along the passage between the two countries.

The Midland Plain of Scotland. This is a rift valley whose floor has been crumpled by folding and broken by the upheaval of volcanic hills. The Highland Line between Dumbarton and Stonehaven is a definite boundary, but the separation from the Southern Uplands is not so clear, since the plain rises here into a kind of shelf. A central downfold has admitted the sea to form the Firths of Forth and Clyde and has permitted a barge canal to join the two estuaries without crossing the 100-foot contour. In this area there are layers of carboniferous rock, but folding and subsequent denudation have reduced the coalfield area to three relatively small districts: Kyle in Ayrshire, Lanarkshire,

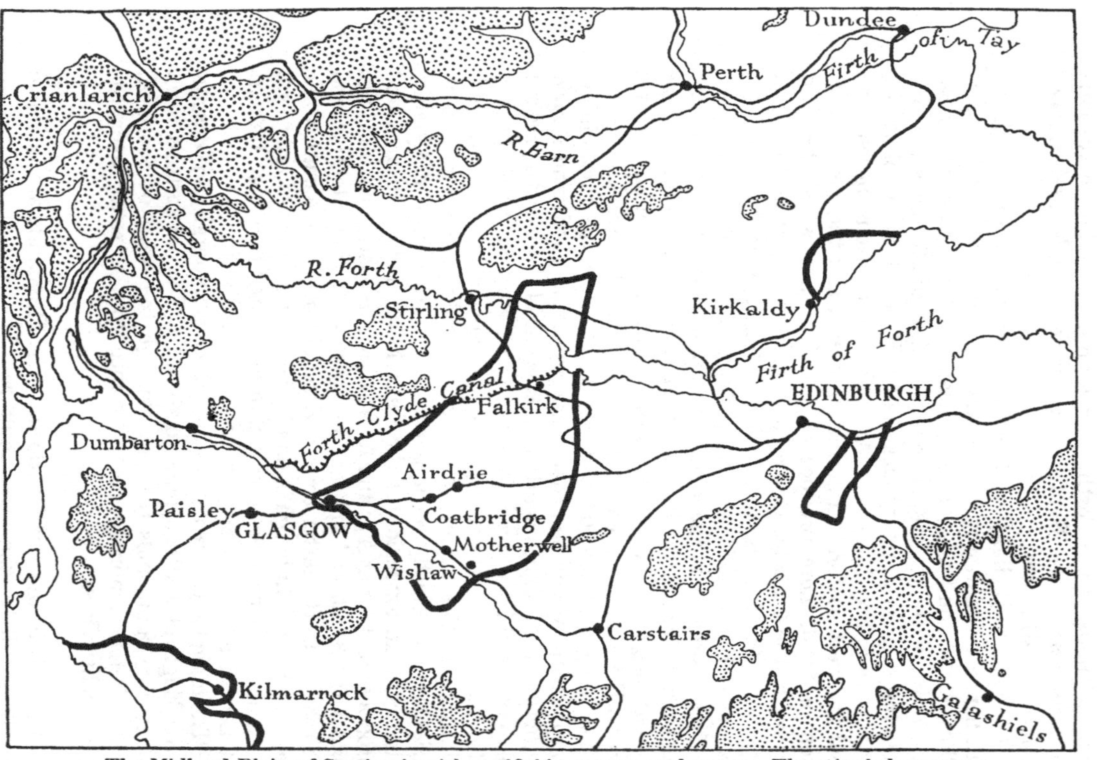

The Midland Plain of Scotland, with coalfields, routes, and towns. The stippled areas are above 500 feet. The coalfields are ringed with thick lines

and the shores of the Forth. Farther north, the line of volcanic hills comprising the Renfrews, Campsies, Ochils, and Sidlaws isolates Strathmore.

The climate is slightly colder than that of the English plain; but here too the west is damper and more equable than the east. Thus, Ardrossan has a January mean of 41° F. and a July mean of 57° F., with an annual mean rainfall of 37 inches, whilst Edinburgh has a January mean of 39° F., a July mean of 59° F., and a rainfall of 25 inches. As a result the east is agricultural and the west pastoral. Fifeshire, Lothian, and the eastern portion of Strathmore produce oats and barley, while the fruit grown in the Carse of Gowrie has given rise to jam-making at Dundee. Ayrshire is, on the other hand, a dairying county, except in the south where sheep are reared.

This subregion is the most important part of Scotland and contains three-fourths of the population of that country. This importance largely depends on the industries which have grown up on the coalfields. The chief town on the Lanarkshire field is Glasgow (pop. 1,000,000 in 1931) which has large shipbuilding yards on the Clyde and manufactures cotton goods and a large variety of miscellaneous articles. It is one of the world's greatest seaports, and is engaged chiefly in trade with America. Paisley also manufactures cotton goods, specialising in cotton thread. Motherwell, Coatbridge, Wishaw, and Falkirk have iron foundries, relying chiefly on imports of Swedish ore. The Ayrshire district is an extension of the Lanarkshire industrial area and produces textiles at Ayr and Ardrossan. The eastern coalfield lies under the Forth and appears on either shore in Midlothian and Fifeshire. In the latter county Kirkaldy makes linoleum, while in the former Edinburgh is rapidly becoming an industrial town, with factories for paper-making, printing, and brewing. Its port at Leith trades with Scandinavia and the Baltic.

The main routes are those which reach Edinburgh and Glasgow from England and continue northwards. The old fortress town of Stirling commands the way through the waist of Scotland, while Perth dominates the entrance to the route over the Drumochter Pass.

The Highlands of Scotland. The Highlands are bounded in the south by the great fault running from Dumbarton to

Stonehaven, and they are separated by the narrow rift of Glenmore into northern and central portions. The eastern hills are humpy ridges with a relatively smooth outline, whilst those on the west tend to form separate peaks, the highest of which is Ben Nevis (4448 feet). The west coast is very ragged, being deeply indented with sea-lochs and fringed with islands ranging in size from Harris, Mull, and Skye to mere stacks and skerries.

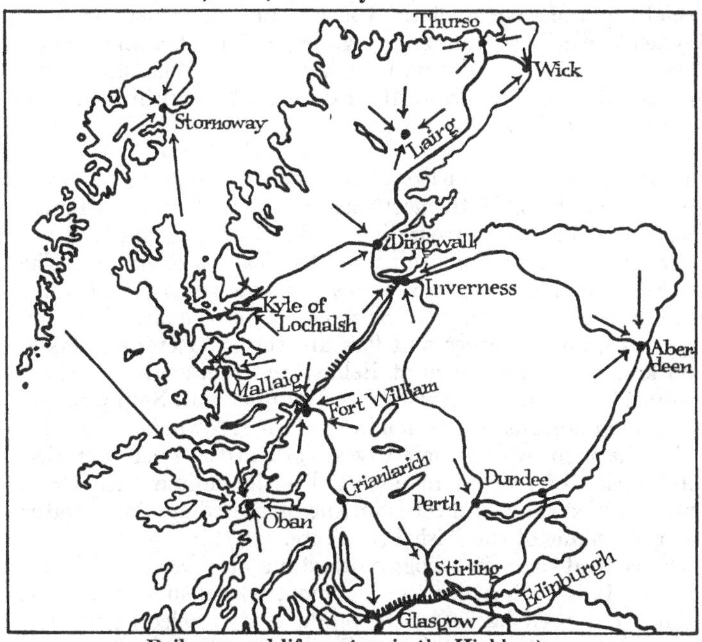

Railways and life centres in the Highlands

The east coast is even and rocky, with few harbours. The climate is on the whole bleak and rainy, but equable on the west, though these conditions improve on the east. Thus, at Fort William the mean temperature in February, the coldest month, is 39° F., and for July, the warmest, 57° F.; whilst at Bræmar the corresponding temperatures are 34° F. and 55° F. The eastern highlands are forested, whilst the western hills have bare slopes patchily covered with heather, poor grass, and bracken. Many peat bogs occur.

The subregion is barren, with the exception of the district of Aberdeen, where oats, beans, and potatoes are grown. Aberdeen, the focus town, is also a fishing port. Elsewhere, the scanty population clusters on small areas of alluvial soil at the head of lochs, where they combine fishing with the raising of a few head of cattle and sheep. The cottagers engage in weaving homespuns, and Fair Isle and Harris are well known for special types of woollen cloth. Tourists and sportsmen give employment to a large number of gillies, guides, boatmen, hotel keepers, etc. A railway from Glasgow leads to Oban, and a line from Edinburgh through Perth crosses the Drumochter Pass to Inverness and Wick.

Ireland. Ireland consists of a central plain bordered by a broken rim of hills. In the northeast, Ulster is the basin of Lough Neagh and the hills surrounding it, of which the chief are the Sperrin and Mourne Mountains and the plateau of Antrim. The latter is formed by an outflow of basaltic lava which disintegrates into a fertile soil. Owing to the moistness of the air, cereals do not thrive, and potatoes and flax are the chief crops. Flax is spun and woven into linen at Belfast and Londonderry, whose factories burn coal from Ayrshire and Cumberland. Ship-building is also an important occupation in the former town, being carried on by the men, whilst their wives work in the linen factories. This portion of Ireland remains politically united with Great Britain, whilst the rest of the island has acquired dominion status under the name of the Irish Free State.

The central plain is a boggy limestone area used mostly for grazing. It is drained by the Shannon, whose sudden drop at Killaloe has been harnessed to produce hydro-electricity. The light and power thus obtained will, it is hoped, be sufficient to supply nearly all the needs of the Free State. Limerick, the focus of a fertile area, is a port, but is handicapped through facing west. On the east coast is Dublin, the capital of the Free State, which derives its importance from its position relative to England. It has various industries, including brewing and the manufacture of poplin. Canals connect Dublin with the Shannon and Barrow. The latter runs through a fertile valley in which barley, wheat, and potatoes are cultivated and dairy-farming practised. In the parallel east-and-west valleys of the southwest similar occupations

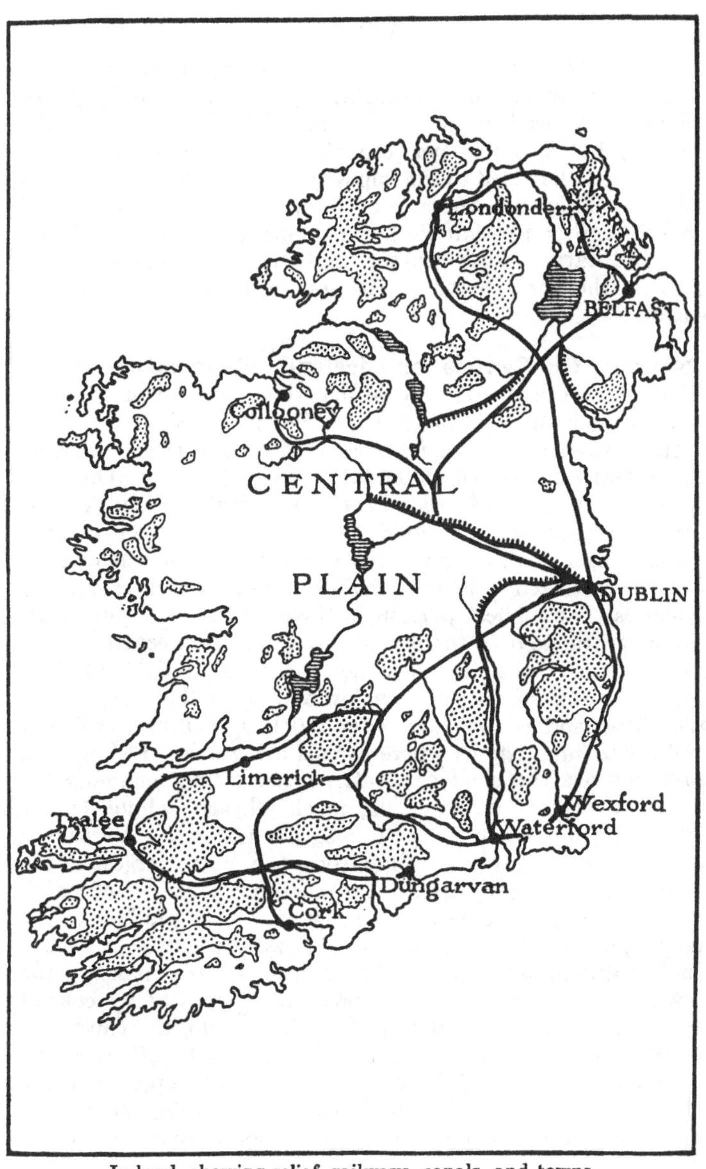

Ireland, showing relief, railways, canals, and towns

are found, the produce being exported through Cork and Wexford. The exports of the country as a whole are live cattle, pigs, and horses as well as dairy produce. The market is normally England, but quarrels between the English and Free State Governments have led to the imposition of a tariff on Irish goods and the consequent decrease of trade. Regular cross-channel services connect Rosslare with Fishguard, Kingston with Holyhead, and Larne with Stranraer, while most of the goods traffic between the two countries passes through Liverpool and Bristol.

Norway, the Færoes, and Iceland. These have a climate which may be described as cold maritime. Bergen has a mean temperature in January of 34° F. and in July of 58° F. Exposure to the southwest winds and a backing of high relief give a heavy rainfall which in Bergen amounts to 81 inches a year. Only the southwestern corner of Iceland comes within the region, the rest of the island consisting of moorland varied with lava fields, sandy expanses, and small ice-caps. Norway comprises a narrow strip of deeply indented fjord coast backed by bleak and ice-covered highlands. The highest portion of these occurs in Jotun Fjeld, Jostedals Bræ, and Hardanger Fjeld. Population centres round the deeply cut dales, of which the chief are Österdal and Gudbrandsdal, and in the narrow lowlands near Trondheim and Öslo. The land is too barren for much agriculture, which is confined to small strips where cereals are grown. Fishing and lumbering are the chief occupations, the former enabling the country to export cod, herring, mackerel, and salmon to the value of £5 million in 1930. It also has fostered a mercantile marine of two and a half million tons, the fourth largest in the world. Lumbering is carried on now on scientific lines and, besides timber, wood pulp is being increasingly produced. The many rapid streams supply abundant power for hydro-electricity, which is accordingly used in the factories and on many of the railways. Communications are poor, being mostly by coastal steamers; but there are railway lines from Öslo, the capital of Norway, to Trondheim and to Bergen, the headquarters of the fishing industry. Narvik in the north is an ice-free port which is connected by rail with the Swedish iron-producing district at Lulea and is used for the export of ore, especially in winter.

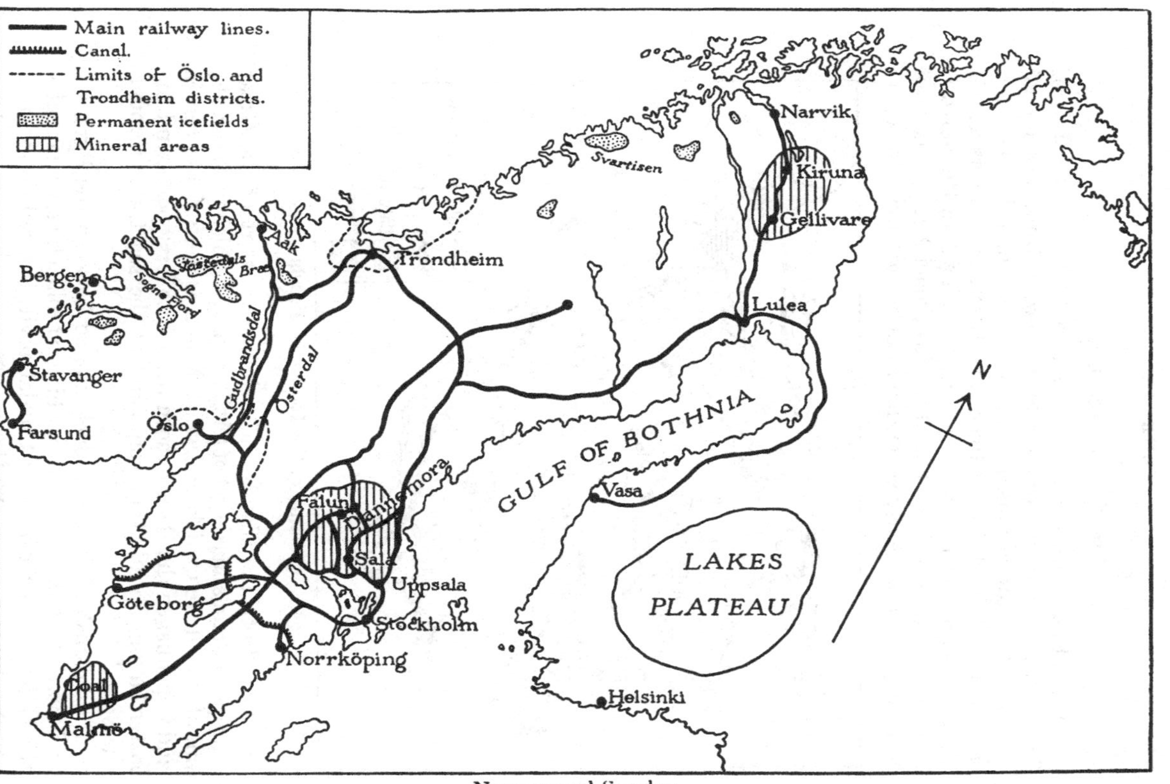

Norway and Sweden

The Transitional Region

A mean temperature in January of about 30° F. and in July of 65° F., with a rainfall régime which shows a total of some 25 inches with a summer maximum, but no dry month—these are the features of the climate of the region. In extent it covers central Europe, the Spanish Meseta, and the Po valley. In the north it includes the Danish Islands and the low, glaciated peninsula of Schönen, which is nearly separated from Scandinavia by a chain

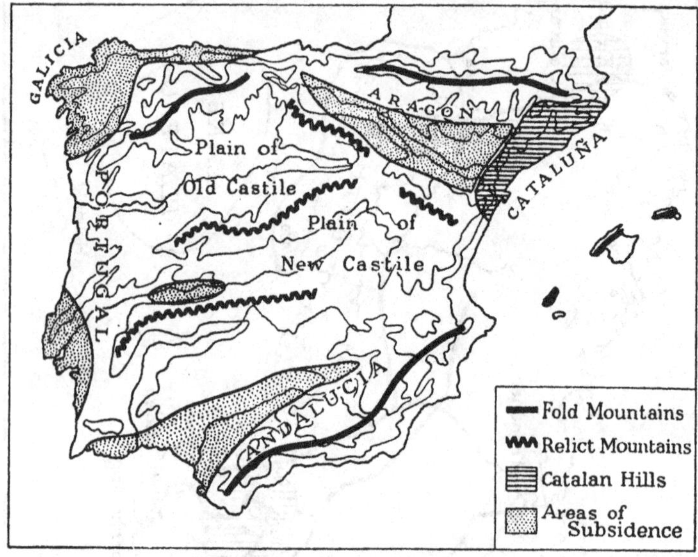

Structure of the Iberian Peninsula

of lakes. Both the islands and Schönen are part of the Great European Plain and consist of fertile lowlands which yield cereals, but above all are used for dairying. Malmö is the Swedish centre and Copenhagen the Danish, both being also ferry ports, and the latter having, in addition, command of the sea traffic through the Sound.

The Spanish Meseta is a block upland between the Pyrenees and Sierra Nevada. Its surface, which is broken by the triangular rift valleys of the Ebro and Guadalquivir, is tilted downwards towards the southeast, and the western edge, therefore, forms a mountain barrier to the winds from the Atlantic. Hence, most of the rainfall is due to convection in summer, and some areas, notably La Mancha and Estremadura, are very dry. A line of relict mountains crosses the tableland from east to west, the highest part being known as the Sierra de Guadarrama. The rivers, which derive most of their volume from the precipitation in these hills, cut deep trenches in the dry areas through which they flow and thus make communication difficult. The three largest are the Duero, Tagus, and Guadiana, all unnavigable except for a short

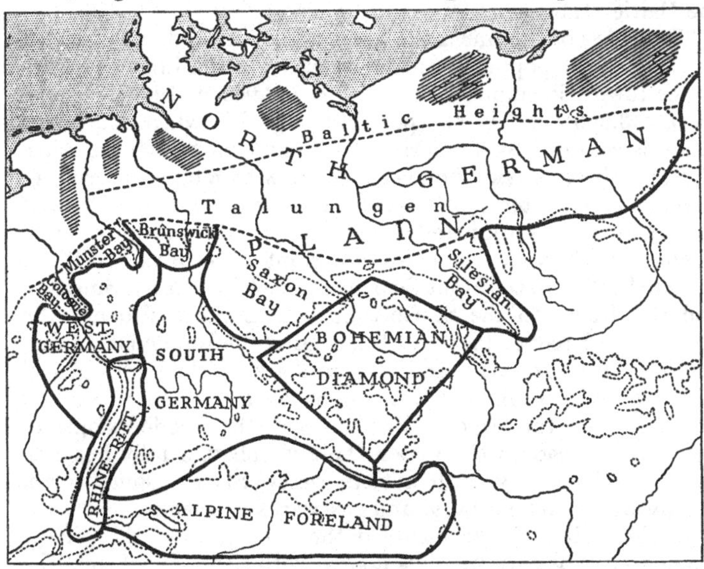

Subregions of Germany

distance from the sea. The Ebro is also unnavigable and shares with the others a tendency to great differences in volume in summer and winter. Most of the tableland is occupied by the plains of Old and New Castile, whose landscapes show a groundwork of bare soil and rock patches with clumps of southern pine and cork oak, or irrigated fields of maize. A number of small towns occur, still contained within their medieval walls. Madrid is comparatively modern and is placed with no geographical advantages other than that of central position.

Central Europe is divided into two parts by the great ridge of the Alps and Carpathians. North of the ridge the relief runs in three belts: (1) the outer ridges and valleys of the great fold ranges, including the valleys of the Aar in Switzerland and the Danube in Austria, the plateau of Bavaria, and many fertile valleys like that of the Inn and Upper Rhine which penetrate deeply into the great mountains; (2) the block uplands, including the Black Forest, Thüringerwald, and Boehmerwald; and (3) the plain of North Germany, whose southern limit is the 600-foot contour and which stretches east and northeast into Poland and the Baltic States.

The valley of the Aar is the heart of Switzerland and contains the bulk of the population of that country and all but one of its larger towns, including Zürich (pop. 250,000 in 1930), Geneva (pop. 143,000), and Berne (pop. 112,000). Elsewhere, the population is confined to the narrow valleys which trench the Alps. Many of these provide holiday resorts, with mountaineering in summer and winter sports from January to March. The chief centres are St Moritz and Pontresina in the Engadine, and Chamonix. The valleys are also important route-ways into Italy, the Arc leading to the Mont Cenis Pass, the Rhone to the Simplon, and the Reuss to the St Gotthard. Important also are the Maloja and Brenner Passes at the head of the Inn and its feeder, the Sill. The Aar valley and its slopes are overlaid with glacial deposits and contain lakes, of which Geneva, Neuchâtel, and Constance are the largest. Farther east, the wooded slopes of the pre-Alps lead to the upper Danube valley, a hilly but rich agricultural country with outlets westwards at Donauwörth, northwards at Regensburg, and eastwards at Vienna.

In this belt must be included the rift valley of the Middle Rhine (see diagram), which is floored with rich alluvial soil. The Rhine enters it at Basle and leaves it at Mainz. The Ludwig Canal connects the Main, a right bank tributary, with the Danube, while the Zabern Gap provides a westward exit from Strasbourg, the French capital of Alsace. Other routes issue from the south through Belfort and Basle. Thus, the valley is one of the chief highways of Europe. At the lower end of the valley the Rhinegau is a wine-producing district, and Wiesbaden and Frankfurt have unusually mild winters owing to their southern aspect.

The belt of blocks consists on the west of broken uplands whose higher ground is wooded and whose valleys are well populated and agricultural. The nature of the ground formerly caused this area to be divided politically into a number of petty German states, which the improvements of modern communications have welded together into Germany. Stuttgart and Nürnberg are the largest towns, the latter being the focus of several main routes. Farther east is the 'Bohemian diamond', whose four sides are formed by the Boehmerwald, the Ore Mountains, the Sudetes, and the Moravian plateau. Within, lies the basin of the Upper Elbe, of which Prag is the centre. It is an area rich in soil, the chief crops being potatoes, wheat, hops, and sugar-beet. It also has coalfields on which have sprung up cotton and chemical manufactures, and it is famous for its beer (Pilsen) and watering-places (Karlsbad and Marienbad). Between it and the Car-

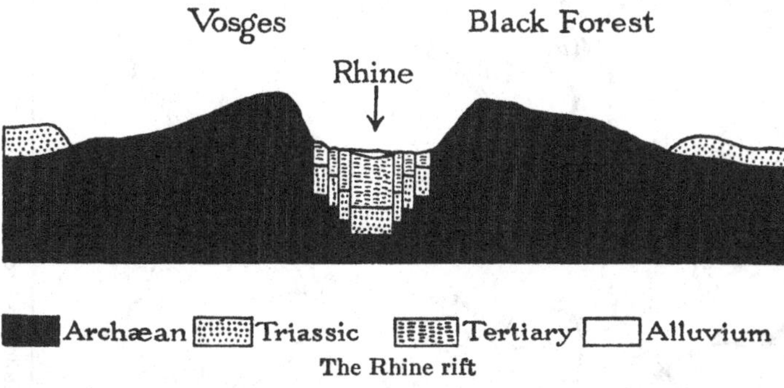

The Rhine rift

pathians is the ill-defined rift of the Moravian Gate, which forms the chief highway from Germany into Hungary and to the Adriatic. Just outside the 'diamond' is the mineral district of Saxony, with large deposits of brown and bituminous coal, and some small deposits of iron, silver, and zinc. Here Chemnitz ('the Manchester of Germany') and Zwickau manufacture cotton and woollen goods, and Dresden makes its famous chinaware.

The North German Plain is part of the Great European Plain and is an expanse of lowland whose slope was originally northwards. During the Tertiary ice ages it was partially covered by an ice-cap which laid down a pile of rock waste across the drainage lines. Although the main streams of the Oder and Vistula have resumed something like their former direction, their feeders have an east-and-west trend which has facilitated the development of

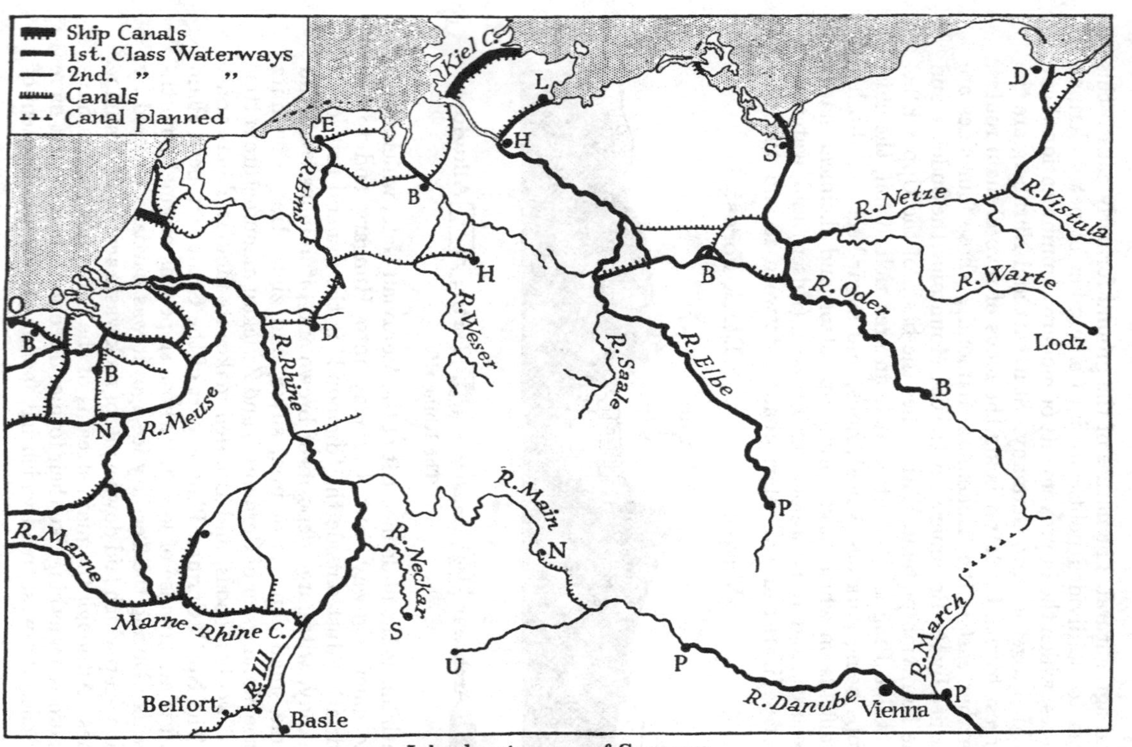

Inland waterways of Germany

NASSAU, GERMANY

The Lahn has carved a steep-sided valley from the ancient plateau, whose generally flat top can be seen in the background. The town was the birthplace of Baron Stein

THE KIEL CANAL

The size can be judged by comparison with the two tramp steamers visible. Note the huge bridge

Plate IV

THE RHINE GORGE

The view shows the point at which the river enters its gorge. The town is Bingen, at the inflow of the Nahe into the Rhine. On the opposite bank rise the vine-clad slopes of the Niederwald

(*Mondiale*)

HUNGARIAN PEASANTS

Note the dress of the man and woman, and the long horns of the cattle

a good system of inland waterways. Furthermore, the deflection of the Elbe towards the North Sea has given a northwestern outlook to most of the plain. The soils are poor and the drainage

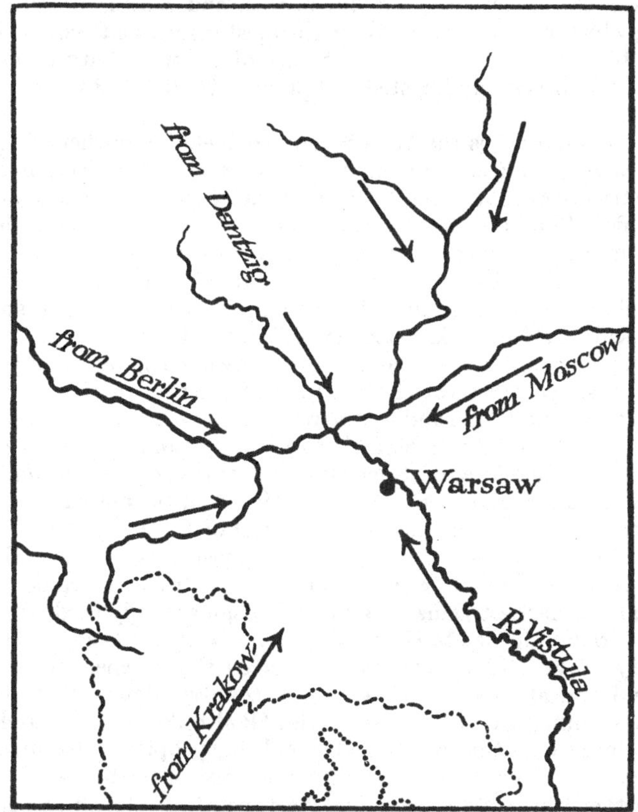

Position of Warsaw. The town is at the focus of the natural routes indicated by the converging rivers

naturally bad, but care and hard work in modern times have enabled much of the area to be used for agriculture. Rye, oats, sugar-beet, and potatoes are the chief crops, the last being used for making alcohol and sausages. Population centres round the coastal inlets and the 'bays' where the rivers from South Germany debouch on to the plain—Dresden, Breslau, Leipzig, Magdeburg, and Hanover in the south, and Stettin, Danzig, Königsberg, and Riga on the coast. Exceptions are Warsaw, the focus of the basin of the middle Vistula, and Berlin, a modern town which has been found to be the most convenient centre for a united Germany. The Baltic States of Estonia, Latvia, and Lithuania are largely forested and produce timber, but they also export flax.

In the far west is the Westphalian coalfield, an outlier of the Franco-Belgian (see map on page 36), with its collection of industrial towns, of which Dortmund, Essen, and Düsseldorf are the chief. Since only a little iron is found nearby (at Siegen), the Lorraine deposits were used before the war of 1914–18, the loss of this source of supply being a severe blow to German industry. Düsseldorf, Elberfeld, and Barmen manufacture cotton and chemicals, Crefeld makes silk, while Essen is famous for its iron and steel goods. Cologne is an old bridge town, being placed at the best crossing of the Lower Rhine, and it is also the junction of the route down the Rhine and that across the Great European Plain.

The coastline of the plain is low, sandy, and fringed with lagoons. It therefore repels maritime enterprise, except at points where the larger rivers enter the sea. Hence, until comparatively modern times the coast has been an object of attack by the sailor folk of the more encouraging shores opposite. Even now the outlets of the North German Plain are through North Sea ports, though farther east Danzig, Riga, and Libau are fully used, since the overland journey to Hamburg and Antwerp is too great. The Treaty of Versailles (1920) re-established the independence of Poland and allotted to her a strip of country along which she might communicate with the sea. This 'Polish Corridor' is purely artificial and has the political defect of being peopled by Germans and of separating East Prussia from the rest of Germany.

Farther south, there have also been political difficulties over the position of Silesia, i.e. the basin of the Upper Oder, between Germany, Poland, and Czecho-slovakia. The coalfields make the area desirable, but it is geographically a unit, and no division can be wholly satisfactory. The present arrangement gives to Germany the industrial portion which is centred about Oppeln, but to Poland and Czecho-slovakia the mineral deposits on which the industry depends.

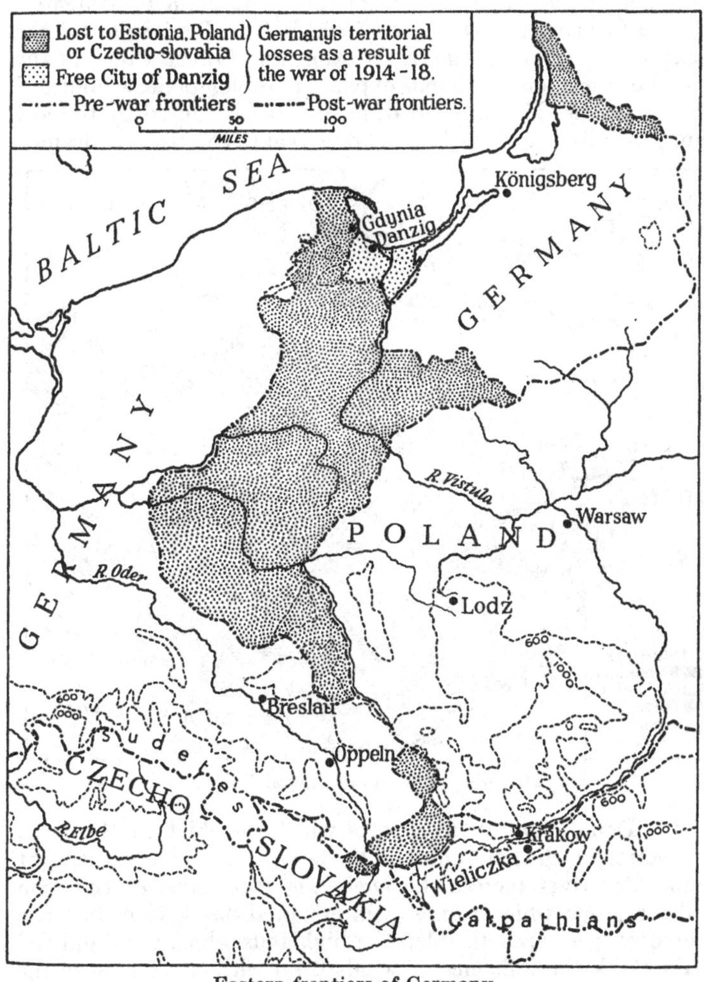

Eastern frontiers of Germany

Germany's considerable territorial losses are shown by stippling

South of the Alps-Carpathians line central Europe comprises five areas of different relief: (1) the great plain of Hungary, (2) the Carpathians, (3) part of the Balkan Peninsula, (4) the eastern valleys and spurs of the Alps, and (5) the valley of the Po. The first is a monotonous expanse of rolling plain surrounded by mountains. The only break in its relief is caused by the low range of hills known as the Bakony Wald. The plain is drained

Historical districts of central Europe

by the Danube and its feeders, of which the chief are the Tisza and March on the left bank and the Sava and Drava on the right bank. The river pierces the encircling mountains at the Iron Gate, a gorge which formerly interrupted navigation, but has now been made practicable. The highlands which surround the plain check the influence of winds from without and cause the climate to be dry, with extremes of temperature in winter and summer. Budapest has a mean temperature in January of 28° F. and in July of 70° F. and a rainfall total of 23 inches, of which

15 inches fall between April and October. The natural vegetation type is grassland, and the plain really forms an outlying portion of the steppes of south Russia. This fact is confirmed by the Asiatic origin of the Magyar population which occupies most of the area. The whole plain is given up to agriculture: rye, oats, and barley being grown on the poorer soils of the north, and wheat, maize, tobacco, and sugar-beet on the fertile lands farther south. These crops lead to flour-milling, sugar-refining, and distilling. The focus of the whole is the twin city of Budapest (pop. 1,006,184 in 1930), which occupies the geographical centre; but the corners of the plain have local centres at Debreczin and Zagreb.

The highland regions around the plain of Hungary were formerly split up into a number of political areas whose local names and customs still distinguish them (see map opposite). The Carpathians, whose main range bears the local names of Beskids in the north, the Forest Carpathians in the northeast, and the Transylvanian Alps in the south, slope sharply down on their outer side, but broaden out into a mass of broken highlands on the inner side of their curve. Here are found the Tatra, which contains the highest peaks in the whole system of ranges, and the Bihar Mountains. The highlands are forested, first with beech and then pines, up to 7000 feet, and the trees yield a large amount of timber. Mineral deposits are plentiful, but are very little worked.

On the west side of the plain the country resembles Austria, a fact which gives significance to the dispute over the possession of the Klagenfurt district of the Drava valley after the war of 1914–18. Main routes from Vienna and elsewhere cross the mountain ridge at either end of the Julian Alps and make for the Adriatic coast. Farther south, the Dinaric Alps are a region of barren karst with a block of uplands to the east of them. The valleys opening out on to the plain of Hungary are fertile and produce grapes which are made into prunes, but the rest of the country is pastoral and backward. A narrow complicated rift which carries the north-flowing Morava and the south-flowing Vardar leads from the Danube to the Ægean. The town of Belgrad stands at the north end and the port of Salonika at the south. A branch rift, forming the valley of the Nishava, leads past Sofia to the valley of the Maritsa and so to Istanbul.

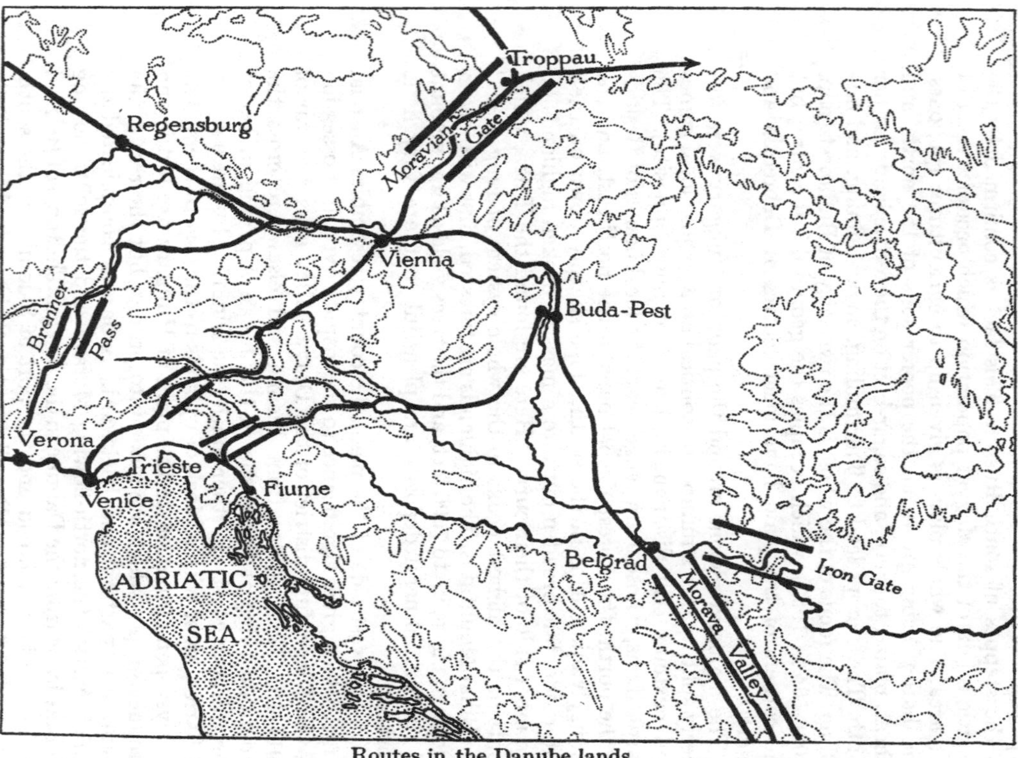

Routes in the Danube lands

The pairs of thick parallel lines show important gaps in the relief

The valley of the Po consists of the southern slopes of the Alps together with the plain of Lombardy. The south-facing Alpine slopes have a mild climate and their valleys are fertile, producing wine and olives. The scenery is beautiful around Lakes Maggiore, Garda, and Como and attracts many tourists every year. Most of the plain of Lombardy is alluvial, the lower end being fringed

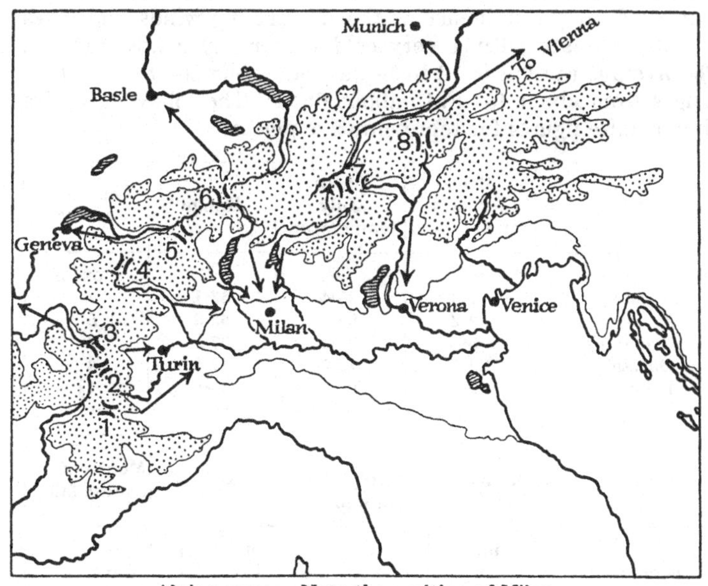

Alpine passes. Note the position of Milan

1. Mont Genèvre
2. Mont Cénis
3. Little St Bernard
4. Great St Bernard
5. Simplon
6. St Gotthard
7. Reschen
8. Brenner

with swamps and lagoons. Owing to the height of the surrounding mountains the climate is little influenced by the winds and bears all the features of that of central Europe. There is, however, no lack of water from the mountains, whose torrents supply hydro-electricity for the growing industries of Turin and Milan. The plain is agricultural as a whole, wheat, maize, and rice being the chief crops. The mulberry thrives here and gives rise to the production of raw silk which used to be exported to France, but is now increasingly manufactured locally. The Alpine passes cause the railways from the north to converge on Turin and Milan, thus adding to the importance of those towns. A gap in the Apennines gives Milan an outlet through Genoa, though the natural port of the Po valley is the old city of Venice.

The Continental Region

As we have seen, its distance from the Atlantic removes eastern Europe from the influence of the moderating winds and gives it a dry climate with a very cold winter and a hot summer. The average rainfall is 16 inches a year, while the temperature ranges from about 15° F. to about 65° F. The following tables give definite figures for representative places:

Place	Mean January temperature	Mean July temperature	Range	Length of winter
	° F.	° F.	° F.	months
Archangel	7·3	60·4	53·1	9
Moscow	12·2	66·0	53·8	5
Orenburg	3·4	70·9	67·5	6
Odessa	25·3	72·7	47·4	4
Kiev	20·8	66·6	45·8	4½

Place	Mean January rainfall	Mean July rainfall	Total	Month of greatest fall
	in.	in.	in.	
Archangel	0·8	2·2	15·3	July
Moscow	1·1	2·8	21·0	August
Orenburg	1·1	1·7	15·2	June
Odessa	0·9	2·1	16·1	June
Astrakhan	0·5	0·5	5·9	June

It will be seen, as might be expected from the great size of the region, that there is a wide variation in the climate, and five climatic subregions may be distinguished: (1) The polar north, where winter lasts for ten months and is broken by a short summer of six or eight weeks during which the temperature is only just warm. (2) The centre and north of Sweden, Finland, and north Russia, where winter lasts between four and six months and is followed by a warm summer during which most of the year's rain falls. This and the polar subregions have no spring or autumn, the change from winter to summer and from summer to winter being abrupt. Icy winds sweep across the plains in winter, giving short periods of the most intense cold. (3) East Poland, west Russia, and Little Russia feel to some extent the influence of the southwesterlies from the Atlantic and have a more equable climate than the other parts, but the annual range of temperature is still 43° F. and the rainfall shows a marked summer maximum. (4) Romania, the northern strip of Bulgaria, and a wide belt of south Russia come under the influence of the Black Sea. Furthermore, the effect of lower latitude makes itself felt. Hence, though the winters are still cold owing to the bitter north winds which prevail in that season, they are somewhat shorter, while the summers are extremely hot. The rainfall is low and has its maximum at the beginning of summer. The plain of Hungary, which for convenience of treatment was described under the head of the transitional region, properly belongs here. (5) Eastwards, this Pontic subregion shades off into the Caspian climate, where the winters last five months and are very cold, while the summers are blazing hot. The scanty rainfall makes this area into a semi-desert. See map on page 74.

Generally speaking, the whole region is one vast plain, but it may be divided into five subregions according to variations in structure and relief: (1) The Baltic platform or shield (see map on page 75) is a much eroded plain of very old rock which has been partly drowned by the Gulf of Bothnia and Finland. It bears many points of resemblance to the Canadian Shield, the Gulf of Bothnia corresponding to Hudson Bay. Its surface has been planed down and smoothed by the Tertiary ice-cap, which, however, has laid down rock waste across the drainage lines in Finland, thus causing a large number of lakes, waterfalls, and rapids in the courses of the streams. (2) Between Finland and the Urals is the Arctic plain, a low-lying area in which the ground slopes northwards to the Arctic and White Seas. The Dvina and Pechora rivers are large, sluggish streams, but are useless for navigation, since their lower courses are frozen over for nine or ten months in the year. The ice dams which block their mouths in spring flood the country around and convert it into marsh. (3) Farther south, is the great plain of central Russia, which includes the central plateau and extends thence eastwards across the Eastern Heights to the Urals. The highest portion of the Central Plateau lies in the so-called Valdai Hills northwest of

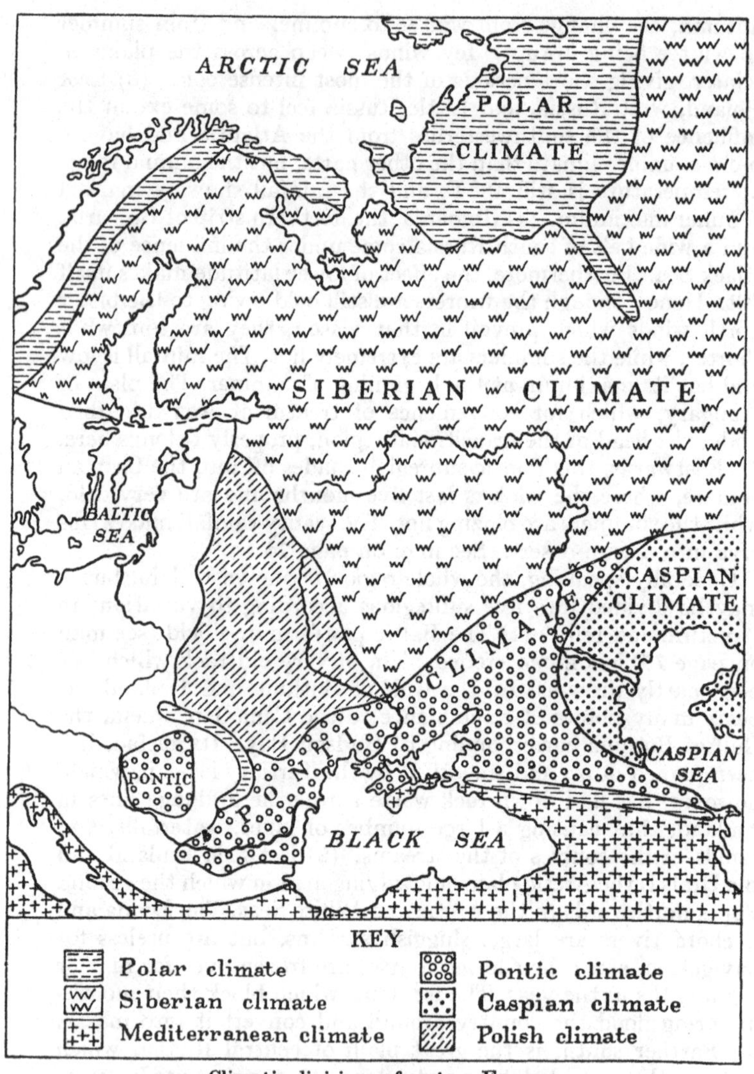

Climatic divisions of eastern Europe

Moscow and rises to 1200 feet above sea-level; but the gradients leading up to it are so gentle that it does not interrupt the general

Relief divisions of eastern Europe

flatness of the region. The same is true of the Eastern Heights and the Urals. The Central Plateau is important because the

Volga, Don, Dnieper, and Duna, the chief rivers of the region, take their rise in it and flow off in all directions. Since the streams rise near each other and have slight gradients, their use as routes has caused the plateau to be a natural focus of movement. It is to this fact that Moscow owes its growth and its status as the true capital of Russia. In modern times its focal position has been improved by canals which join the several rivers and by railways which converge from all directions. (4) To the west of the plain of central Russia is an undulating area consisting of some outliers of the Central Plateau in the north and the wide marshes of the Pripet in the south. The latter district is of little human importance owing to its bad natural drainage. (5) The Black Sea plain, which stretches from the Carpathians to the Caspian, is far from being uniform in relief and is cut into two parts by the Donetz plateau. This upland district rises to 1400 feet and is a mass of old, mineral-bearing rock of quite different formation to that of the surrounding lowlands. Southeast of it, the ground is very low, a portion of the plain actually lying some 80 feet below sea-level at the north end of the Caspian Sea, while another part has been drowned by the Sea of Azov. At one time there was communication between these two seas through the depression now occupied by the River Manich, but the fall in level of the Caspian owing to a decrease in rainfall over most of its basin has interrupted the connexion. Southwards, the depression gradually rises to meet the spurs of the Caucasus. To the west of the Donetz plateau the Pontic depression is a low coast strip which finds its greatest expanse in Walachia. In Podolia and Bessarabia upland shelves lie against the Carpathians and form an area of broken surface whose most valuable parts are the valleys of the southern Bug and Dniester.

The coasts of the region are all marked by drowned valleys, though the type varies considerably. The Baltic coast is rocky and fringed with islands, and the inlets usually form good, though often small, harbours. The Arctic shores are low and uninviting. The river mouths are drowned and even a portion of the plain has been covered to form the White Sea. The Black Sea coast is also low, and its estuaries are blocked by sandbanks and shoals formed mostly by the sediment brought down by the Danube. The great river itself has a delta at whose head is the little port of Galatz, while the Dnieper has its port at Odessa, not at its mouth, but some miles to the east.

The rivers of the region are of great importance. In Finland and Sweden, where the gradients are steep and broken, they float timber to the sea and provide hydro-electricity. Elsewhere, they are means of communication, being navigable in summer and offering a smooth, hard, icy surface in winter. Large quantities of sturgeon (from whose roe caviar is made) are caught in the Volga. Unfortunately, this great river empties into an inland sea, and the Danube, Dnieper, and Don into the Black Sea which, for political reasons, is nearly as isolated. Frost, which binds the water in the soil, and a winter rainfall minimum cause the volume of the rivers to be least in winter. The release of melt-water in spring is followed by floods, since all the streams are frozen over at their mouths; and the volume is maintained by the summer rains.

Differences of climate cause differences in the natural vegetation. Along the Arctic coast, where the soil is frozen for ten months in the year, ordinary trees cannot grow and plant life consists of dwarf birches, willows, and larches, of mosses and lichens, and of a number of flowering annuals. Man finds life difficult in such tundra country, but a few Lapps and Samoyedes manage to exist with the help of their reindeer. The only settlements are on the mouths of rivers, and the Russians have established railway connexion from Moscow to Archangel and Murmansk with the object of using these places as summer ports. The rest of the region, except the Black Sea plain, is forest clad. In Sweden, Finland, and north Russia the coniferous forests still hold their own, though clearings have been made for the cultivation of flax. Most human occupations here are of the forest type and are chiefly lumbering and fur hunting. Farther south, the trees are leaf-shedders, the birch being the predominant species; but large clearings have been made and the forest now survives in patches, the one occupying the marshy tract between the Bug and the Dniester being the most extensive. The chief crops are flax, rye, barley, oats, and potatoes. A coalfield around Tula has given rise to textile manufactures, though the output is of low grade and expensive, owing to the workmen's lack of skill. Leningrad and, to a much less extent, Riga are the outlets for this area.

The forests of the region have played a great part in the historical development of the Russian nation, since they formed

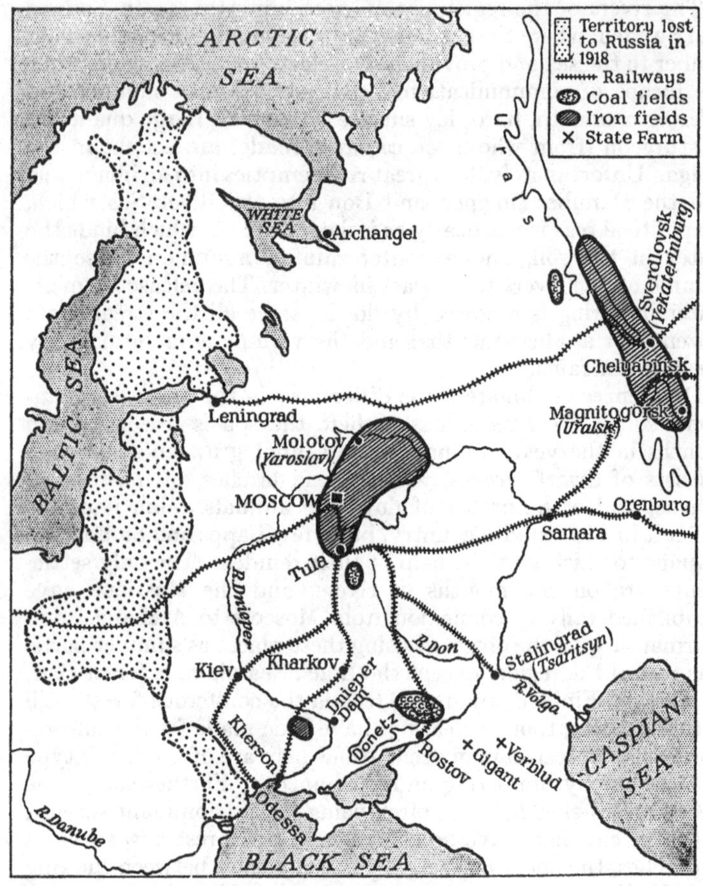

The new Russia

an impenetrable refuge in former times against the incursions of the mounted invaders who swept across from the plains of Asia. These attacks were successful in the south, where the wide grasslands of the Black Sea plain are continuous with the steppes of

Asia. Much of the Ukraine is covered with rich, black soil which yields good grain crops, and before 1914 was one of the great wheat-producing areas of the world. Kiev and Kharkov were the chief centres, and Odessa was the outlet. The Russian Revolution destroyed this trade, which however has survived in Wallachia, where the grain is shipped from Constantsa. East of the Donetz plateau the steppes quickly become arid and saline and are patchily covered with poor grass and scrub. The Caucasus and Urals are forest clad.

The population of the region is mostly rural, industrialism, which causes the growth of large towns, being but poorly developed. Stockholm (pop. 500,000 in 1932), the capital of Sweden, is a beautiful city built on the islands and shores of the so-called Lake Mälar, an inlet which provides a good harbour at the focal point of central Sweden. Leningrad (pop. 2,780,000 in 1930), first built by Peter the Great and till the Revolution the capital of Russia, is now the chief port of that country, but like Stockholm and most of the Baltic ports it is frozen up in winter. The inland towns in Russia are local agricultural centres or else route foci, and they are not as large as one might expect. Bucharest, the capital of Romania, is the natural focus of Walachia.

The railway systems are wide-meshed and undeveloped. In Sweden a coastal line connects Stockholm with the iron mines of Gellivare in the north. In Russia Moscow is the chief centre, but secondary foci are at Leningrad, Samara, Orenburg, and Nijni Novgorod (now Grodno). The gauge used is larger than that of the neighbouring countries, but the system joins those of Germany and Poland, so that overland journeys are possible from Berlin via Königsberg to Leningrad or via Warsaw to Moscow. From either of these towns it is possible to travel to Transcaspia or to China by the long-distance trains running into Asia.

The people of eastern Europe are of several races. The majority are Slavs and are closely related to the Alpine stock which inhabits the uplands of central and western Europe. They occupy nearly the whole of Russia and are divided into three branches: the Little Russians of the Ukraine (capital Kiev); the White Russians (capital Minsk) occupying the Pripet marshes; and the Great Russians centring round Moscow. The Swedes are Nordics, as are also the numerous German immigrants who are found in

Hungary and throughout Russia, but especially in the Donetz district. Patches of Asiatics occur in Finland, Lapland, the Baltic States, and Hungary, while the Bulgars and the Kalmuck and Khirghiz tribes of the Caspian have the same origin. Large numbers of Jews are scattered throughout Russia and Poland, where their thrift and unrelenting pushfulness has made them unpopular; and in Walachia there are distinct traces, including the use of a neo-Latin language, of the Roman colonists who were planted in the country two thousand years ago. The Poles are of Alpine race and for some reason are more cultured than their eastern and northern neighbours.

The Mediterranean Region

Between latitudes 30° and 45° on the west coast of every continent there occurs a transition belt between the temperate climate and the tropical desert, a belt which lies in winter within the path of the westerly winds, but in summer is beset by desert conditions. Geographers have given it the type-name of Mediterranean region, after the best known example which surrounds the Mediterranean Sea. But no two individuals of the regional type are exactly alike, and the Mediterranean region of Europe has several peculiarities which are due to the existence of a great continental sea penetrating two thousand miles into the land. The corresponding regions in North and South America, being not only without an inlet to carry the influence of the sea far inland, but even having a lofty mountain range parallel to the coast, are limited to a comparatively narrow coast strip. But other effects arise from the nature of the Mediterranean Sea.

The Mediterranean Sea is divided into eastern and western basins by a submarine ridge across the Strait of Tunis, while branching peninsulas and large islands mark off a number of secondary marginal seas, of which the Black, Ægean, Adriatic, and Tyrrhenian are the chief. Such a structure prevents active circulation of the water and produces a kind of stagnation which is increased by the fewness of large inflowing rivers and the presence of a sill some 200 fathoms deep across the Straits of Gibraltar. Hence, the waters of the sea below the 200-fathom contour, having a mean temperature of about 55° F., form a vast storehouse of warmth which keeps the winters mild along the coast. The tendency of mountains to run parallel to the shore restricts that influence to a narrow coast strip, except in Spain, the south of France, peninsular Italy, and Greece.

Except where large valleys reach the sea, the coasts are rocky and barren, and natural features like Gibraltar and Monte Pellegrino near Palermo are not uncommon. The rivers, most of which are of the *wadi* type, have wide stony beds that scar the landscape in summer. In winter the dark green vegetation makes a fine contrast with the red or brown rocks, but, when the heat of summer has scorched the grass and shrubs and a film of dust turns the leaves grey, the view is drab and lifeless.

The Mediterranean portion of the Iberian Peninsula includes the south of Portugal and the south and east of Spain below the scarp of the Meseta. The richest part is the valley of the Guadalquivir, whose triangular rift is enclosed between the dark, oak-clad slopes of the Sierra Morena and the snow-topped peaks of the Sierra Nevada. The summer rains of the Meseta, the winter rains of the valley itself, and the melt-water of the Sierra Nevada keep a fairly constant volume in the river and enable irrigation to bring out the full wealth of the soil. The lower end is swampy, and the old Phœnician port of Cádiz is built partly on a rocky peninsula some way from the river mouth. Seville, a river port and the chief focus of the valley, is clear of the swamps and can be reached by small ships. Córdoba, once the seat of the Moorish caliphate, is the focus of the upper valley. Rice, wheat, and maize are cultivated, but the characteristic products are the fruits grown in the tributary valleys which pierce the Sierra Nevada. The valley of the Genil is famous for the wine named *sherry* after the centre of the industry at Jérez. Farther north-east is the famous Vega of Granada, whose orange orchards surround the romantic town of Granada and the fortress-palace of the Alhambra. The coast strip is narrow elsewhere, and the uplands are barren; but the lowlands connected with Murcia and Valencia are, together with the Balearic Islands, pleasant fruit-growing areas whose crops are improved by irrigation. The most important town, however, is Barcelona (pop. 990,000 in 1931), where industries of wool and cotton cloth have sprung up and where sardine and tunny fishing have centred for some time past. The headquarters of Catalan culture, it nurses political separation on the grounds that its industries are made to pay for the relative idleness of the rest of Spain.

Mediterranean France consists of the alluvial lands along the coast from the frontier of Spain to Marseille and the Riviera as far as Mentone (see map on page 25). It runs up the Rhone valley as far as Valence and up the Aude to Carcassonne. The coast is fringed with lagoons and does not encourage seafaring, except at the eastern and western ends of the silt-formed area. West of the Rhone is Languedoc, which produces a large quantity of low-grade wine, while Provence on the east bank is a silk-producing country. The drought of summer, which parches the vegetation, forces the people to lead their flocks and herds up the slopes of the Pyrenees, Cévennes, and Alps. This practice of *transhumance* is found throughout the Mediterranean region and especially in Spain, where special sheep trains are run in early summer from the plains of Castile to the Pyrenees. Mediterranean France, Rome's earliest province, preserves more marks of Latin civilisation than any other place outside Italy. Nîmes contains a well-preserved temple, an amphitheatre, and a cemetery, and the old ship basins of the port of Arles can still be seen, though they are now completely silted up and the town itself is twenty miles from the sea. Marseille (pop. 800,000 in 1931) has had a revival since the opening of the Suez Canal and is now one of the world's greatest ports. Through it passes most of France's trade with her north African and Indo-Chinese possessions. Farther east, the Riviera is a popular international health and pleasure resort on account of its mild climate and beautiful scenery.

Peninsular Italy consists of the backbone of the Apennines and a narrow coast strip on either side. It is a hilly and infertile country except in its western valleys, where the Arno, Tiber, and Volturno have cut deeply into the mountains and reach the sea through small plains. Florence is the inland focus of the first and has its port at Leghorn; Rome, the capital of Italy because of its central position in the peninsula, is the focus of the Tiber basin and originally owed its growth to its defensive site on its seven hills near the limit of boat navigation on the river; and Naples, on its famous bay, is at once an important town and a great port. Sicily contains a number of valleys whose wealth has in times past caused the island to be regarded as a prize to be won or has enabled it to become the headquarters of a state. Here Palermo is the chief town. Corsica and Sardinia are block uplands whose rough, infertile surfaces keep the islands in a backward condition.

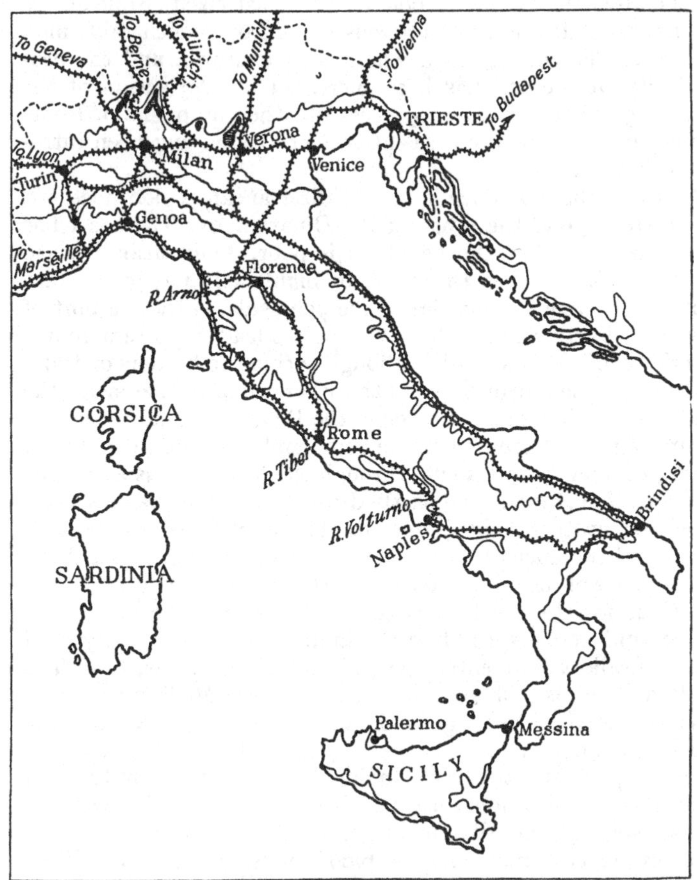

Italy: routes and focal points

The concentration of the inhabitants in more or less isolated valleys formerly caused the peninsula to be divided into a number of petty states, and it is only within the last eighty years that Italy has become a political unit. The hilly backbone forces the main routes on to the coast, a fact which makes Italy's communications vulnerable in time of war. This weakness is increased by the absence of good harbours on the east coast, while across the narrow Adriatic are numerous bases from which raids may be made. The country as a whole is agricultural, and the productivity of the soil has been increased by the action of the present government, swampy areas like the Campagna of Rome having been drained and scientific methods having been introduced into farming.

In the Balkan Peninsula the Mediterranean region does not penetrate beyond the range of the Dinaric Alps. The coast has many harbours but the backland is poor. Only near Scutari, where a rift breaches the chain of highlands, and gives communication with the interior of the peninsula, is there a port of any size. Farther north, however, passes lead important routes to Trieste and Fiume, which, though within the borders of Italy, are yet Yugoslavian outlets. At the other end of the peninsula the Gulf of Corinth cuts right through the Pindus Mountains, nearly separating the Morea from the mainland. A canal capable of taking coastal steamers cuts through the low isthmus and leads to the Gulf of Ægina. The partly drowned fold valleys of Greece, separated by high ridges, were the cradle of Greek civilisation of old and the ancient city states. Central position and a slight convergence of the inlets have made Athens and its port at Piræus the focus for the south of Greece. Sheltered water backed by fertile land and a sprinkling of innumerable islands early bred a race of sailors of an enterprising, if not adventurous, type. The modern Greek is still the sailor of the eastern Mediterranean.

To the north of the Ægean the valleys are larger, those of the Vardar, Struma, and Maritsa ending in small plains. By running back deeply into the Macedonian block, they introduce an element of civilisation into this backward area. On the uplands, especially in Albania, the peasants still wear strange, if picturesque, costumes, carry on blood feuds, engage in banditry, and take no part in the life of the world at large. They are mostly engaged in pastoral pursuits, the village being the social unit. The absence of good roads has in the past precluded good government, and to-day there is little sign of improvement. Salonika at the southern end of the Vardar-Morava rift is an important seaport.

The double straits of the Dardanelles and Bosporus, the latter only nineteen miles long and three-quarters of a mile wide at its narrowest point, form an important highway between the Black Sea and Mediterranean. Its importance is increased by the crossing of the overland route from Asia to Europe, and history records many struggles for the possession of this strategic, military, and economic position. The city of Istanbul (formerly

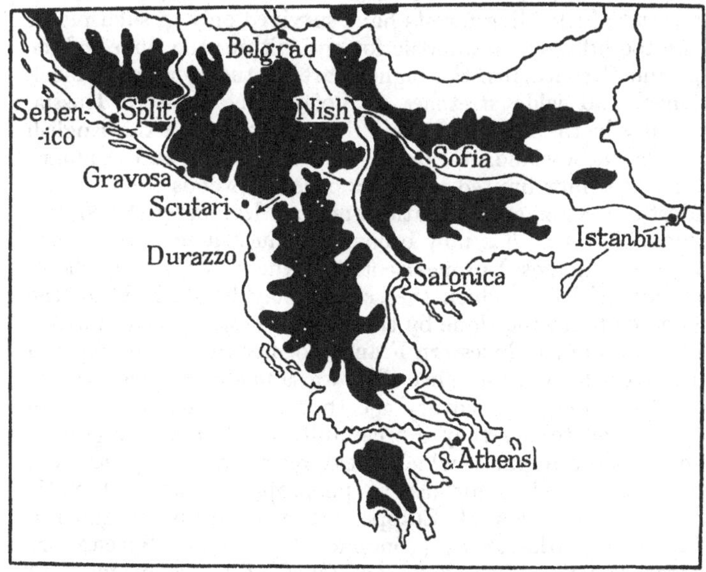

The Balkan routes

Constantinople) occupies a little peninsula between the Golden Horn and the Sea of Marmara and thus controls these routes.

MAN'S USE OF NATURE

Forest Products. Nature supplies man with certain materials, like plants and animals, which are renewable and others, like minerals, which are unrenewable. Man's control of nature in Europe is demonstrated by his mastery of the vegetable and mineral world. In the early Middle Ages the whole continent outside the Mediterranean region and the natural grasslands was covered with vast expanses of forest, but gradually the trees have been replaced by crops and to-day the forest areas have been greatly reduced. In the north, where the summers are too short for crops to ripen, the pines and firs still hold sway and spread through northern Russia, Finland, Sweden, and Norway, though even here large areas have been cleared. Farther south, the vast leaf-shedding forests have survived only in such places as are too hilly or too infertile for agriculture: the slopes of the Alps and Carpathians, the higher parts of the Hercynian block uplands, and wide stretches of Poland and western Russia, especially in the region of the Pripet marshes. In the English plain and the lowlands of France, the Netherlands, and Denmark the forest is now limited to copses or large 'woods'.

Most of the last are plantations more or less under the control of man. Forestry has now become as much a science as agriculture, and in most European countries there are training schools and universities in which the science may be studied. In the past much harm was done by the reckless destruction of forests. The Landes of southwestern France, for instance, were exposed to the overwhelming advance of blown sand and became a desert of shifting dunes. A century ago they were reclaimed by Bremontier, and to-day they are carefully planted with regularly spaced southern pines which yield resin until they are old enough to be felled for use as pit-props. Under systematic care the woodlands of Europe are treated as ornamental 'wilds' and parks or as economic plantations. Species are selected and improved, trees are felled at the right moment and are replaced by seedlings so as to maintain the area under wood.

The leaf-shedders yield hardwood which is used for making furniture, implements, handles for tools, etc. The most commonly used are oak, beech, walnut, elm, and ash. The willow which grows near running water bears long, pliable twigs which are plaited into baskets and wickerwork. The conifers on the other hand give softwood which is used as timber for the construction of houses, window and door frames, posts, and the cheaper forms of woodwork. Lumbering as an occupation exists in areas of coniferous forest (see map on page 16), the chief

European timber countries being Russia, Finland, Sweden, and Norway.

Timber production in 1931

Country	Acres under forest (millions)	Value of timber exported (million £)	Value of pulp and paper exported (million £)
Russia	405,000	15,560	—
Sweden	53,740	9,540	13,650
Finland	62,400	7,365	9,140
Norway	18,532	1,080	5,560

Inferior trees, which were formerly useless except as firewood, are now reduced to matchwood or pulp and cellulose from which newsprint, artificial silk, linoleum, and a growing number and variety of other articles are made. Young birches which are too small to be sawn into boards are planed spirally into thin sheets of plywood, which is being increasingly used for light, strong construction. In the forests of Bohemia, Thuringia, and south Germany generally, the practice of making wooden toys and musical instruments has long existed, having come into being as a means of whiling away the long hours of winter. Hereditary skill thus acquired has facilitated an extension of the industry to metal articles of the same kinds.

Forest products include some of animal origin. Hunting and trapping still survive in the coniferous forests of Russia, whence are sent the pelts of bears, martens, ermines, and foxes to supply the demands of the fur markets of the world. The beech and oak forests farther south provide mast and acorns for feeding pigs, and these animals have become the chief part of the meat diet in Germany, where pork is eaten mostly in the form of sausages. In the arid west of Spain the cork oak, which is cultivated primarily for the cork which is cut from its thick bark, feeds numerous pigs which form the only wealth of the poorer folk of the province of Estremadura. In most countries, however, the rearing of pigs has outgrown the forest industry stage, and the animals are kept in sties and fed on maize, potatoes, and specially prepared foods. The chief pig-keeping countries were, in 1931:

	Million pigs
Germany	22·815
Russia	13·300
France	6·398
Poland	5·841
Spain	5·102
Denmark	4·886

Pastoral Products. The pasture lands of Europe are either natural areas of grass or else parts of the forests which have been cleared. The former include the steppes of south Russia and Hungary, the summer pastures above the tree line in the Alps, and the dry chalk or limestone areas in England and France. The pasture animals of Europe are seven in number: the sheep, cattle, horse, goat, mule, ass, and reindeer. The last is restricted to the tundra and is of no general importance. The sheep, cow, and horse are found everywhere, except in the habitat of the reindeer, but the ass and mule thrive best in the southern countries and are not found in considerable numbers away from the Mediterranean. Spain, with 1,175,000 mules and 1,004,000 asses in 1931, and Italy, with 455,000 mules and 870,000 asses in the same year, have the greatest number of these animals. The chief pasture animal in Europe is the sheep, which, originally a mountain animal, has never become thoroughly at home on damp lowlands, but thrives best on grassy uplands like the Pennine moors, the southern uplands of Scotland, or the Cévennes, and is most numerous on the dry pastures of the Mediterranean coasts. It is bred either for wool or for mutton or for both. The following table shows the chief sheep-rearing countries in 1931:

	Million sheep
Russia	84·900
Great Britain	26·412
Spain	20·046
Romania	12·356
Italy	10·043
France	9·845

The goat, which is a hardy animal and can feed on dry leaves and twigs, replaces the sheep in more barren areas, especially in the Mediterranean *garrigues*, though it is found as far north as Finland. Russia, Greece, and Spain have the greatest number of these animals, which yield hair, skins, and milk, and are usually the property of the poor. Sheep are often replaced on the lowlands by cattle, the second in importance among pasture animals. These animals are bred for dairy purposes, meat, or draught. Dairying is carried on chiefly in the British Isles, Denmark, Czecho-slovakia, and Switzerland. In the last-named country milk preserved in a condensed form and cheese of fine quality are exported. In the other countries dairy-farming is combined with the keeping of poultry and pigs, the latter being fed on whey. Large quantities of butter, cheese, bacon, and eggs are thus produced. Great Britain does not, however, produce enough to meet her own demands, and further supplies are imported from Ireland and Denmark. The cheapness of beef produced on the great grass plains of America and Australia confines the European production to the high-grade article obtained from the slaughter of male cattle which are required neither for stud nor for draught purposes. The chief cattle-rearing countries in 1931 were:

	Million cattle
Russia	53·000
Germany	19·122
France	15·434
Poland	9·457
Great Britain	7·591
Italy	7·022
Czecho-slovakia	4·500

Horse-breeding has decreased since the introduction of the motor vehicle, but the animal is still used for draught purposes on farms, for sport, and for military needs. Russia, with 32,800,000 head, is by far the greatest producer, followed by Poland (3,938,000 head), Germany (3,393,000 head), France (2,920,000 head), Romania (1,988,000 head), Yugoslavia (1,169,000 head), and Great Britain (1,067,000 head). A great many horses were formerly bred on the grass plains of Hungary and are still reared on the Russian steppes, where the Cossacks and Tatars have scarcely yet settled down to a sedentary life. But improvements in the implements and methods of agriculture have turned or are turning the plains into corn-lands, since these crops give far greater returns than herding does.

Agricultural Products. The spread of agriculture not only to the grass plains, but also to areas formerly regarded as uncultivable, is due to: (1) the greater demands for food owing to increased population; (2) a deeper chemical and biological knowledge of plant life, which has led to the use of manures and the lightening of heavy, clayey soils by mixture with sand and ash; (3) improved methods of tilling, among which are the use of a rotation of crops and the breeding of plants to secure greater yields or better health; (4) more efficient ploughs and other implements; and (5) the perfection of drainage and irrigation systems.

The crops produced in Europe fall into four classes: (*a*) cereals, (*b*) other food crops, (*c*) fruit, and (*d*) raw materials for industry. The cereals are wheat, barley, oats, rye; maize, millet, and rice. The last three are confined to the south, since they need greater warmth than can be had in northern Europe. Rice is cultivated chiefly in the swampy alluvial plain of the lower Po basin, but small quantities are also grown in Spain in the similar country around the mouth of the Guadalquivir and on the coast strip from Barcelona to Valencia, and in the neighbourhood of Salonika in Greece. In 1932 the total area under rice in Italy was 333,000 acres, from which were produced three million bushels of grain. Millet, which is a hardy plant and thrives where other cereals will die, is grown chiefly in south Russia, the area devoted to it amounting to close on ten million acres. Maize stands cold better than either rice or millet and flourishes in the Plain of Hungary and in south Russia as well as in the Mediterranean coast lands. It is interesting to note that it is used as food for poultry in the Plain of Hungary and that this region exports annually a vast number of eggs. The following table shows the area under the grain in 1932 in the countries where it is most cultivated:

	Million acres
Romania	11·75
Russia	8·00
Yugoslavia	6·00
Hungary	5·00
Italy	3·25

Wheat thrives best between latitudes 45° and 55° N., but it is successfully cultivated north of that belt in Russia, the Baltic States, and Sweden; and south of it in Spain and Italy. These extraneous areas, however, need a special variety of wheat to suit local conditions; thus, in southern Italy a hard wheat is sown, while farther north a softer variety is used. In western Europe, and especially in Great Britain and France, the grain is made into white bread; but in central and eastern Europe, where black bread is eaten, wheat is not a popular crop and is replaced by rye. In the famous black earth region of south Russia, however, where the soil is particularly suitable for wheat, the grain is grown in large quantities for export. Before the Revolution of 1917, this district was one of the granaries of the world. The following table indicates the chief wheat-growing countries in Europe. It will be seen from the last column that the English and German farmers get the greatest yield from their soil, while the Russians and Romanians obtain least.

Country	Area under wheat (acres)	Yield in 1932 (qrs.)	Yield per acre (qrs.)
Russia	93,150,000	102,000,000	1·08
France	13,389,000	41,000,000	3·06
Italy	12,236,887	34,000,000	2·80
Germany	5,700,610	23,000,000	4·03
Spain	11,240,113	23,000,000	2·04
Romania	8,562,049	8,000,000	0·90
Yugoslavia	5,287,047	7,000,000	1·30
Hungary	4,011,001	7,000,000	1·70
Czecho-slovakia	2,059,576	7,000,000	3·40
Great Britain	1,363,883	6,000,000	4·40
Poland	4,265,482	6,000,000	1·40
Bulgaria	3,078,224	6,000,000	1·90

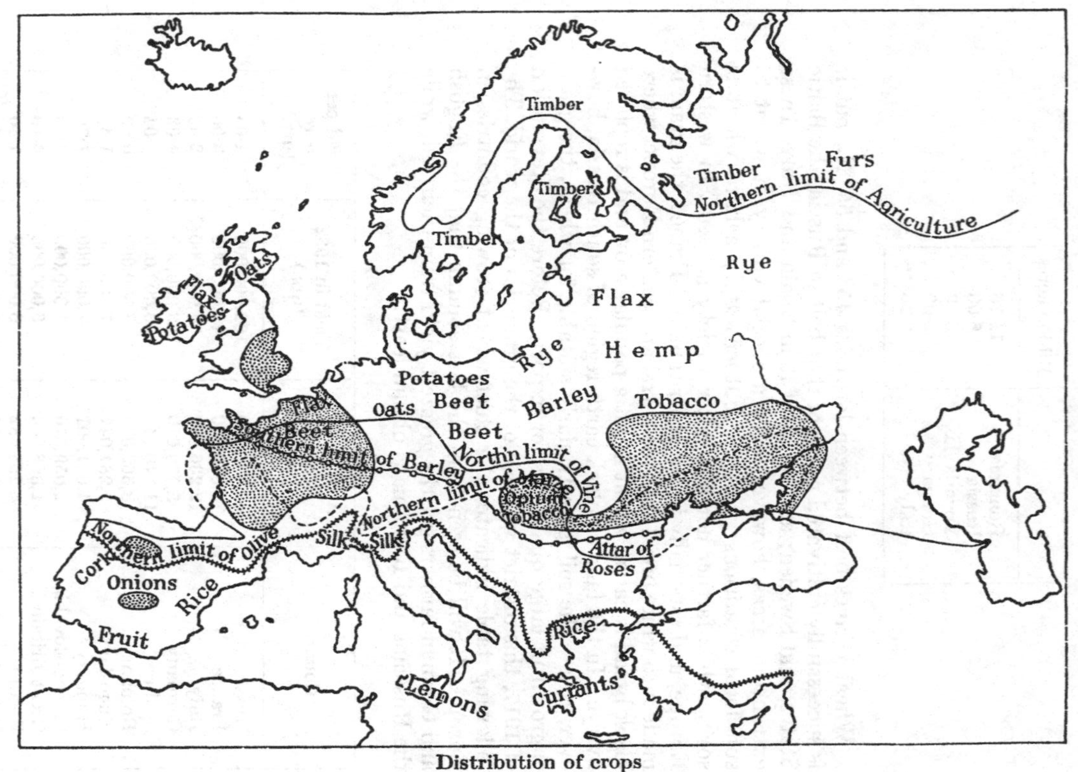

Distribution of crops

Barley will grow farther north than wheat, but not so far south. It covers wide areas in the British Isles, Russia, Germany, Spain, Romania, Poland, Czecho-slovakia, France, and Denmark, where it is used mostly for making beer and alcohol. Oats and rye do not require a great deal of sunshine and will grow on poorer soil than either wheat or barley. The countries of central and eastern Europe are the largest rye producers in the world, while oats form the chief crop in Scotland and Saxony and are grown to some extent in Spain and in most European countries away from the Mediterranean.

Among other food crops the potato is the most important. Introduced from America some three centuries ago, it was slow in being accepted in Europe; but now it forms, next to wheat and rye, the commonest of vegetable foods. It does not flourish in the Mediterranean region or in the extreme north, but in Ireland and Germany it has become the chief crop, forming in the latter country one of the ingredients of the national sausage. Beet, from which white sugar is made, is a principal crop in Germany, Russia, France, and Czecho-slovakia, and it is increasing in Poland and England. Other root crops are turnips, carrots, onions, and swedes, which with beans are usually grown in market gardens around large towns, though onions are one of the chief field crops in Spain. In the Mediterranean region, where the warm climate deprives meat of its fat, the place of animal fats in the diet is taken by olive oil, and the olive tree is therefore widely cultivated. The habit of pouring quantities of the oil over one's food, however disagreeable to a northern European, is necessary in the Mediterranean lands.

Fruit crops are of two kinds: (1) the hard fruit of the north, and (2) the soft fruit of the south. The chief southern kinds are the orange, lemon, peach, apricot, cherry, fig, pomegranate, and, above all, the grape. The orchards of the Granadine valleys, Valencia, and Murcia in southern Spain are especially noted for the bitter, or Seville, orange from which marmalade is made. Oranges and other rind fruit ripen in winter along the Mediterranean. Sicily specialises in lemons, though Spain also produces a great many. Greece cultivates sultana raisins and the small seedless grape known as currants (after Corinth, the centre of their production). The grape, however, is usually devoted to wine-making. In Greece the local wine is flavoured with resin and is drunk by the peasants only. Italy produces a good deal of wine, the acreage under vineyards in 1932 being 2½ millions; but unfortunately the same care is not taken in making the beverage as in other countries, and the liquor does not improve with keeping. Special kinds are Marsala, which is made in Sicily, and Chianti, which comes from Tuscany. Spain and Portugal are famous for sherry and port respectively, while France produces many well-known brands from the vineyards of Champagne, the Côte d'Or, Languedoc, and the Bordeaux district. The latter country is the world's chief producer, having 3½ million acres under grapes in 1932 and a vintage of over 500 million gallons. The slopes of the Moselle, Rhine, and Danube are also celebrated for their wine, the last producing the famous Tokay.

The northern fruits are apples, pears, and plums. Apples are chiefly grown in the sheltered valleys of England (Kent, Isle of Ely, Severn valley, Devon) and Normandy, where they are used for making cider and jam. The Shumadia district of Yugoslavia is the land of plums, but large quantities are also grown in Bulgaria, Czecho-slovakia, and Hungary, where they are dried and exported as prunes. Small fruit, chiefly strawberries and raspberries, are also cultivated for jam-making, especially in England.

Of the plants which supply raw materials for industry the most important are hemp, flax, and the mulberry. Hemp is used for making rope and coarse cloth and is grown chiefly in Germany, Spain, Czecho-slovakia, Russia, and Hungary. Flax supplies the raw fibre for linen, and its seeds yield linseed oil. It is cultivated in the same countries as hemp, with the addition of the Baltic States and Romania. The mulberry, which is typical of the Mediterranean, is grown for its leaves, on which silkworms are fed. The Po valley and southern France are the chief areas in which it is cultivated, but there are also large plantations in Spain, Bulgaria, Hungary, and Greece.

Other plants which come under this head are hops, the flower of which is used to give a bitter taste to beer and which are therefore cultivated near barley; tobacco, which, though it does not really thrive in Europe, yielding only an inferior leaf, is cultivated chiefly in Russia and Italy. and also to a less extent in Germany, Spain, Czecho-slovakia, Romania, Bulgaria, Hungary, and Greece; roses, from which a concentrated perfume, or attar, is extracted, in Bulgaria; and the poppy, which is cultivated for its opium, in Hungary.

Mineral Products. The only unrenewable products in Europe are minerals. Man's progress has enabled him to renew some gifts of nature which were formerly unrenewable. Thus, the fur-bearing animals of the north, which are gradually being exterminated in their wild state, are now being kept on farms in England. The fish in the North Sea and elsewhere are not inexhaustible, and among the measures taken to conserve the supply is an experiment in Lijm Fjord in Denmark, where fish are bred in tanks. Of course, oysters have for years been grown in artificial beds.

Europe was never very rich in precious metals, and most of the gold and silver mines are now worked out. But the former metal is still found in Transylvania and the southern Urals as well as in many places in smaller quantities. In 1932 the Russian mines yielded 1¾ million ounces, or nearly one-twelfth of the world's total output. The chief silver mines are at Mansfeld and Freiburg in Saxony and at Sala in Sweden. The useful metals, which man learnt to handle at a far later date, are still plentiful. This is especially true of iron, which is found all over the continent wherever old rocks occur. It is never found pure, but always mixed with carbon and rock. The mines at Dannemora, Gellivare, and Kiruna in Sweden, together with those at Bilbao in Spain and near Barrow-in-Furness in England, yield the purest ore, but iron of high quality is mined in Lorraine, Styria, Elba, the Urals, south Russia, Romania, and Hungary.

Copper is obtained chiefly from the Harz Mountains, Falun in Sweden, the Banat of Romania, and in the Isker valley in Bulgaria; but most of the European supplies are now imported from Chile and Canada. Quicksilver is mainly found at Almaden in Spain and at Idria in the Istria peninsula at the head of the Adriatic. The tin mines for which Cornwall was once famous have now been practically worked out, and supplies are imported from Malaya, Bolivia, and the Dutch East Indies. Lead is got at Sala in Sweden and in the Banat, while platinum is one of the many metals of the southern Urals.

Other minerals of importance are building stone, sulphur, salts, coal, and oil. Among the best known kinds of stone are the

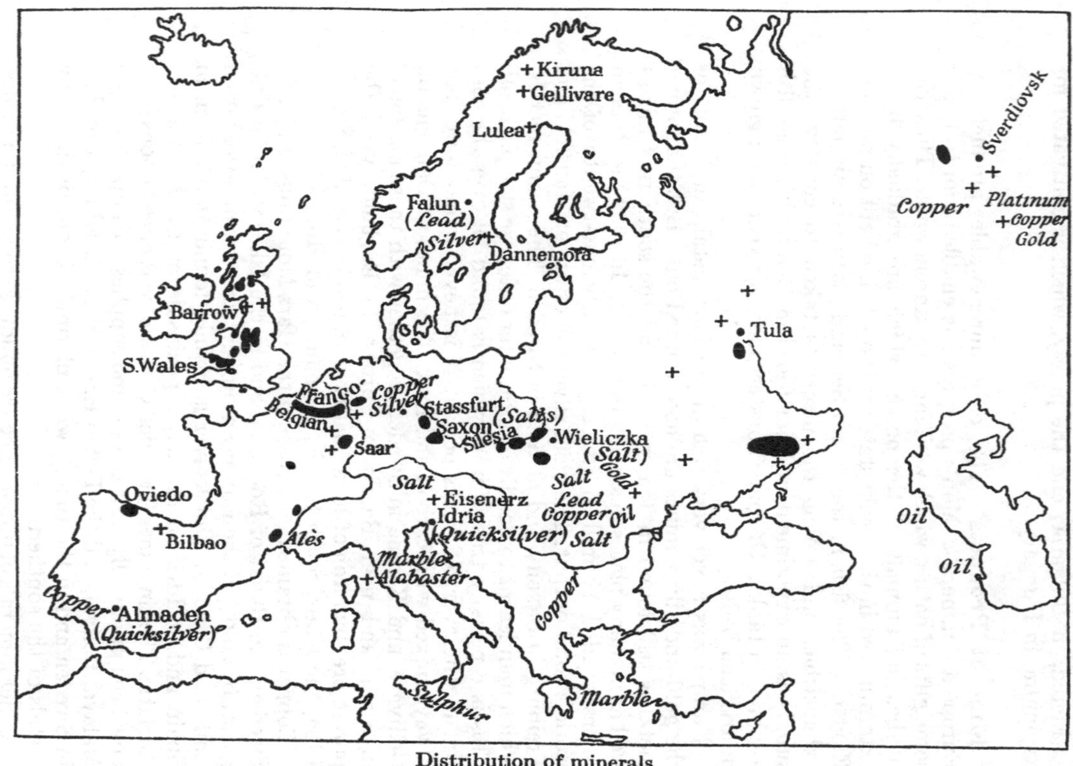

Distribution of minerals

DRAMMEN, NORWAY

The old town lies at the head of a long fjord. The view looks up the ice-carved valley

A TIMBER MILL AT JESSHEIM, NORWAY

Plate VI

(*Canadian Pacific Railway*)

BENARES

Bathing at Mani Kainika Ghat, the most sacred on the Ganges

Portland stone quarried in Devonshire, slate in Wales, granite in Cornwall, red granite in the Urals, and so on. Pure white marble is got from Paros in the Ægean, and alabaster has long been quarried near Florence. Some minerals are used for chemical purposes: thus, sulphur, which is found in Sicily and at Cerna in Romania; and various kinds of salts. The latter are mined at Stassfurt in Germany, in Salzburg, in Austria, at Wieliczka near Krakow, in Transylvania, at Slanic in Romania, and in Cheshire. More important are the fuels, oil and coal. These, unlike most minerals, are found in newer rocks. Oil occurs chiefly in Caucasia and in the districts of Campina and Buchtenai in the Carpathians. The plentiful supply in Caucasia has led the Russians to replace coal by oil, since coal is not plentiful in eastern Europe.

Coal is the basis of modern industry, since it is used for driving the engines of factories, trains, and steamships, and the most densely peopled areas are now on coalfields. The chief coalfields are in the Midlands and north of England, in South Wales, the Midland plain of Scotland, Belgium, the Ruhr, the Saar basin, Bohemia, Silesia, and Pecs in Hungary, while smaller ones are found in Schönen in Sweden, near Oviedo in Spain, around Tula and in the Donetz district of south Russia, and in the Banat.

Nowadays, coal is being ousted by hydro-electricity. Mountainous countries like Italy, Switzerland, Norway, Sweden, and France are rapidly developing this new driving force. It has great advantages over coal, for besides being clean and free from soot, it can be distributed cheaply and easily over a wide area, thus avoiding the need for crowding factory workers together in gigantic towns where they must be less healthy than in the country.

Industry. Europe is the most industrialised of the continents. Since the Industrial Revolution, which began in England towards the end of the eighteenth century, all the larger towns have established manufactures of various kinds. But the advantages of abundant fuel offered by the great coalfields have caused the growth during the past hundred years of industrial areas in which sprawling, untidy towns have sprung up with such ill-defined boundaries that the name 'conurbation' has been proposed for indistinguishable groups like Burslem, Hanley, and Stoke or the Enneperstrasse of the Ruhr district.

There are eleven important industrial areas in Europe based on coalfields, namely, (1) the Midlands and north of England, (2) the Midland Plain of Scotland, (3) the Franco-Belgian, stretching from Béthune in north France through Belgium to the Ruhr in Germany, (4) Saxony, (5) Czecho-slovakia, (6) Silesia, (7) the Lyon district in France, (8) the Lodz district in Poland, (9) the Tula district in Russia, (10) the Donetz area in Russia, and (11) Styria and Salzburg in Austria. Two others depend on hydro-electricity, viz. the upper Po basin and the Aar valley in Switzerland. Besides these, there are a number of areas of less importance scattered all over the continent. Thus, in Spain there are the Barcelona and Oviedo-Bilbao districts; in Holland there are the famous potteries at Delft, textile manufactures at Utrecht and the working up of colonial produce at Amsterdam and Rotterdam; around Krakow and Lwow in Poland there is much industry based on the salt mines at Wieliczka and the oil fields at the foot of the Carpathians; in Sweden there are the production of pig-iron and match-making; and in France Rouen is fast developing manufactures based on raw material from the New World. Even in the more backward countries the larger towns, at any rate the capitals, engage to some extent in industry, and the whole continent is becoming industrialised. This means that a large proportion of the population live in towns and work at some form of productive industry or transport, depending for food on supplies imported from the new countries overseas. The attractions of town life are many, and a district engaged in industry can carry a far larger population per square mile than one engaged in agriculture or pastoral pursuits. Thus, a population map of Europe indicates fairly accurately the distribution of industry. But the crisis in world trade which began in 1931 has shown how unwise it is for a nation to depend for vital supplies on foreign countries.

To become important, an industrial area needs driving force—coal or electricity—near at hand; it must be able to get raw material easily; it must have good means of transport and ready markets; and it requires an intelligent population with adequate numbers of skilled workmen. The coalfields of Europe are only fully worked in the west, since this part of the continent can

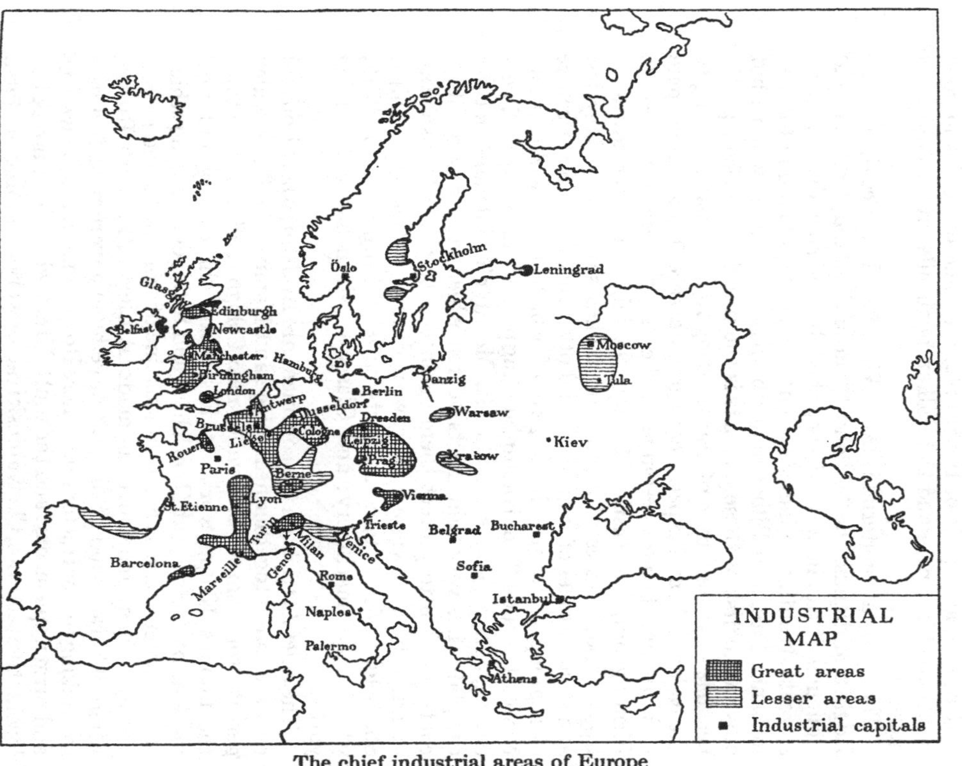

The chief industrial areas of Europe

easily draw supplies of raw material from abroad and export its manufactures. But to some extent it is due to the industrious and progressive spirit of the people, for some countries in good positions, like Spain and the Balkan peninsula, are backward in industry.

The chief raw material is iron. When found near coal, as in the English Midlands, the Midland Plain of Scotland, Brabant, and Bohemia, it can be worked very cheaply, and such districts have a great advantage in industry. The purest ore is found in Furness in Lancashire, near Bilbao in northern Spain, and in Sweden; but only in the first of these districts is the iron fully worked, partly because of the expense of transporting fuel and partly owing to the lack of skill and experience in manufacture. The mines in Spain and Sweden being accessible to the sea, the ore is shipped to Birmingham, Newcastle, Merthyr Tydfil, and other towns in England and Wales. Birmingham makes small arms, bicycles, and small articles of iron, steel, and other metals. It is the centre of a group of towns, the largest of which is Wolverhampton. Newcastle-upon-Tyne is a shipbuilding port with satellites at Tynemouth, Jarrow, and South Shields. Merthyr Tydfil manufactures tin plate, i.e. very thin sheets of iron which are covered with a coating of tin to protect them against rust. Other British iron-manufacturing towns are Barrow-in-Furness, which builds ships; Sheffield, which is famous for its cutlery; Stirling and Falkirk, which make machinery; and Glasgow, which has shipbuilding yards.

The chief districts on the continent are Brabant, where iron and coal are found together near Liége and Namur; the Ruhr, which gets its iron from the Sieg valley near Cologne and which contains the famous Krupp works at Essen and the cutlery and steel implement factories at Ramscheid and Solinger and along which the Enneperstrasse is a continuous line of iron works and forges; Saxony, where machinery is made at Chemnitz; south Russia, where Kharkov and Stalingrad have huge government factories for making motor tractors and where Rostov manufactures wagons and farm implements; and the south Urals, where Sverdlovsk has locomotive factories, Chelyabinsk tractor works, and Magnitogorsk forges for turning out heavy machinery. Other districts of less importance are Prag in Czecho-slovakia; Eisenerz in Austria; Nancy and le Creuzot in France, and Leningrad and Molovsk in Russia.

One aspect of the iron industry is shipbuilding. The chief yards are at Newcastle-upon-Tyne, Glasgow, and Belfast in the British Isles; Havre, Bordeaux, and Marseille in France; Genoa in Italy; and Stettin and Hamburg in Germany.

Next in importance are the textiles. Raw cotton is imported from tropical or subtropical regions and can therefore be manufactured cheaply at places moderately close to the sea. The Lancashire district, with a population long experienced in manufacture, is still the chief world-centre of the industry. In spite of the trade depression, the value of the cotton goods exported from Great Britain in 1932 was £32 million. Most of the goods were made in Lancashire or in Paisley. Manchester is the great cotton market and is surrounded by a number of typical industrial towns, of which Bolton, Preston, Blackburn, and Wigan are the chief. Its great port is Liverpool, which conveniently faces west towards the cotton plantations of the southern United States; but a ship canal allows ocean-going vessels to reach the great city itself.

On the continent Ghent is the centre of cotton manufacture in Flanders, but the Ruhr district is more important. Düsseldorf is the chief cotton town, but Barmen and Elberfeld have many factories. Bremen, which is the cotton market of Germany, is connected by a canal with Dortmund; and Duisburg is a large river port on the Rhine. The chief German cotton area, however, is the Chemnitz-Zwickau district in Saxony. Under the Soviet régime efforts are being made to extend the industry in the Tula district to the south of Moscow, the raw material being derived from Transcaspia. Lodz in Poland, Prag and Pilsen in Czecho-slovakia, Barcelona in Spain, and Vienna in Austria all manufacture cotton, as do also most of the large towns in eastern Europe, such as Moscow, Leningrad, Warsaw, and Riga.

The raw material of woollen manufacture is a product of the temperate belt, and local suitability for sheep-rearing has fostered the industry in districts like Yorkshire, Brabant, and Upper Silesia. But the enormous growth of the demand for wool has greatly exceeded the local supplies, and large imports are derived from Australia, New Zealand, and South Africa. The chief woollen-manufacturing districts in the British Isles are the Yorkshire area, of which Leeds is the centre and Bradford and Halifax important towns, and the south of Scotland, both at

Hawick in the Tweed valley and at Paisley and Glasgow in the Midland Plain. On the continent Verviers in Belgium and Aachen in Germany are important centres, the industry here having grown out of the sheep pastures of the Ardennes. The industrial area of Saxony also turns out a good deal of woollen cloth, especially at Zwickau, but the most important woollen-manufacturing district outside the British Isles is in Upper Silesia and along the northern foot of the Carpathians, whose slopes form good natural pasture for sheep. Görlitz, Liegnitz, and Troppau in Silesia and Krakow and Lwow in Poland are all centres of the industry. The manufacture extends through Moravia, where the centres are Brno and Iglau, into Bohemia, where the chief town is Reichenberg.

Silk is manufactured chiefly in the south, since its raw material is largely derived from the Mediterranean region. The Po valley and the slopes of Dauphiny are the largest producers of raw silk. The chief manufacturing district in France is the neighbourhood of Lyon and St Étienne, the latter town specialising in ribbons. Much of the raw silk used there is imported from the French colonies in the East and enters through Marseille. The recent efforts to develop industry in the plain of Lombardy have included silk manufacture, and there are factories at Milan and Como. Besides these, there are smaller centres at Zürich and Basle in Switzerland, at Budapest in Hungary, at Vienna, and at Krefeld in the Ruhr.

Certain industries are tied to particular districts by the occurrence there of some necessary mineral. Thus, salts mined in Cheshire have given rise to the manufacture of chemicals and dyes at Warrington and St Helens. Similarly, at Stassfurt and Leipzig in Saxony local deposits of pure rock salt and potash salts have led to glass-making and many industries involving the use of chemicals, such as dyeing and the preparation of pigments, fertilisers, explosives and photographic material. Chemicals are also made at Barmen and Elberfeld in the Ruhr and in the Belgian towns of Verviers and Liége, which also make glass. At one time the best glass was produced in Bohemia, but English glass-makers are now unsurpassed. Pottery and earthenware are made where suitable clay is found. The best china is made of kaolin, the English supply of which is found in Cornwall, whence it is shipped to Worcester, Derby, and Stoke-on-Trent for manufacture. These towns also produce ordinary earthenware from local clay. Among the best known potteries in the world are those at Delft in the Netherlands, Sèvres in France, and Dresden in Saxony. The centre of the Sèvres porcelain-making is now at St Cloud outside Paris, and that of Dresden china at Meissen, about twelve miles from Dresden. Brick-making is carried on in all clay plains, since building stone is necessarily unobtainable locally and expensive to transport. The largest works are near Bedford in England and around Ghent in Belgium.

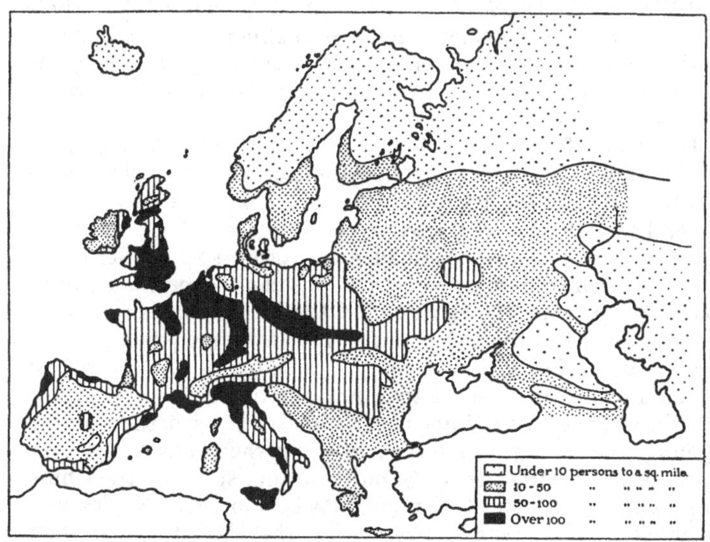

Distribution of population

Other industries depend on certain plants and are carried on where these are grown. Thus, the potato crop of Germany, Poland, the Baltic States, and western Russia supplies raw material for distilleries of alcohol, while the barley of England, Germany, Czecho-slovakia, and Austria is used for making beer. The brands of Münich and Pilsen are famous. A common and easily grown crop in central Europe is beet, which yields most of the white sugar sold in Europe. Then there is the soap which is manufactured in Italy from the local olive oil. Nürnberg and other towns in the forested parts of south Germany are noted for their toys and wooden musical instruments.

An interesting type of manufacture is known as 'Swiss'. In Switzerland and other countries where little work can be done during the long winter, the people take to amusing themselves with making articles needing great skill and patience to manufacture. Such, for instance, are the watches and clocks of the Jura, the gloves of Valais and the Tirol, the embroidery of Basle and Berne, and the lace of the Swiss canton of St Gallen. In Paris and Vienna, the two leading centres of fashions in clothes, there is a similar industry which consists of making novelties in dress, leather bags, silk-covered pincushions, and a hundred and one other articles which can only be seen in the shop windows of those two cities.

The stress of modern life is so great that there is an increasing need for holiday resorts and other places of recreation. This gives rise to a regular tourist trade. In some cases the resorts have mineral springs and are visited for health cures. Thus, Bath and Harrogate in England, Wiesbaden and other spas in Germany. Others are winter resorts where invalids can escape the rigorous cold. Of this type the Riviera and the Mediterranean coast in general is probably the best example. Then there are purely holiday resorts, as, for instance, the many seaside towns in the south of England or on the west coast of France. Switzerland is largely given up to the entertainment of tourists who visit the country to enjoy the mountain scenery, to climb the peaks, or to take part in winter sports. Other favourite beauty spots are the Rhine gorge and the fjords of Norway. Paris, Florence, Naples and other cities of Italy and France use the attractions of their ancient buildings, their museums, theatres, etc. to allure numbers of visitors. This development of travelling gives rise to many hotels and boarding houses and allows swarms of people to earn a living as guides, porters, drivers, and the like.

Commerce. The high industrial development of Europe results in a large exchange of products with other continents and between the various parts within her own borders. Europe exports manufactured goods of every description, but especially cotton and woollen cloth, chemicals, machinery, tools, and metal utensils and implements. The cheapness of mass-production methods and machine-made articles has enabled the European trader to undersell the hand-worker of calico and brass in India and to induce the natives of Africa and South America to exchange primitive digging-sticks for the hoe and spade. India imports annually from Europe cotton goods to the value of £10 million, machinery to the value of £15½ million, and metal instruments and apparatus to the value of £3 million. In 1932, the total value of the imports into India of European manufactures was £68½ million. Similar figures could be given for all parts of the world outside Europe, except the United States and Japan, which now have a fully developed industrial system of their own.

On the other hand, Europe imports vast quantities of cereals, meat, and fish from temperate lands overseas, and raw materials and foods of various sorts from tropical countries. These imports can best be shown by a table (p. 106) which gives a list of the chief kinds of goods and the countries from which they are mainly got.

A great deal of this overseas commerce is indirect, that is, the goods exchanged do not travel straight from the country of origin to the districts in which they are to be used. The central position of the ports of the British Isles, Belgium, and the Netherlands with respect to the distribution of land on the earth's surface has caused these ports to develop what is known as an entrepôt trade. Thus, the wool exported from Australia and South Africa and intended for the markets of Germany, France, and the United States is shipped first to London and thence re-exported to its final destination. Similarly, a good deal of the cotton imported into Liverpool and Manchester from the U.S.A., India, and Egypt is passed on to the various cotton-manufacturing districts of the continent. On the other hand, European exports often go through London; e.g. French silks which are sent on to Australia. In fact, it has been estimated that one-fifth of the exports into Britain are intended for re-shipment. Antwerp, Rotterdam, and Amsterdam perform similar functions, though to a less extent.

There is also an enormous trade between the various countries and districts of Europe. The agricultural areas exchange their products for those of the manufacturing towns. Sometimes whole countries, like Denmark, Bulgaria, Romania, Yugoslavia, and Hungary, are almost purely agricultural, whilst others have large

Class of goods	Sub-class	Kind	Countries of origin
1. Foodstuffs	(*a*) Cereals	Wheat	Canada, India, Argentine, Australia, North Africa
		Maize	United States, Argentine
		Rice	India, Indo-China
	(*b*) Tropical foods and drinks	Tea	India, Ceylon, China, East Indies
		Cocoa	West Africa, Brazil, West Indies, Venezuela
		Sugar	West Indies, India, East Indies
		Spices	East Indies
	(*c*) Temperate foods	Meat	United States, Australia, New Zealand, Argentine, Brazil
		Fish	Newfoundland, Canada, United States
2. Raw materials		Rubber	Malaya, East Indies
		Cotton	United States, India, Egypt
		Wool	Australia, Argentine, South Africa, New Zealand
		Silk	Japan, China
		Cabinet woods	West Indies, Honduras, Burma, West Africa
		Tobacco	United States, India, Nyasa, West Indies
		Jute	India
3. Minerals	(*a*) Fuel	Petroleum	United States, Venezuela, Persia, Mexico, Iraq, Burma, Ecuador, Peru
	(*b*) Metals	Gold	South Africa, Canada, Australia, United States
		Tin	Malaya, Bolivia
		Copper	United States, Chile, Canada, Japan
		Nickel	Canada
		Lead	United States, Australia, Mexico, Canada
	(*c*) Fertilisers	Nitrate	Chile

industrial areas and are not self-supporting in food; e.g. England, Belgium, Czecho-slovakia, France, and Germany. This leads to a natural exchange of products. An examination of the export tables of Hungary and Austria shows that the most important items exported from the former are livestock, cereals, and poultry; whilst the imports are machinery and textiles. In the Austrian tables the items are reversed, and one is not surprised to see that these two neighbouring countries trade very largely with each other, since they are in an economic sense complementary.

There is also an exchange of regional products. Thus, the timber and wood pulp of Sweden, Finland, Norway, Russia, and Poland are sent to the British Isles, Denmark, and the Netherlands. Mediterranean countries export fruit, flowers, wine, and silk to the more northerly lands; the countries with grass plains, i.e. Hungary, Romania, and south Russia, supply cereals and meat to their neighbours; and the coastal areas are the source of fish. Other trade depends on the distribution of minerals. Thus, one of the chief exports from Great Britain is coal, while iron bulks largely on the Swedish and Spanish lists. These natural impulses to trade expose the futility of all attempts to make any one country self-supporting and to throttle trade by tariff walls.

Commerce seems to group itself into five areas: (1) The Baltic, in which the primary products of the north (timber, pulp, iron) are exchanged for the foodstuffs and manufactured goods of the south. It should be noticed that Great Britain falls within this group, since she is the chief customer of Sweden and Finland. Denmark sends most of her dairy products to Germany, but also exports large quantities to Great Britain and Sweden. The Baltic trade is very old and dates from the time of the medieval Hansa towns. (2) The Danube lands, in which internal trade moves along the Danube and its feeders and consists of the exchange of cereals and meat for manufactured goods and minerals. Surplus produce finds its way (*a*) up the Danube to Switzerland, France, and the Rhine valley, (*b*) through the Moravian Gate into Poland and Germany, (*c*) down the Danube and through the Mediterranean to Great Britain, and (*d*) through the Alpine passes to Italy. This group includes Hungary, Austria, Yugoslavia, Romania, and parts of Czecho-slovakia and Bulgaria. (3) The Eastern, including Russia and Poland, is mostly agricul-

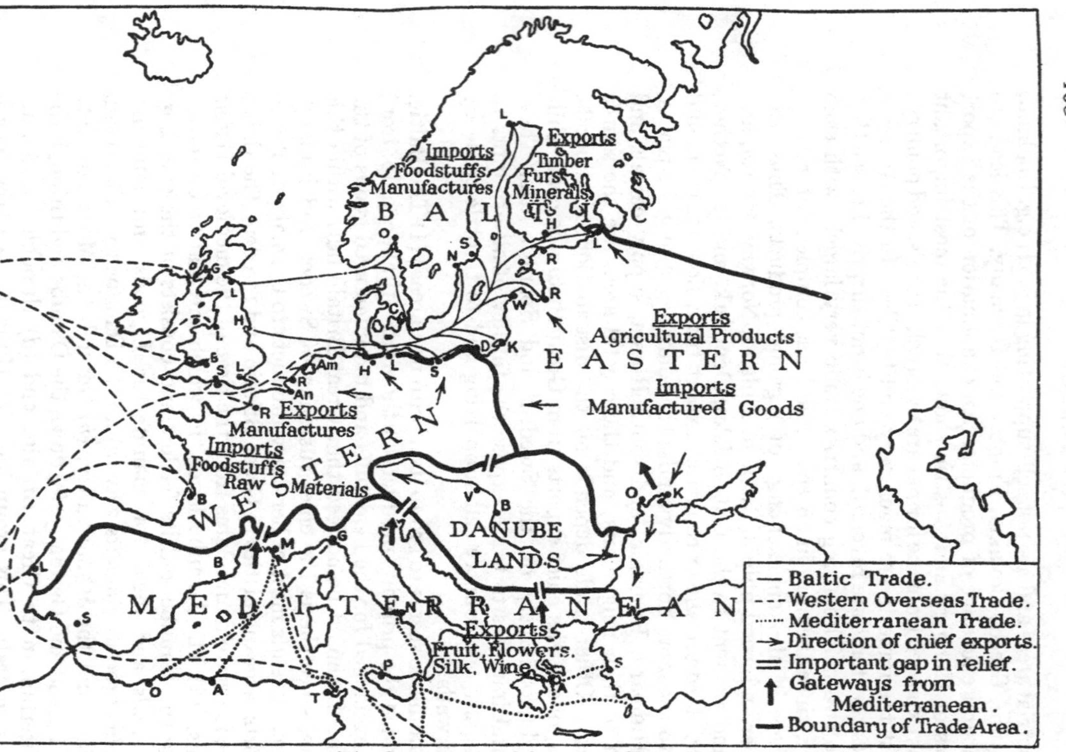

The five commercial areas of Europe

Thick arrows point to gaps in relief which concentrate the passage of trade; thin arrows show the direction of export trade. The most important trade centres and ports are marked by an initial letter. Reference to an atlas will show the full name in each case

tural and tends to exchange the products of the land for manufactured goods from neighbouring countries, viz. Germany and Czecho-slovakia; but Russia has a large trade with Great Britain and Poland with France. (4) The Western, in which the trade is largely overseas, either with the United States, or with colonial possessions, or with various other parts of the world. The chief countries included are Great Britain, France, the Netherlands, Belgium, Norway, Portugal, and Germany; but every part of Europe joins more or less in this western commerce. Thus, the United States is the second largest customer of Russia and Poland, the third largest customer of Sweden and Greece, and the chief customer of Italy. The entrepôt trade of Great Britain and the Netherlands plays a great part in the commerce of this group, and there is also considerable local exchange of goods between the countries of the western seaboard and those farther inland. Thus, Great Britain, Germany, and France are each other's chief customers. (5) The Mediterranean, in which there is considerable trade between the European shores on the one hand and the African and Asiatic coasts on the other. But a great deal of regional trade takes place between the Mediterranean region and the regions farther north. This is the oldest commerce in Europe and dates back to the days of the Phœnicians and Greeks. Traffic northwards from the Mediterranean passed through four main gates, which are historically of the utmost importance, since through them went, along with trade, the culture of the south.

The Nations of Europe. Europe is divided politically into five Great and twenty-two Lesser Powers, as shown in the table on page 110.

The peoples of these nations are not racially homogeneous, but are composed of mixtures of the Nordic, Alpine, and Mediterranean types. The first of these types centres round the western Baltic and the North Sea and is mainly represented in Sweden, Norway, Denmark, northwestern Germany, the Netherlands, Flanders, northwestern France, and England. The Mediterranean type occupies the shores of the great inland sea, while the upland belt running through the middle of the continent is the home of the Alpine type, the Slav branch of which spreads over most of Russia, Poland, and Yugoslavia. With these there are various Asiatic intrusions in the Balkan peninsula, Russia,

Hungary, the Baltic States, and Finland. Except in certain areas, racial purity is rare, the three types having mingled to a very great extent, especially along the main overland routes and in the centres of trade and culture.

Country	Capital	Area in sq. miles	Population in 1930	Kind of government
I. GREAT POWERS				
Great Britain	London	131,000	47,500,000	Kingdom
Germany	Berlin	182,000	63,000,000	Republic
France	Paris	213,000	40,000,000	Republic
Italy	Rome	120,000	42,000,000	Kingdom
Russia	Moscow	1,492,000	108,000,000	Union of Soviet Republics
II. LESSER POWERS				
Albania	Tirana	20,000	1,000,000	Kingdom
Austria	Vienna	31,000	6,000,000	Republic
Belgium	Brussels	11,000	7,000,000	Kingdom
Bulgaria	Sofia	40,000	5,000,000	Kingdom
Czecho-slovakia	Prag	54,000	14,000,000	Republic
Denmark	Copenhagen	56,000	4,000,000	Kingdom
Finland	Helsinki	150,000	3,000,000	Republic
Greece	Athens	49,000	7,000,000	Republic
Hungary	Budapest	36,000	8,000,000	Republic
Netherlands	The Hague	12,000	7,000,000	Kingdom
Norway	Öslo	124,000	3,000,000	Kingdom
Poland	Warsaw	150,000	27,000,000	Republic
Portugal	Lisbon	35,000	6,000,000	Republic
Romania	Bucharest	160,000	18,000,000	Kingdom
Spain	Madrid	196,000	21,000,000	Republic
Sweden	Stockholm	173,000	6,000,000	Kingdom
Switzerland	Berne	16,000	4,000,000	Republic
Turkey	Ankara (in Asia)	10,000	2,000,000	Republic
Yugoslavia	Belgrad	248,000	13,000,000	Kingdom
Baltic States:				
Estonia	Reval	18,000	1,000,000	Republic
Latvia	Riga	41,000	2,000,000	Republic
Lithuania	Kovno	20,000	2,000,000	Republic

The frontiers between the states vary in quality from those of Great Britain and Spain on the one hand to those of Belgium, Poland, and Russia on the other. The absence of good frontiers and the desire of nations for areas which are fertile in soil or rich in minerals has led to many wars in the past. Thus, the Middle Rhine valley and the iron and coal deposits of Lorraine and the Saar basin have long been a bone of contention between the Germans and French. The map on page 112 shows the readjustments which followed the victory of France in 1918. Germany also lost small areas in Holstein, Silesia, and East Prussia. The peace treaties which followed the Great War recognised differences of race and aimed at the creation of ethnographical boundaries where no definite natural frontier existed. Holstein, which had been wrested from Denmark in 1866, and which was inhabited by many Danes, was subjected to a popular vote, as a result of which a slice of territory was returned to Denmark. The same principle of deciding boundaries by popular vote was used in Silesia, and the district was divided between Germany, Poland, and Czecho-slovakia. The treaty of Versailles arbitrarily instituted the 'Polish Corridor', a strip of territory without a vestige of natural boundary leading from Poland to the Baltic near Danzig. This 'Corridor' had the fatal defect of passing through German territory.

Also as a result of the War of 1914–18, the old empire of Austro-Hungary was split up into the three independent republics of Austria, Hungary, and Czecho-Slovakia, while large slices of its territory were handed over to Romania, Yugoslavia, Italy, and Poland. The division was based on racial principles, except on the Italian frontier, where military strategical considerations led to areas inhabited by Austrians being handed over to Italy. The Klagenfurt district was left to the decision of a popular vote.

After the Russian Revolution, the Baltic States, Finland, and Poland became independent, and Bessarabia was included in Romania. These losses of territory to Russia are shown in the map on page 78. Readjustments have also been made in the frontiers of the Balkan States as a result of the Balkan War of 1912, the Great War of 1914–18, and negotiations since the latter date. The most important political factor here is Italy's fear of the now powerful Yugoslavia being left in possession of the naval ports at the head of the Adriatic. Hence, contrary to geographical principles, the peninsula of Istria has been given to Italy, though commercially Trieste and Fiume are Yugoslavian ports.

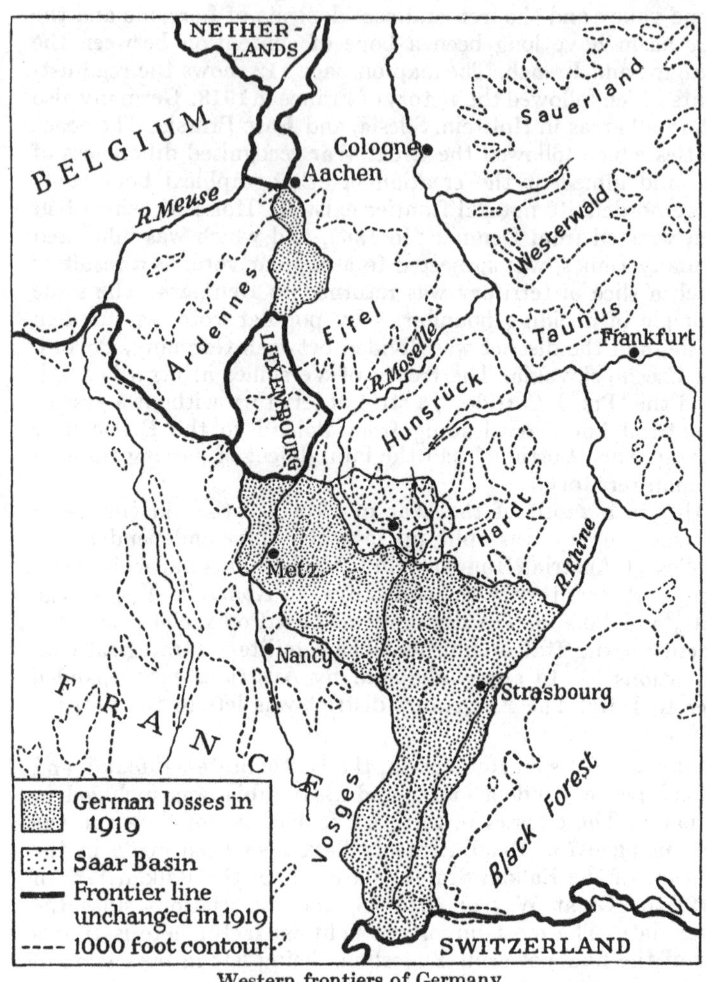

Western frontiers of Germany

As a result of the war of 1939–45 further changes are in progress. Russia has reabsorbed the Baltic States and Bessarabia, acquired part at least of East Prussia, and has moved her frontier further west in Poland. In compensation the Polish frontier has been advanced westwards to the line of the Oder at Neisse.

Highways in and out of Europe. Europe has six great overland highways which have been in use from early times and

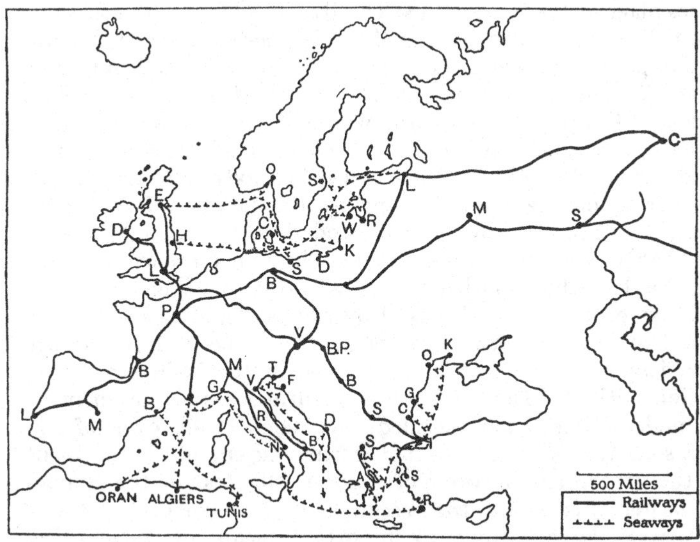

Railways and local seaways

have played an important part in the history of the continent: (1) The first leads from northwestern Europe to the Mediterranean through the natural corridor of the Rhone valley, reaching the sea at Marseille. It is now followed by a railway which, beginning at Marseille and passing through Lyon and Paris, reaches the Straits of Dover at Calais and Boulogne. Cross-channel steamers continue the route to Dover and Folkestone, whence railways lead to London and on to Edinburgh and Dublin. (2) The second also leads from northwestern Europe, but uses the Rhine and Danube valleys to reach Vienna, Budapest, and Belgrad. From this city it ascends the Morava and Nishava valleys to Sofia and thence passes down the Maritsa valley to Istanbul. (3) The third route follows the Great European Plain from the Straits of Dover through Brussels, Cologne, Hanover, Berlin, and Warsaw. At that point it forks, one branch going through East Prussia to Leningrad and so on across Siberia to Vladivostok. The other branch runs through Moscow to rejoin the first branch at Omsk. Russian trains have a gauge which is slightly larger than the standard, and a change of trains takes place at the frontier. (4) The Baltic-Adriatic route ascends the Oder, passes through the Moravian Gate, crosses the Hungarian plain, and reaches Trieste after threading the Semmering Pass. (5) The Iberian route starts from Paris and runs through Orleans, Bordeaux, and Irún, where it crosses the Pyrenees. Thence, it ascends the Meseta and at Medina sends one branch to Madrid and another to Lisbon. (6) The sixth route crosses the Alps from northwestern Europe by the Simplon Pass, under which the railway now tunnels, and descends to the plain of Lombardy, whence one branch goes through Florence to Rome and Naples, while another turns eastwards to Venice.

These routes are all followed by railways and roads, but the first two have waterways as well. The Seine and Rhone are joined by the Canal de Bourgogne, while the Main has a connexion with the Danube between Nürnberg and Regensburg. In modern times good metalled roads connect all towns of any size, so that it is possible to motor from one end of the continent to the other. The network of roads is slightest in the Balkan peninsula, Russia, and the Scandinavian peninsula.

Airways have become an important means of transporting passengers, mails, and light goods. Before the outbreak of war in 1939 all the principal towns in Europe were connected by regular services (see map on opposite page). These services have not yet been fully resumed, but will eventually follow the same routes. The British Overseas Airways Corporation has organised services between Great Britain and (1) Canada and the United States, (2) Brazil and the Argentine via Lisbon, (3) South Africa, and (4) through the Mediterranean to India and Australia.

Local European seaways run: (1) from the North Sea ports (Leith and Hull) to Scandinavia and the Baltic; (2) from Marseille to North Africa (Algiers, Tunis, Oran); (3) in the Adriatic from Venice and Trieste; (4) in the Black Sea from Odessa, Kherson, etc. through Istanbul. The Kiel Canal considerably shortens the Baltic route. It takes nine hours to pass through.

Great ocean routes carry the trade of Europe to distant lands. The most important are: (1) from northwestern Europe to Canada and the U.S.A.; (2) to India, Australia, and the Far East through the Suez Canal; (3) to the east coast of South America; (4) to the West Indies and, through the Panama Canal, to the west coast of North and South America and to New Zealand; (5) to South Africa and Australia.

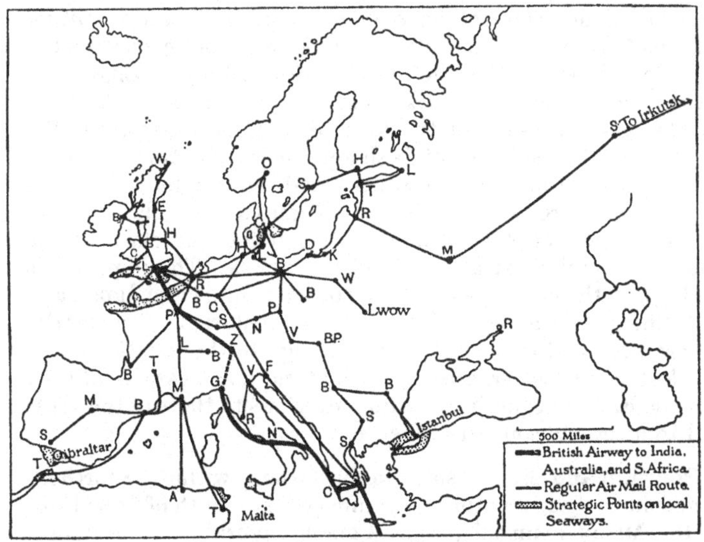

Airways and strategic points in 1939

Several of these routes run through straits whose shores are strategically important, since whoever holds them commands the passage. The Sound and the Great and Little Belts, which form the natural outlet from the Baltic, give Copenhagen its importance, and Istanbul is the key to the Bosporus. The Straits of Gibraltar are guarded by the famous rock fortress which has been held by the English since 1704, while much of the trade of London and Antwerp is due to the Straits of Dover.

PART II

ASIA

IN a sense Asia is the oldest of the continents, for it has been the cradle of the great civilisations, and the earliest chapters of the history of man were written within its borders. It is supposed that civilisation first sprang up on the banks of the Tigris and Euphrates and that culture was carried thence to India and China. Whether this be so or not, the fact remains that the three great civilisations of the Hindus, the Chinese, and the Semites have had their origin and have grown up in Asia. Nilotic culture was an offshoot of the Babylonian, and Greece but carried on the torch laid down by the Phœnicians. Rome, France, England—these three in succession took up and improved the work of the Greeks, and this shift of the focus of civilisation towards the northwest into the hands of fresh and vigorous peoples has kept it from stagnation. On the other hand, geographical isolation sapped the energy of Chinese and Hindu culture. The study of the geography of Asia to-day centres round the clash of the young, forceful West and the old, fixed East. The impact has to-day overthrown the whole social order of India and Japan and cast China into a state of chaos and revolution. Democratic government and industrialism are foreign to the Asiatic, whose intellectual outlook transcends the material and seeks the infinite. But the East is beginning to adjust its views to those of the West and to assert its counter-influence.

Position and Size. Asia lies almost wholly within the northern hemisphere. The Equator passes ninety miles south of Cape Busu in the Malay Peninsula, while Cape Chelyuskin, the northernmost point, reaches to the 78th parallel of north latitude. The 90th degree of longitude east almost divides the continent into halves. On the northeast Asia approaches within sight of Alaska, though the breadth of the Pacific separates it from North America farther south. The archipelago of the East Indies joins, rather than severs, it from Australia. In the west Africa is actually united to it by the Isthmus of Suez, while Europe is merely a peninsula of the great continent, from which it is divided by cultural and conventional, rather than physical, frontiers.

Asia is the largest of the continents, having a surface area of 17 million square miles, as compared with the 11½ million square miles of Africa and the 8 million square miles of North America. Its greatest length from Cape Chelyuskin to the southern point of Malaya is 5350 miles, while its greatest breadth, from Chosen to Asia Minor, is 6000 miles. It is very irregular in shape, having several large peninsulas which jut out towards the east, west, and south, while many groups of large and small islands festoon its coasts.

Build and Relief. The surface relief of Asia is best studied in three separate divisions: (1) the vast plain of the northwest, (2) the mass of highlands which run through the centre and occupy the east, and (3) the two crustal blocks of India and Arabia which are separated from the mountains by broad alluvial plains.

The northwestern plain consists of a huge expanse of horizontal layers of rock, undisturbed by earth movement from the Yenisei River and the Tien Shan to the Ural Mountains. South of the Urals the plain is uninterrupted and extends through Russia and north Germany to the Atlantic. In the Asiatic portion a slight rise of ground stretching from the neighbourhood of Lake Balkhash to the south of the Urals forms a watershed which divides the basin of the Ob from that of the Caspian. The former is gentle in gradient and drains into the Arctic Sea through the sluggish streams of the Ob and its feeders. Owing to the block caused by the melting of the winter ice in the upper course before the lower course is free, the wide flood plain is swampy or even flooded in summer and a frozen marsh in winter. The Caspian basin falls away some 200 feet below sea-level. It now drains into the two shallow salt lakes known as the Caspian and Aral Seas, whose margins are edged with saline marshes or salt-encrusted mud and whose levels are subject to considerable, irregular variations.

The highland area is seamed by great lines of fold mountains which continue the European system right through to the Pacific. The lines enter Asia in the Taurus Mountains of Asia Minor and in the Caucasus. The Elburz branch of the former unites with the latter to form the Hindu Kush, while the Zagros Mountains sweep round to the south to return to the main line through the Kirthar and Suleiman Mountains. The junction of the Hindu Kush and Suleimans forms the Pamirs, 'the Roof of the World'. Soon, however, the folds diverge again, the Tien Shan running northeastwards, the Kunlun and Altyn Tagh eastwards, and the Karakorums and Himalayas southeastwards. The last named contain the world's highest peaks. Mt Everest (29,000 feet) and Mt Kinchinjunga (27,000 feet) have never been climbed, in spite of the frequent determined efforts of organised expeditions to reach their tops. In the Karakorums Mt Godwin-Austen towers up to 28,000 feet, and beside these there stand many other giant peaks of scarcely less height. After a hair-pin bend due to the resistance to folding of the plateau of Yunnan, the Himalayas continue southeast through the Arakan Yoma, the Andaman Islands, Sumatra, Java, and the Sunda Islands. Probably, they are carried still farther east in the mountains of New Guinea, while a branch which turns through the Banda arc runs northwards in the Philippines, Taiwan, and Japan. These great folds are all recent, geologically speaking; but in the centre and east the Altai, Yablonoi, Stanovoi, Khingan, and most of the ranges of China are of older date and are consequently far more worn.

Between the folds are pinched up great blocks of topland which for the most part are areas of inland drainage. Beginning with the comparatively small toplands of Anatolia and Armenia in Asia Minor, we pass eastwards to the extensive tableland of Iran. Beyond the Pamirs the Lop Basin lies between the Tien Shan and Altyn Tagh, with Tibet farther south. The former is continued in Gobi, the latter in the plateau of Yunnan.

The southern tablelands of Arabia and the Deccan are said to be remains of the former continent of Gondwana Land and are thus akin to Africa, from which indeed Arabia is not wholly separated by the rift valley that contains the Red Sea. In each of these peninsulas the land surface slopes gently down towards the east, leaving the western rim, notched by erosion, to stand up as a range of hills. In India this edge of hills is known as the Western Ghats.

The eastern portions of the great lines of fold still show much instability in the Earth's crust. The islands off the east and southeast coast are often blasted by volcanic eruptions and rocked by severe earthquakes. The island of Java alone has fifty active volcanoes, while there are even more in Japan. The most terrible eruption occurred in 1883 in the small island of Krakatoa, hurling thirteen cubic miles of rock into the air and causing the island almost to disappear. But the most destructive have taken place in Japan, where in 1912 buildings were damaged merely by air

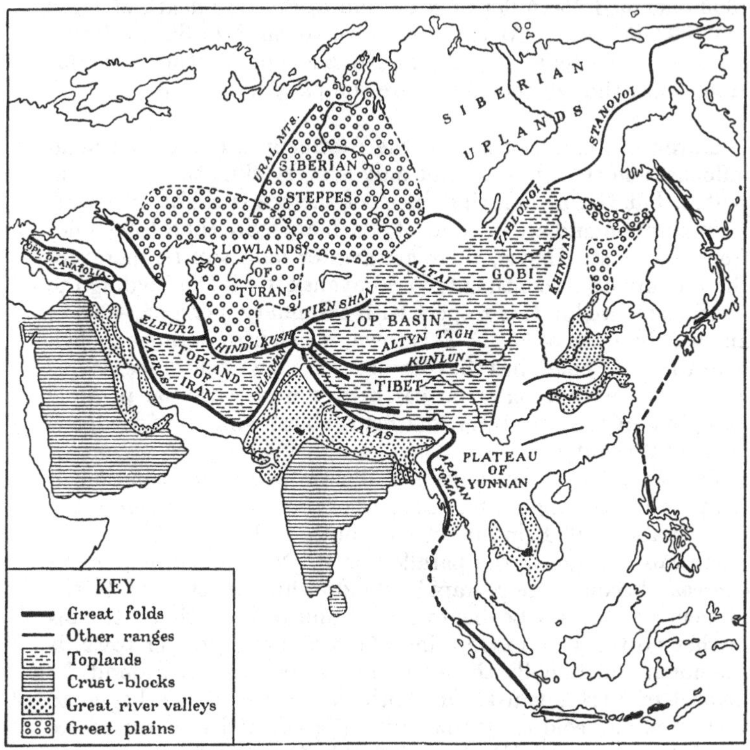

Structure and relief of Asia simplified

concussion within a radius of eleven miles from the volcano of Asama-yama, while in 1913 the solid matter ejected by Sakura-shima was calculated to be enough to cover an area of fifty square miles to a depth of 100 feet. The earthquakes which are connected with this volcanic activity are even more destructive. Tokyo and Yokohama were destroyed in 1923 by a shock which completely altered the configuration of the floor of the Gulf of Tokyo. No fewer than 170,000 houses were destroyed in the capital alone and altogether nearly 200,000 persons were killed. These earthquakes are often followed by seismic tidal waves which cause great havoc along the coast. One such wave which occurred in 1896 damaged 250 miles of coast and caused the death of more than 30,000 people. At the western end of the fold lines volcanic activity also occurs in the Ægean Sea and in the Caucasus Mountains, though the effects are comparatively mild.

Climate. The vast size of the continent is one of the main influences on the climate. Not only is every climatic type represented, but the moderating effect of the sea is slight or absent from many inland areas, some of which are more than 1500 miles from the coast. Hence, there are greater ranges of temperature than in any other continent, Verkhoyansk in Siberia holding the 'world's record' in this respect with a January mean of −60° F. and a July mean of 60° F. This town, which is sometimes spoken of as the 'cold pole', has the coldest recorded winter temperatures and is the coldest place on earth in winter with the possible exception of the Antarctic continent, about which knowledge is at present scanty.

The set of maps on page 121 illustrates the distribution of temperature during the four seasons of the year. In January the mean is about 80° F. in the extreme south, with a gradual decrease northwards as far as the parallel for 40° N. From this line the decrease becomes more rapid inland, while on the coasts the influence of the sea begins to be felt and to push the isotherms farther north. This moderating effect becomes greater towards the north, until finally the isotherms enclose a space around the district of Verkhoyansk. In April the continent has begun to warm up, the cold centre has disappeared, and a warm centre has formed over the Deccan. By July the vast land mass has become thoroughly heated, a large warm centre exceeding 90° F. has settled to the northwest of India, and only the northern coast fringe has a mean of less than 50° F. It will be noticed, too, that the influence of the sea has been reversed and that the coastal areas are cooler than those farther inland. October has an isotherm pattern similar to that of April, though the rapid cooling of the land mass renders the temperatures everywhere lower.

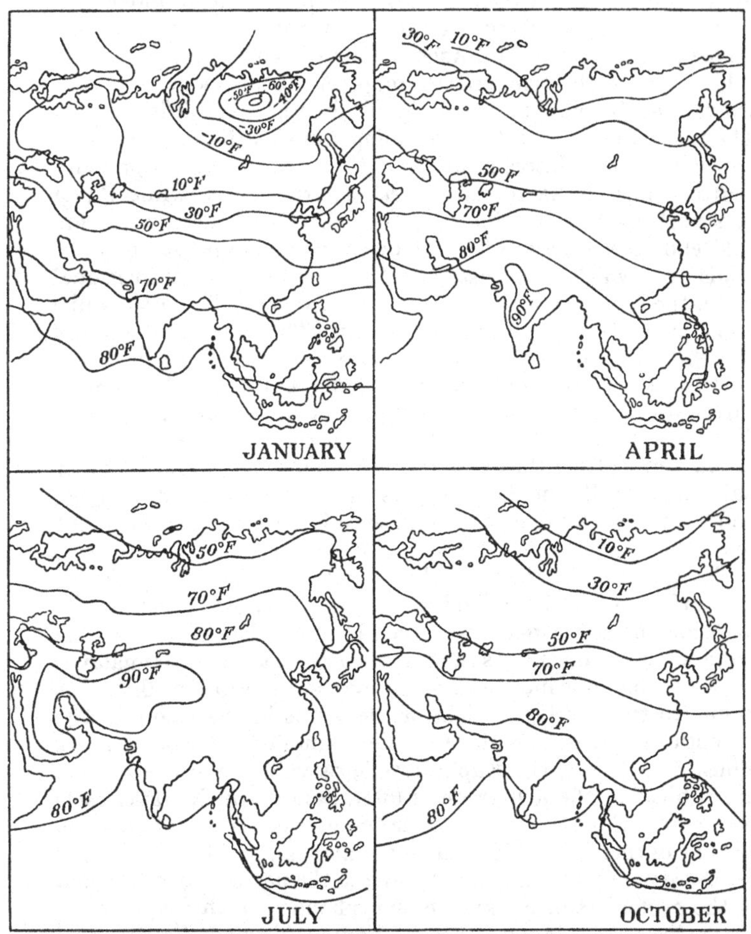

Distribution of temperature in Asia
Note the 'cold pole' district around Verkhoyansk

The relation of pressure to temperature systems is brought out by a comparison between the set of maps described above and those on page 123. In the latter it is seen that a vast high-pressure centre occurs over central Asia in January, corresponding to the cold centre of the temperature map. At this season, therefore, winds tend to blow outwards in all directions towards the sea. The point from which the prevailing wind reaches a given area depends on the relative positions of the area to the high-pressure centre. Thus, India receives northeasterly winds, while China gets northwesterlies. Whatever their direction, these winds are cold and have a great effect in lowering the temperature in the lands over which they pass. The April map shows a transitional state, the high-pressure centre moving away to the north, while a low-pressure centre forms in the south. The winds are light and variable. In July a vast low-pressure centre has formed to the northwest of India, corresponding to the centre of high temperature noticed in that month. Winds now blow inwards from all directions, those in India originating in the southwest, those in China coming from the southeast. October shows a similar isobar pattern to April, but the high-pressure centre is more developed, and the low-pressure centre has moved farther away towards the south.

The direction of the prevailing winds and the distribution of the surface relief of the land are the chief factors in the rainfall of Asia. The set of maps on page 124 shows that in January all but a few areas on the east and west coasts and in the Himalayas have less than one inch of rain and may be described as dry. This is due to the overland origin of the winds in that month. The exceptional areas owe their greater rainfall to various causes. Thus, Asia Minor, the Euphrates-Tigris valley, and the Punjab are subject to the low-pressure disturbances which travel eastwards from the Atlantic in winter and give convection rain along the line of their path. The cause of winter rain in China is similar, for there precipitation is mostly due to the passage over the land of the cyclonic storms, known as typhoons, which frequent the China Seas. In Japan, Chosen, Annam, Taiwan, Ceylon, and the northern Philippines the prevailing winter wind crosses the sea and absorbs moisture before it reaches these places and is thus able to yield rain. Malaya and the East Indies, like all other countries on or near the Equator, have an abundant convection

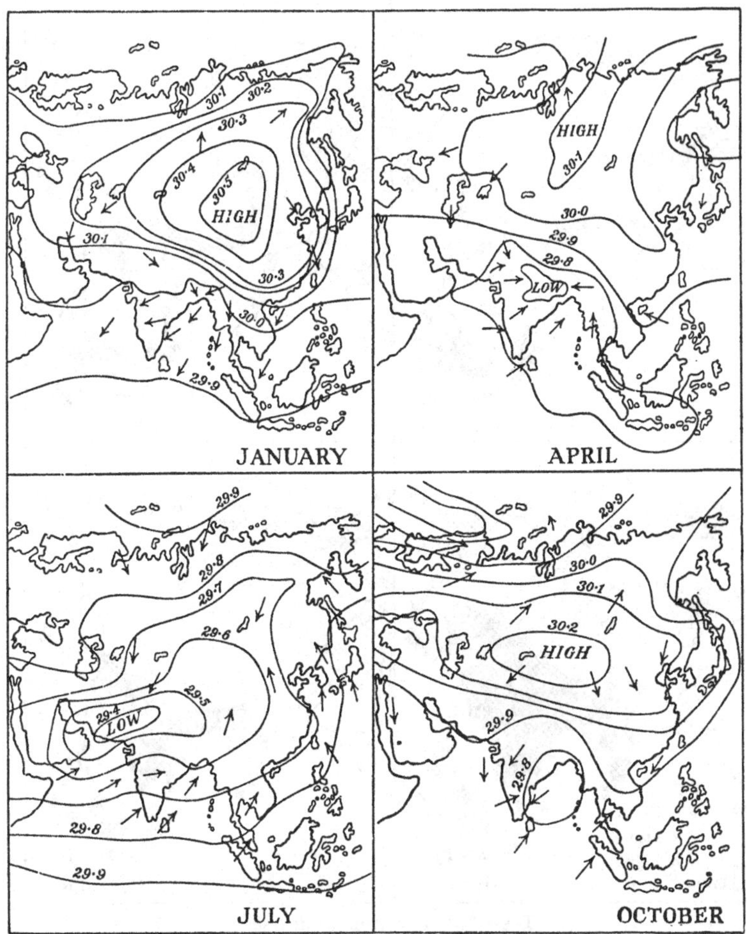

Distribution of pressure and winds in Asia

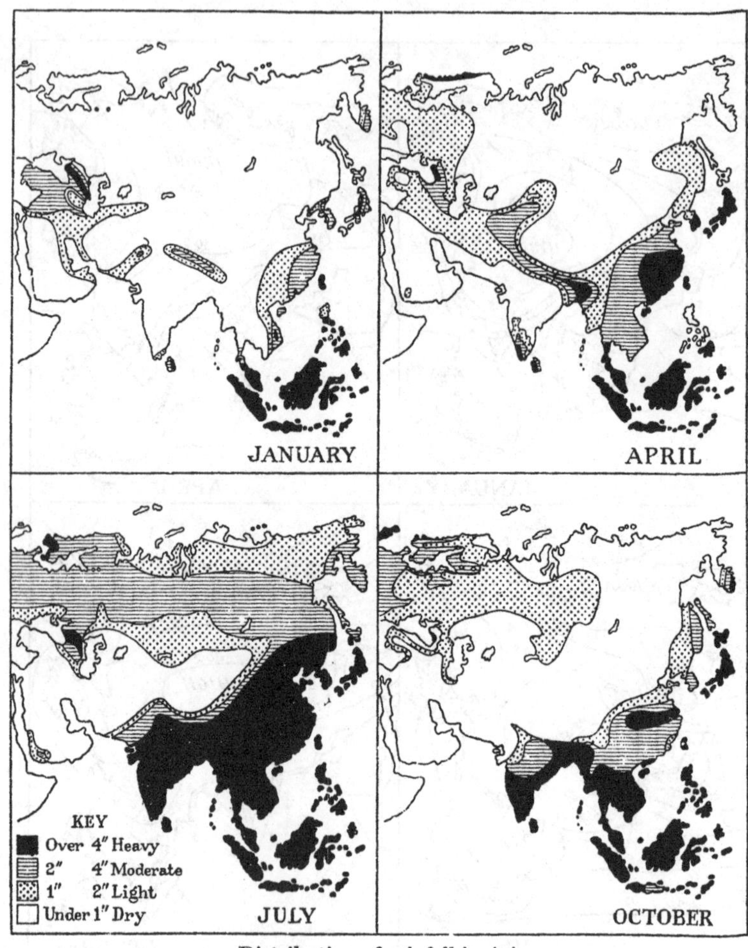

Distribution of rainfall in Asia

Distinguish the areas with rain (1) in summer, (2) in winter, (3) throughout the year

rainfall well distributed through the months. In April, as the map shows, summer conditions have begun to form, but only coastal and highland regions have more than light rains and a mean total of more than two inches. The July map shows summer conditions at their height. Heavy relief rains fall in southeastern Asia from India to Japan, while north of about 40° there is a moderate rainfall mostly due to convection. A vast tongue of drought extends eastwards from the Mediterranean and Red Seas right into the heart of the continent. This is due partly to the Horse Latitude conditions of pressure and wind, partly to the overland origin of the wind, and partly to the rain shadow effect of the Himalayas and other ranges. October represents the transitional phase between summer and winter conditions.

The absolute reversal of the direction of the prevailing wind in summer and winter, together with an alternation of heavy summer rainfall and winter drought, forms what is known as a monsoon system. This name is derived from the Arabic *mausim* (=season). The periodic reversal of winds blowing across the Arabian Sea was known to Arab traders in the Middle Ages and was used by them on their trading voyages between Zanzibar and the Malabar Coast. In due course the knowledge of it was passed on to the Portuguese, who brought it to Europe. So long as wind remained the motive power of ships, the navigational importance of the monsoons was great, and the routes followed by ships varied according to season and were arranged so as to take advantage of the prevailing wind.

But the monsoons have a more general human importance. The conditions of great heat and moisture which prevail in summer are extremely favourable to plant growth and enable crops to give an immense yield. On the other hand, the cool, dry winter period checks luxuriance at that time and permits clearings to be made and maintained easily. Hence, the lowlands and especially the great river valleys of monsoon lands are intensely cultivated and support dense populations. The standard of life among these swarms of agriculturalists is very low, but an assured existence has sapped their vigour and courage and has rendered them in the past liable to the depredations of their hardier neighbours, who have ended by establishing themselves as the despotic rulers of the farmers.

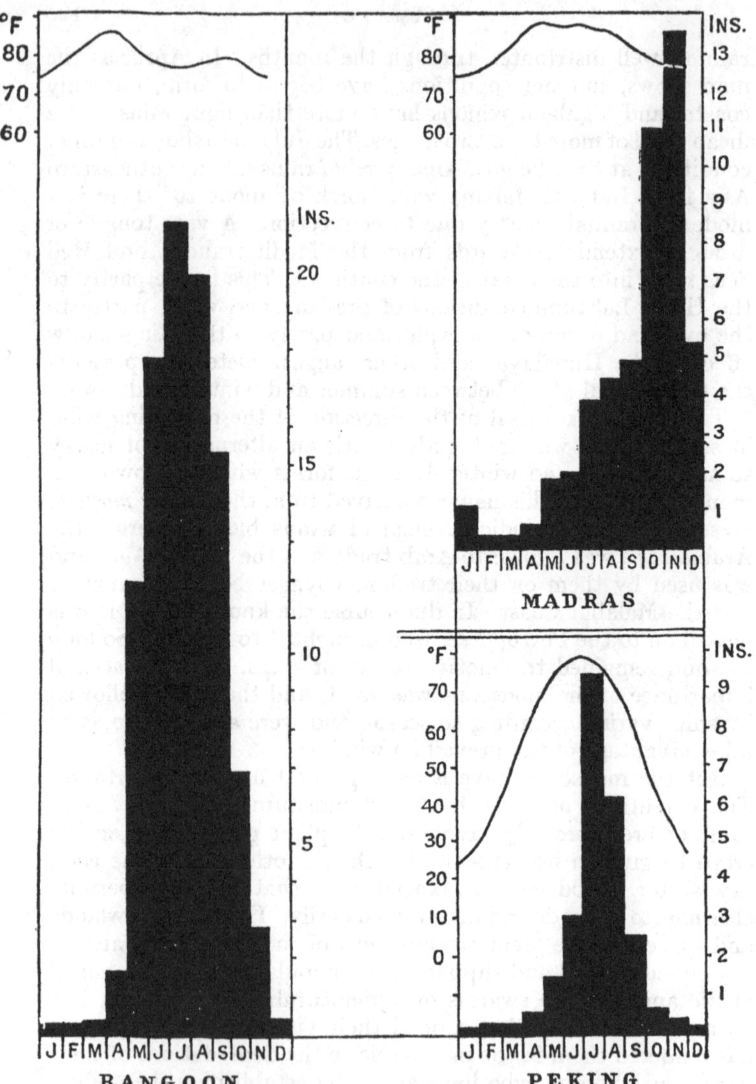

Rainfall pillars of monsoon Asia

Illustrating (1) the ordinary tropical system (Rangoon), (2) the régime due to the retreating monsoon (Madras), and (3) the temperate system (Peking)

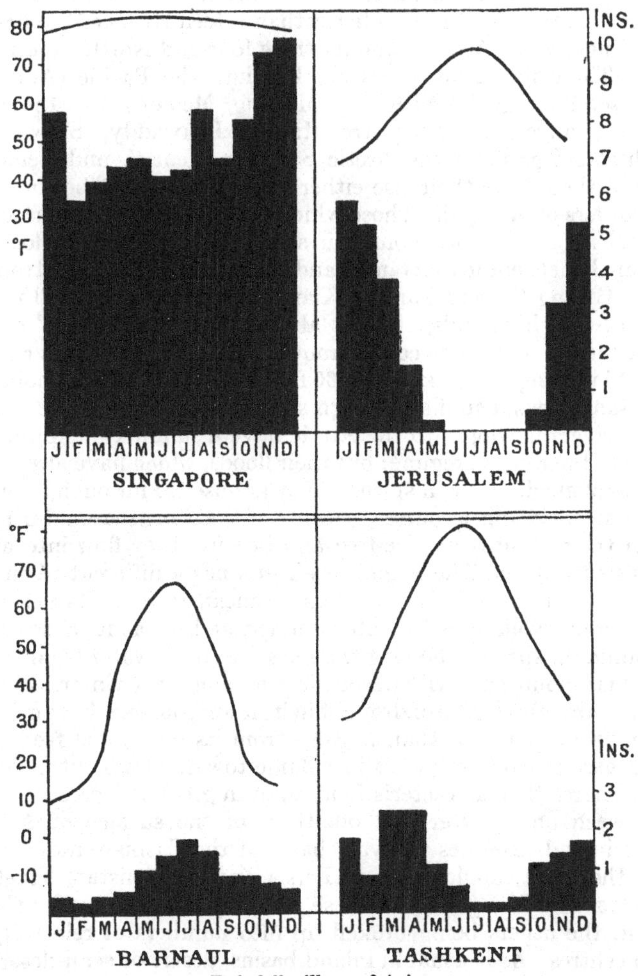

Rainfall pillars of Asia

(1) Singapore—equatorial system, (2) Jerusalem—Mediterranean system, (3) Barnaul—extreme continental, and (4) Tashkent—mountain-foot

Drainage. Asia is a continent of large rivers. Not counting secondary streams, it has no fewer than fifteen rivers of the first magnitude, while Africa can only show four and North America three. These first-class rivers all flow into the Pacific (Amur, Yangtse, Hwang Ho, Si-Kiang, Mekong, Menam), the Indian Ocean (Ganges, Brahmaputra, Indus, Irrawaddy, Salween, Euphrates-Tigris), or the Arctic Sea (Ob, Yenisei, and Lena). Most of them take their rise either in southeastern Tibet or on the borders of Mongolia. Those which pass through the monsoon region have a maximum volume in summer, which leads to floods that are beneficent in the Ganges and Mekong, but often disastrous in the Hwang Ho and Yangtse Kiang. Since these rivers all rise in snow-clad highlands, there is also a spring 'maximum' due to the melt-water which comes from the snow. The Yangtse rises 40 feet in spring at Hankow and 60 feet in July at the same point. The Ganges has a similar, though smaller, range of volume.

The rivers flowing into the Arctic have a summer maximum, owing to the rainfall régime; but their floods, which have already been explained, occur in spring. They are useless for navigation, except in their upper courses, because of the short season during which their mouths are ice-free and because they flow into an ice-obstructed sea. The Euphrates-Tigris has a different régime from the rest, since it is in a Mediterranean region. Its period of minimum volume is the late summer, and it has its absolute maximum in spring following the release of melt-water from the Armenian mountains, with a secondary maximum in winter. Since it flows through a region of drought in its lower course, it loses more water by evaporation than it gains from its occasional feeders and consequently diminishes in volume towards its mouth. The Indus shares this characteristic in an even greater degree.

Between one-quarter and one-third of the surface of Asia drains inland. Besides the vast basin of the Caspian and Aral Seas, there are smaller ones in Persia and Afghanistan, Tibet, Sinkiang, and Mongolia. Of these, the Lop Basin is by far the largest, the others being broken up into numbers of relatively small centres. As is usual in inland basins, desert or semi-desert conditions prevail everywhere.

Vegetation and Animals. The types of vegetation and their distribution will be seen from the map on page 131, and their

Plate VII

(*Mondiale*)

RICE FIELDS IN YUNNAN

The system of terracing is clearly shown

Plate VIII

(*Canadian Pacific Railway*)

TYPICAL NATIVE HOUSE, SUMATRA

The features of interest are the steep roofs, the minute carving and decoration of the walls, and the open ground floor

relation to the rainfall and temperature should be sought by a comparison of this map with the climatic map-sets given above. It will be seen that the equatorial lowlands in Malaya and the East Indies are clad in the tangled forest typical of that region,

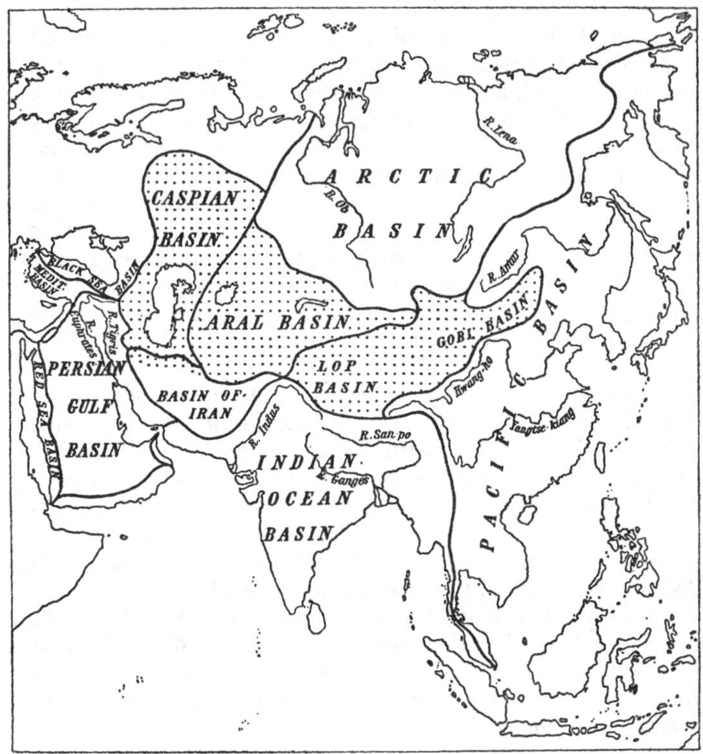

Drainage basins of Asia

The stippled areas are inland drainage basins

while the highlands of Sumatra and Borneo degenerate into savana where they are sheltered. Monsoon tropical forest, containing many leaf-shedders like the teak, is found in Burma, Siam, and India, though cultivation has restricted the limits to the mountain areas. It is replaced by savana in the rain shadow areas of the Deccan and central Burma and on the Indo-Chinese plateau. In China the natural vegetation is subtropical and similar to the magnolia forests of the corresponding region in the United States; but the Chinese hatred of trees has caused the removal of the woods except the clumps of two or three trees which grow on the family burial grounds.

In the latitude of the Tropic of Cancer outside the monsoon area there are the usual deserts and semi-deserts. The Thar in Rajputana is small, but practically the whole Arabian peninsula is barren. Northwards, the deserts are succeeded on the west by the Mediterranean region and in the interior by cold deserts and semi-deserts which cover the basins of inland drainage. Farther north again the semi-desert improves into temperate grassland with wide expanses in western Siberia and in Manchuria. This belt is followed by the *taiga* or great forest of Siberia. In the south this forest consists of mixed deciduous and coniferous trees, but gradually the latter prevail towards the Arctic. Finally, beyond the July isotherm for 50° F., the trees disappear and are replaced by tundra.

The animals of Asia offer a wide variety. In the tropical forests and jungles of the hot belt there are tigers, buffaloes, elephants, and other large species. In Burma and India the elephant has been tamed and used for various purposes, especially for that of moving logs in the forests. The water buffalo is used for ploughing and drawing carts in the wetter parts of India, while the zebu replaces it in the drier regions. The grass plains and semi-deserts of Asia are the original home of the horse, but this animal is not now found in its wild state. Sheep and cattle are plentiful in these regions too, while the camel is the chief beast of burden in the sand deserts. The topland of Tibet has the yak, a beast of burden which yields meat and a warm coat of hair when dead and milk and dung-fuel when alive.

People. A line drawn from the southern end of the Caspian Sea along the Hindu Kush and Himalayas to the Bay of Bengal divides the people of Asia roughly into two different types. To the southwest of the line the races are brown-skinned and have black, wavy hair and straight-set eyes. Many varieties occur with these characteristics, from the almost black, short, lithe Tamils of southern India to the tall, light brown Sikhs of the Punjab. Arabs, Persians, Jews, Armenians, and Afghans all differ from each other clearly; yet they are still more clearly different from the people to the north and east of the racial frontier which we have described above. These other people are distinguished by a sallow skin, black lanky hair, beardless faces, and a flat face with a sunken nose and apparently obliquely-set

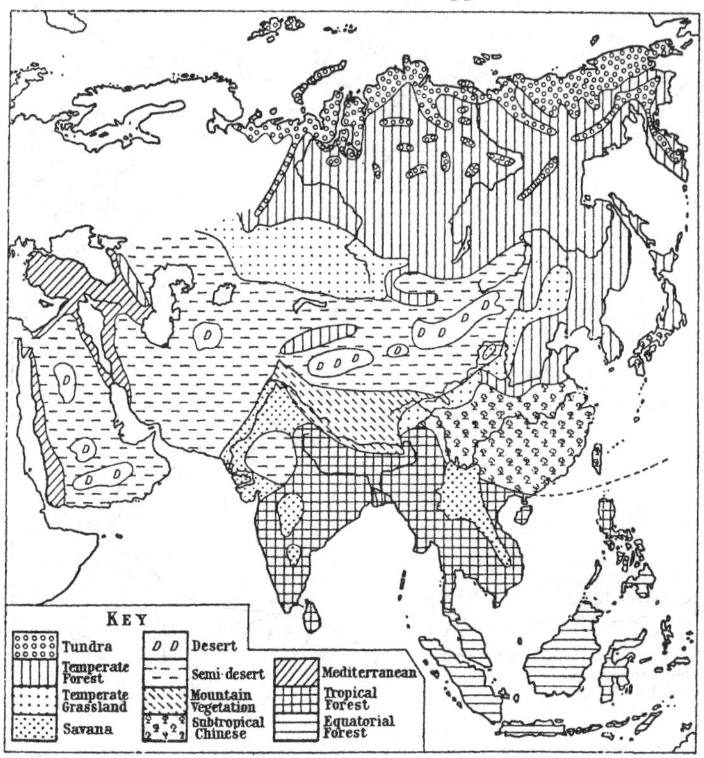

Distribution of vegetation in Asia

eyes. Two subtypes may be observed: the yellow, or Mongolian, of China and the north; and the brown, or Malay, of Indo-China, the East Indies, and Japan. In this type too there are many variations. The Malays are dark, short, and lithe, while many of the northern Chinese are moderately tall, thickset, and almost fair of skin. It should be noticed that the Turks are an intrusion of the yellow subtype among the Mediterranean peoples of Asia Minor. The two types differ in language, the peoples to the south-west of the racial line speaking inflexional tongues of the Indo-European family, those to the north and east using 'agglu-

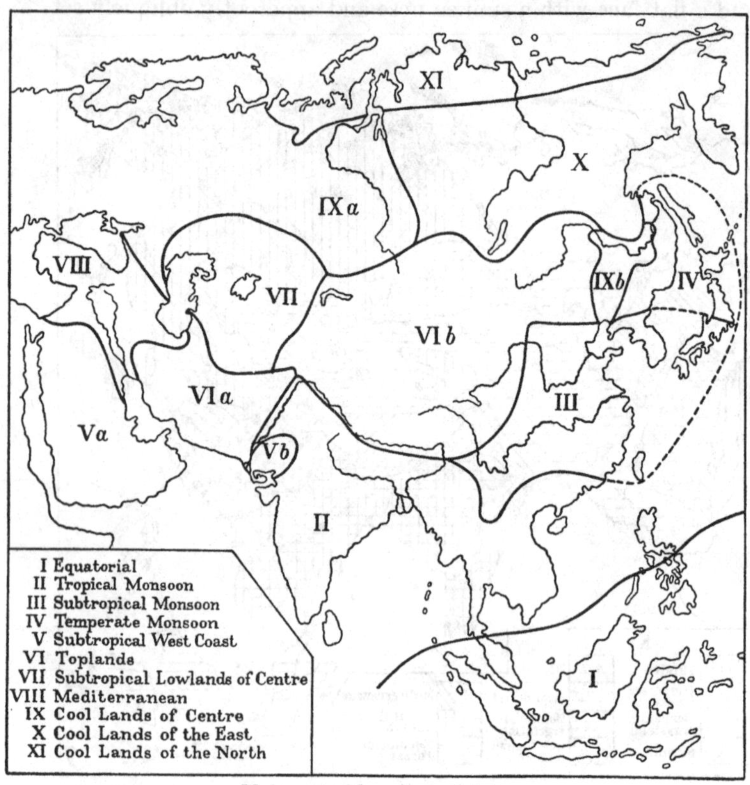

Main natural regions of Asia

tinative' languages which differ fundamentally in system from the Indo-European.

The Natural Regions. The peculiar conditions of the monsoons and their reaction on human life mark off the area which they influence from the rest of Asia. But certain climatic features which are not characteristic of monsoons cause variations, and 'Monsoon Asia' may be subdivided into (1) a Monsoon Equatorial Region, (2) a Tropical Monsoon Region, (3) a Subtropical Monsoon Region, and (4) a Temperate Monsoon Region. The whole area covered by these regions may be defined as those parts of south and eastern Asia which are shown in black in the July rainfall map on page 124. Outside the monsoon regions, there are seven regions whose numbers in the following list will, for convenience of reference to the map on page 132, be given as continuous with those of the monsoon area. They are (5) the Subtropical West Coast, (6) the Toplands, (7) the Subtropical Interior Lowlands, (8) the Mediterranean Lands, (9) the Temperate Interior, (10) the Temperate East Coast, and (11) the Cold North Coast. These will now be dealt with in detail.

THE REGIONS

I. Monsoon Asia

The Equatorial Region. The area covered by this region, whose exact boundaries are shown in the map on page 135, includes the Malay Peninsula and the East Indian Islands. Sumatra, Java, and Borneo stand upon the continental shelf of Asia, and are the undrowned highlands of the now submerged 'Sunda-land'. Old lines of fold run through Malaya, the islands of Banka and Billiton, and Borneo to the Philippines, but most of this part of the region consists of the remnants of a former peneplain. Tertiary earth movements have thrown up recent folds on the west and south of Sumatra and Java and in the Philippines and have caused corresponding trenches to be formed in the sea bottom in the Java and the Tuscarora Deeps. These recent lines of folding are marked by active volcanoes and by seismic activity. Where the folds curve round in the Banda Islands, the ridges are gradually being raised above the sea, as is proved by the presence of coral reefs a thousand or more feet above sea-level. In spite of these recent developments, the Wallace Line remains the structural division of the Asiatic from the Australasian portion of the region.

The climate is equatorial, but influenced by the monsoons. The temperature has the usual even record: e.g. Batavia has an annual mean of 78° F. and a range of 2° F. The diurnal range is 5° F. But the winds are monsoonal and are controlled by the alternating high- and low-pressure centres in Asia and Australia. In July they blow from the south with an eastern component in the southern hemisphere and with a western component in the northern. In January these directions are reversed. Sea influence helps to steady the temperature, which is unusually equable. Owing to the high relief of the land, the rainfall is very heavy. The annual mean for the region as a whole is 100 inches, but among the hills of Java and the Philippines it rises to 250 or 300 inches. Baguio near Manila holds the world's record of 46 inches for precipitation during a single day. The weather hardly varies from day to day. The mornings are bright at first, but begin to cloud over by 9 o'clock. In the afternoon a heavy shower falls, after which the sky clears again. The nights are clear and windless.

The great heat and damp of the climate causes a luxuriant growth of forest throughout the region. As in the Amazon and Congo regions the 'storey' arrangement is found, and the trees, which include large numbers of sappy species, harbour dependent suites of lianas, epiphytes, and parasites. In certain parts of eastern Java and in Borneo, where a rain shadow causes a dry season, the forest gives way to savana, and on the western slopes of the mountains of Java there is a deciduous monsoon forest of teak. The animals are those of the mainland of Asia within the hot belt.

The interior of Luzón in the Philippines and the forests of western Sumatra contain very primitive races, but the basis of the population consists of Malays. These show a great range of culture, from the most primitive in remote parts to the high civilisation of the Javanese. Indians, Arabs, and Chinese have mingled with the coast tribes, and in the fifteenth century Hindu influence caused the rise of the states of Palembang and Malacca. The first suffered gradual decay, but the latter was destroyed by the Portuguese in 1511. There are a number of Dutch officials, plantation managers, and traders who are concentrated mainly in Java.

Malaya, which is part of the British Empire, was formerly divided politically into the Crown Colony of the Straits Settlements of Singapore, Wellesley, Penang, Malacca, and Labuan;

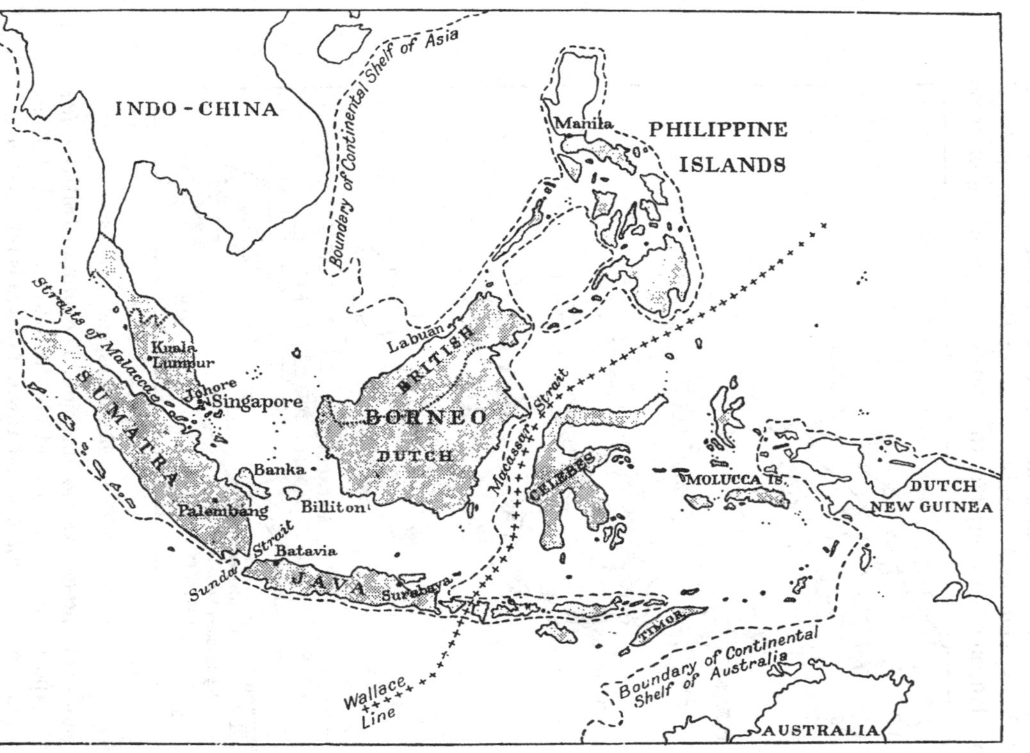

The Equatorial Region of Asia

The area occupied by the region is shaded. The names given are those mentioned in the text

the Protected States of Johore, Kedah, Kelantan, Trengganu, and Perlis; and the Federated States of Perak, Selangor, Negri Sembilan, and Pahang; but the status and grouping of these

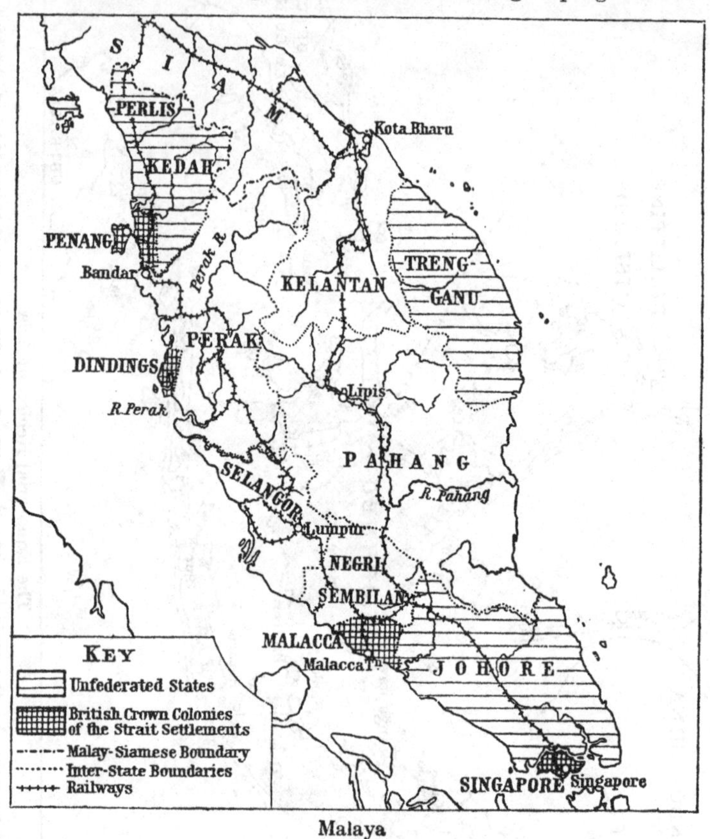

Malaya

Political divisions, chief towns, and railways

divisions is in process of modification. The administrative centre is Singapore, on a small island off the south coast, but Kuala Lumpur is an important native town. Malaya is relatively one of the wealthiest parts of the Empire. It is the largest rubber-producing area, exporting 283 tons in 1926, and its tin production, valued at £16 million in 1931, was the greatest in the world. It also exports mineral oil, rattans, copra, pepper, and sugar, while rice is grown in large quantities for local consumption as well as export. The natives make poor labourers, and in consequence coolies are imported from China and India. Numbers of Chinese have also immigrated and have become the

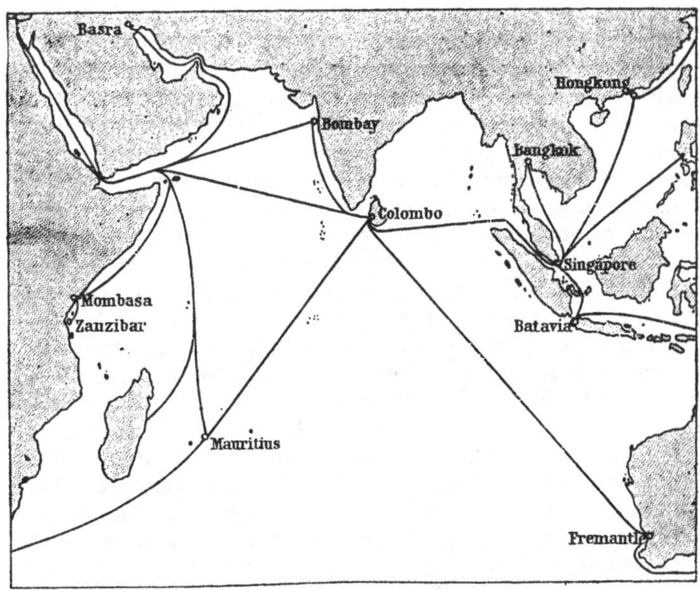

Positions of Singapore and Colombo

Note the focal situation of each

retail traders throughout the peninsula. The great wealth of the exports has enabled a good system of main motor roads to be built, and railways run from Singapore along the west and east coasts to the frontier, where they connect with the Siamese line to Bangkok. The position of Singapore at the gateway of the Straits of Malacca, which connect the Indian Ocean to the Pacific, gives the port great economic and political strategic importance and makes it one of the chief entrepôts of the Far

East. Its half-million inhabitants belong to a diversity of races, and the streets offer a picturesque scene of varied costumes.

Except a portion of Borneo and the Philippines, the rest of the region belongs to the Dutch, the headquarters of whose administration is at Batavia in Java. This port has similar advantages of position to those of Singapore, but has them in a less degree. It shares with Surabaya the export trade of cane-sugar, of which Java is the world's chief producer. This island is extremely fertile owing to the volcanic soil and in 1928 was able to maintain a population of 38 million persons. Besides the equatorial crops of rubber and quinine, both of whose trees have been transplanted from South America, tropical crops of rice, sugar, tobacco, coffee, and cocoa are produced by intensive agriculture and the terracing of the slopes. Mineral oil is being raised in increasing quantity.

Sumatra, Borneo, and the southern Philippines are less fertile, their soil having been leached by the constant heavy rainfall. Sumatra produces rice on its swampy eastern coast as well as tobacco, rubber, coffee, and tea farther inland. It has a good deal of coal and there are some deposits of iron, silver, and gold which are still practically untouched. The picturesque old native city of Palembang is the chief town. The neighbouring islands of Banka and Billiton yielded 35,000 tons of tin in 1927. Dutch Borneo is still largely unexploited, but British Borneo, consisting of the Crown Colonies of Brunei and British North Borneo and the Protectorate of Saráwak, exports rubber and copra together with one-third of the world's supply of sago. The deposits of mineral oil are only just beginning to be worked. Labuan, which yields enough coal to supply the needs of the British Navy on the China Station, is politically one of the Straits Settlements. Saráwak had the interest of being ruled by an English hereditary rajah, but has now become a Crown Colony.

The Philippines belong to the United States, having been wrested from Spain during the Spanish-American War of 1898. Properly speaking, the northern two-thirds of the island of Luzón lies outside the region, but will be treated here for convenience sake. The southern islands, of which Mindanao is the largest, produce rubber, while Luzón exports tobacco and hemp. Rice and sugar are cultivated throughout the group, and small quantities of gold and iron are worked, especially in Luzón. The chief town is Manila in the island last named.

The Tropical Monsoon Lands. These comprise India and Indo-China together with a strip of southern China.

India

India is clearly divided by its relief into three parts: (1) the peninsula of the Deccan, (2) the Indo-Gangetic plain, and (3) the mountains of the north and northwest.

The Deccan is an old crustal block with such geological likenesses to Africa and Arabia as to give rise to the belief that these three land areas once formed part of the same continent to which the name of Gondwana Land has been given. Its worn surface, which has broadly speaking been reduced to the state of a peneplain, though the rejuvenation of river erosion has caused the streams to cut deeply into the rock in many places, stands up as a triangular tableland tilted downwards towards the east. Its western edge, irregularly carved by fluvial erosion, assumes the appearance of a mountain range and is known as the Western Ghats. The low eastern edge has been eaten away widely in parts by the rivers, but its remains form the low Eastern Ghats. The drainage is eastwards into the Bay of Bengal, the largest rivers being the Mahanadi, Cauvery, Godávari, and Kistna. Owing to the large quantities of rock waste brought down by these streams, their mouths are marked by big, swampy deltas. This coast, which is known as the Coromandel Coast, is low, harbourless, and surf-beaten, but has a wide strip of lowland; while the west, or Malabar, coast has little lowland and is crossed by short, fast-flowing streams.

In the southern extremity of the peninsula, where the western edge is highest, the rivers have been most successful in cutting gaps and breaking up the highlands. Thus, the Palghat Gap completely pierces the mountains, separating the Nilgiri Hills from the Cardamom Hills. In the north the plateau edge is marked by the Vindhya Hills, but beyond this is a shelf of upland which is broken in the northwest by the worn stumps of the Aravalli Hills. The drainage of the north Deccan runs either towards the Ganges through the River Son or else westwards through the deep trenches of the Nárbada and Tapti. Between these two rivers the so-called Satpura Range is in truth an isolated part of the tableland.

The soil of the Deccan varies in quality. Between Bombay and Nagpur is a wide area which has been overlaid by a basaltic flow which when decomposed forms a dark, fine-grained earth known as 'black cotton soil' and of great fertility. Elsewhere, much of the surface is covered with a porous, clayey rock called

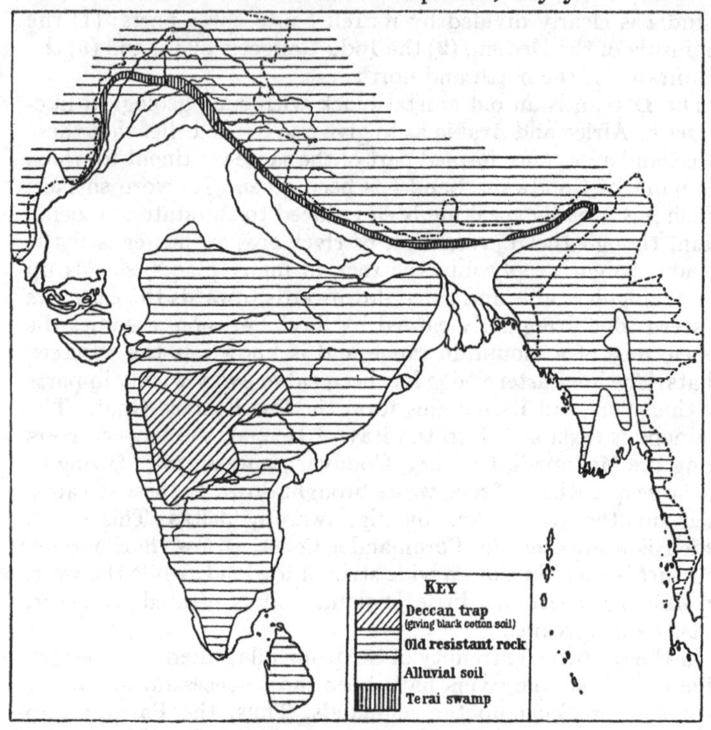

Geological map of India (simplified)
Note the correspondence with the relief

laterite which is infertile, but fortunately only comes to the surface on the higher parts of the tableland. The valley bottoms and other depressions are covered with fertile alluvial soil.

The Indo-Gangetic plain is a vast alluvial covered area 300,000 square miles in extent with a breadth that varies between 90 and 300 miles. It consists of three parts: (1) a deltaic area at each end, (2) a low valley floor following the windings of the main rivers and liable to annual flood, and (3) the higher valley out of the reach of floods and rising sometimes as much as 200 feet above the level of the river. On the Himalayan side of the plain lies a pebbly, porous strip along the foot of the mountains. Lack of water causes this area to be abandoned to forest. On the valley side of this strip runs another known as the *Terai*, a swampy belt also given up to forest. The deltas consist of low, swampy ground scarcely, if at all, above the reach of the tide and covered with dense jungle. The soil of the plain is generally very fertile and bears the great majority of the population of India.

The two great rivers of the plain, the Ganges and the Indus, rise near together beyond the Himalayas and enter the plain through gorges in the mountains. The Ganges, fed by melt-water from the Himalayas and by the heavy rains, is one of the world's largest rivers and is navigable over most of its valley tract. It receives many feeders both from the Deccan and the Himalayas, the largest of the latter being the Jumna which joins it at Allahabad. It shares its delta with the Brahmaputra, which after a long upper course beyond the mountains, where it is known as the Tsangpo, is brought to the plain through the Dihang Gorge. The Ganges valley is separated from that of the Indus by a mere rise in the ground, whose highest point is 925 feet. The Indus is chiefly fed by melt-water from the Himalayas, since it flows through a region of scanty rainfall. Its principal feeders are the Sutlej, Ravi, Chenab, and Jhelum, the rivers of the Punjab.

The great mountain ranges that bound India comprise the Himalayas in the north and the Suleiman and Kirthar Ranges in the northwest. The Himalayas rise in three increasingly high ranges, the main line of peaks being so lofty that none of the passes is lower than 17,000 feet. In the northwest, however, the mountains are far lower, and the Kabul River has cut through a gorge which forms the celebrated Khyber Pass. At the junction of the Suleiman and Kirthar Hills another pass, the Bolan, offers a route way across the mountains. Since there are no practicable routes elsewhere, these two passes have played an extremely important part in the history of India.

Climate. The monsoons play so large a part in the climate of India that many people have heard the term only in connexion with that country. But, immense though the monsoonal influence is, it cannot have the effect of giving a uniform climate to so large an area as India. Hence, while the periodic reversal of winds is a marked condition throughout and heavy summer rains occur nearly everywhere, important local modifications place wide differences between the climate of, say, Madras and that of the Punjab.

There are three seasons of the year: (1) the cold season from October to March, (2) the hot season from March to June, and (3) the rainy season from June to October. It is a strange fact that, though the term 'monsoon' was first applied to the periodic reversal of winds by the Arabs and Europeans who knew the climate chiefly over the sea, to-day it refers rather to the season of rains.

In March the midday sun is overhead at the Equator and the land mass of Asia begins to warm up. The mechanism which produces the summer indraught has not yet started to work, however, and the absence of moist onshore winds causes clear skies, which in turn allows of maximum insolation. At this time parts of the Indo-Gangetic plain have a daily mean temperature of 100° F. By June the indraught has begun and the southwest winds commence first in the south and then gradually farther and farther north. The wall of the Himalayas deflects much of the wind and forces it to flow northwestwards up the Ganges valley. Meanwhile, the high relief of the Western Ghats causes the condensation of the moisture in the warm winds, producing a heavy rainfall. But a rain shadow effect protects the immediate lee of the highlands and causes a moderate fall there. Farther east convection plays its part and gives precipitation in the eastern Deccan and the Ganges valley. In Assam, where the winds are caught in a funnel and forced upwards to great heights, the rainfall is the heaviest in the world, the famous station at Cherrapunji in the Khasi Hills showing a recorded mean of 430 inches a year. Only in the northwest is the rainfall scanty at this season, the cause being that the winds which sweep into the low-pressure centre blow here not from over the Arabian Sea, but from off the dry tableland of Iran. As a result, parts of Sind and Rajputana form the Thar, or Indian Desert. The season of rains, though accompanied by cloudy skies, unpleasant dampness in the air, and thunderstorms, is nevertheless essential to the power of

India to support its vast population. In parts of the Deccan and in northwestern and central India the rains do not always fall

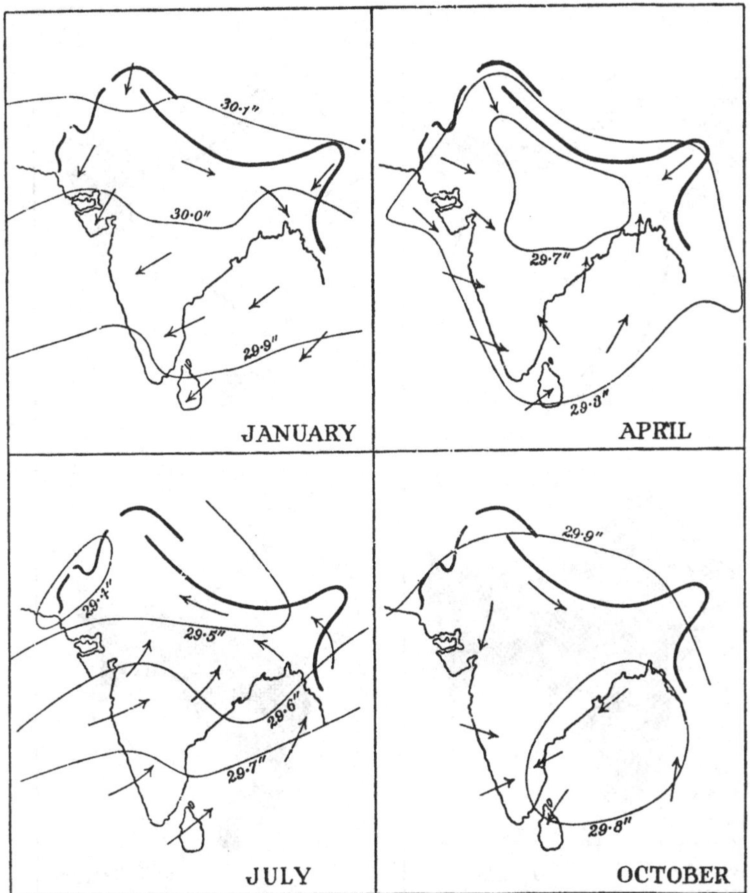

Distribution of pressure and winds in India

in their normal quantity, and famines occur as a result. One of the achievements of the British administration of India has been the establishment of irrigation and transport systems which compensate for the failure of the monsoon or carry food from the areas of plenty to those of dearth.

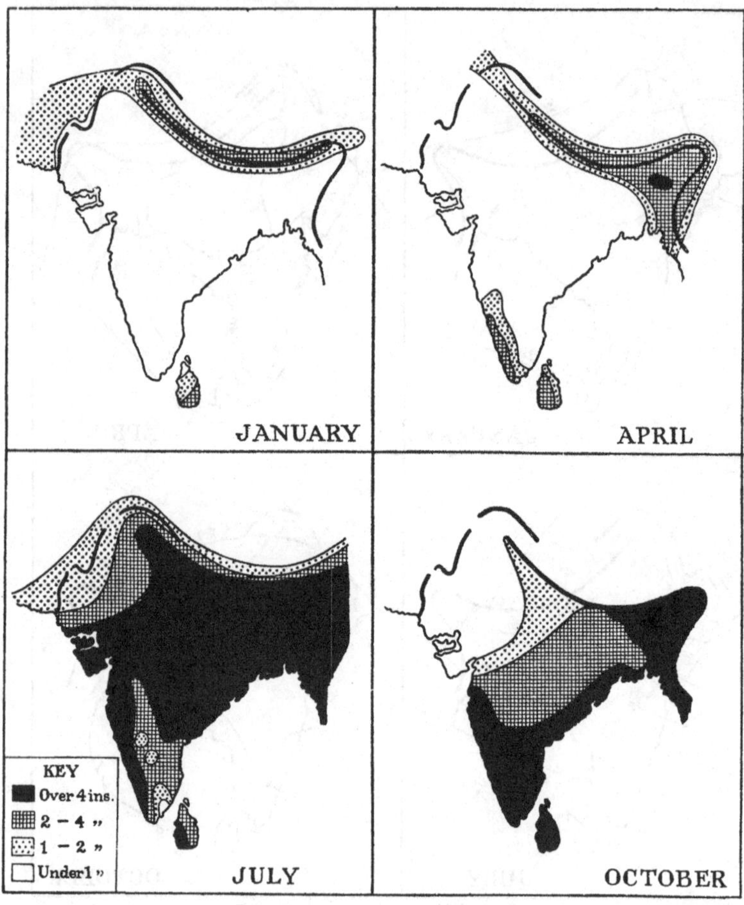

Distribution of rainfall in India

As the sun moves back to the Equator, the low-pressure centre shifts south over the Deccan, and by November this 'retreating monsoon' causes an indraught from the Bay of Bengal along the east coast. It is at this season that Madras gets its maximum rainfall. Elsewhere in India there are clear skies and a cool, dry northeast wind which reduces the temperature to a level bearable to Europeans. The January mean is as low as 50° F. at Peshawar, 60° F. at places like Benares in the Ganges valley, and 65° or 70° F. in the Deccan. Practically no rain falls at this season, the Bombay mean for the five months from December to April being 0·2 inch. The only exception to this winter drought is found in the northern Punjab, where the low-pressure disturbances, which

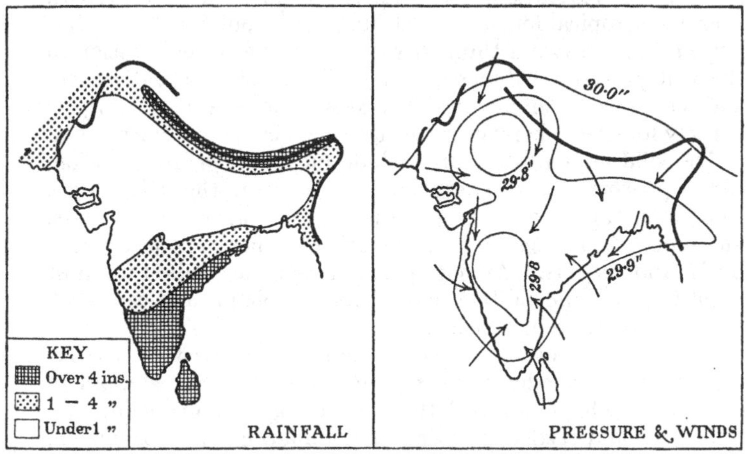

Conditions of the retreating monsoon in India
Note how the southward movement of the low-pressure centre causes onshore winds on the Coromandel Coast

in a weakened form make their way from the Mediterranean, give a secondary 'maximum' in February.

The local peculiarities mentioned above enable India to be divided into a number of climatic regions: (1) the seaward slopes of the Western Ghats which have a heavy monsoon rainfall whose period decreases from south to north, (2) the Coromandel Coast which has a mean temperature of 78° F. and a rainfall maximum in November, (3) northeastern India where the temperature varies from 60° F. in January to 80° F. in June and where the summer monsoon brings heavy rains, (4) central India with a

January mean of 55° or 60° F. and a June mean of 100° F. and a moderate rainfall from July to October, (5) the drought area of Sind and Rajputana, and (6) the district with winter rains in the northern Punjab.

The vegetation of India is as varied as the climate. The slopes of the Himalayas are clad in forest which is composed of temperate species on the higher parts and tropical species lower down. The Ganges valley is mostly cultivated, except in the *Terai*, while the Indus valley is either cultivated or sparsely dotted with xerophilous vegetation. The Western Ghats are covered with evergreen tropical forest of great luxuriance, but the Deccan has large areas of savana through which run lines of cultivation in the valleys and deciduous monsoon forests of teak, satinwood, and sandalwood in favourable places. The east coast strip is densely forested wherever it has not been cleared for cultivation.

The wild animals of India include monkeys, great cats (lion, tiger, panther, etc.), hyænas, dogs, bears in the Himalayas, elephants, tapirs, wild asses in Cutch, wild cattle, sheep, goats, antelopes, deer, pigs, and many others. There are also many birds and insects. Among the domestic animals the most important are the buffalo, zebu, horse, elephant, and camel, all of which are used as beasts of burden.

India is a land of many races of man. In ages past the warlike people of the northwestern mountains and of the steppes of central Asia have invaded the country again and again, conquered it, and settled down as its dominating class. Hence, in the more desirable areas society is divided horizontally into castes which are rigidly exclusive. The chief castes are the Brahmins, from among whom come the priests and scholars; the Warriors, including the princes; the Farmers; and the Pariahs or 'untouchables'. Actually, there are some 1200 subdivisions. Members of one caste may not marry or eat with those of another, so that the class distinctions can never disappear. But, naturally, the waves of conquerors did not attempt to occupy the remote or less fertile areas, and remnants of earlier peoples, like the squat, hairy Bhils of the eastern Deccan and the small, almost black, Tamils of the south, still survive undisturbed. The mixture of races in the Indo-Gangetic plain is illustrated by the existence of some 220 different languages. There are also many religions ranging from the most primitive paganism to Hinduism and

Islam. The caste system is an essential part of the Hindu religion. The Muslims, or followers of Islam, number only 70 millions, as compared with some 220 million Hindus; but they are the descendants of some of the later invaders and are on the whole the more vigorous people. The mutual hatred of Muslims and Hindus is one of the problems of the Government of India.

There is great disparity in the standard of life of the upper and lower classes. Among the former there is wealth and luxury, the arts flourish, and literature is encouraged. But the poor live in mud huts, wear little clothing, and have barely enough to eat. The ordinary farmer, or *ryot* (as he is called in Bengal), tills a few acres with the help of his family and lives on the crops. The chief occupation is agriculture, and more than 260 million acres were under cultivation in 1931, of which 228 million acres were sown more than once. As in all monsoon lands, rice is the chief crop, occupying 34 per cent. of the area under cultivation. Swamp rice is grown wherever the ground can be flooded in the rainy season, i.e. chiefly in the Ganges valley and on the east coast of the Deccan. In drier parts an inferior species of upland rice is grown. In 1932 sufficient rice was produced to supply local needs and to leave an amount valued at £13½ million for export. The chief centre of rice growing for export is Patna in Bengal. Where conditions are unfavourable to rice, millet and other grain crops as well as pulses are cultivated, and these occupied 42 per cent. of the cultivated land in 1931.

Next in importance is wheat, which occupied 10 per cent. of the land cultivated in 1931 and is grown chiefly in the Punjab. Large irrigation works have been established in that district. Wheat is used as a winter crop, other food plants being sown in the warmer seasons. A railway has been built from Lahore to Karachi to take the grain to the sea for export. Jute, cotton, and other fibres rank next in the area they occupy. Jute is grown chiefly on the delta and lower valley of the Ganges where the annual floods restore to the soil the large amount of plant food used up by the crop. Formerly, the fibres were sent to the British Isles, especially Dundee, for manufacture; but in 1932 two-thirds of the raw material was manufactured in the mills at Calcutta. In 1930 the value of the raw jute exported was £20 million and that of jute manufactures £39 million; but the world depression reduced these figures to less than half in 1932.

Cotton is grown chiefly on the black cotton soil in the north-western Deccan. The plant yields fibres and seeds, which latter give oil and seed-cake. The fibres are shipped in bales as raw cotton to England. In 1930 the quantity so shipped was valued at £50 million, but in 1932 this had fallen to £17 million. A good

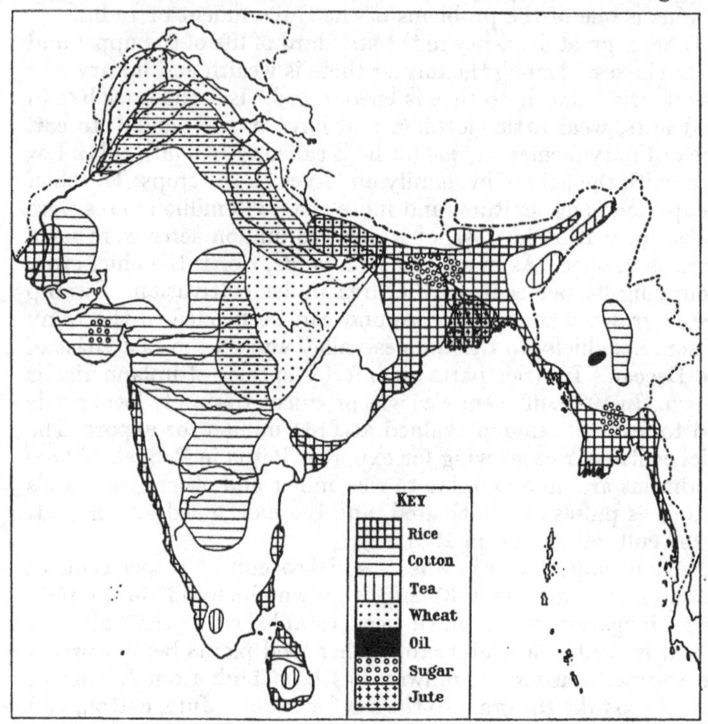

Distribution of the chief vegetable products of India

deal of cotton manufacture is carried on now at Bombay and other towns in the cotton area, but local demands are so far from being satisfied that cotton goods to the value of over £11 million were imported in 1932, almost entirely from the United Kingdom. Other important crops are tea, sugar, and tobacco. Tea, which needs a great deal of moisture, but will not thrive on wet ground, is cultivated on the Himalayan slopes around Darjeeling, in

Assam, and in the Nilgiri Hills. The value of the export in 1932 was £14½ million. Sugar and tobacco are grown on the wet lowlands, and the crops are mostly used locally.

In recent years manufactures have been growing rapidly. Besides the jute and cotton mills already mentioned, there are iron and steel works, oil refineries, printing presses, and other factories. Bombay and Calcutta are the chief centres, but Cawnpore supplies the leatherwork and boots for the British Army in India, Patna has rice mills, and Madras refines oil. Trade is chiefly with Great Britain, 40 per cent. of the imports coming from that country and 25 per cent. of the exports going to it. Other countries with which considerable trade is done are Japan, the United States, Germany, China, the Dutch East Indies, and France. The imports consist chiefly of cotton goods and machinery. The chief ports through which this trade passes are Bombay, Calcutta, Madras, and Karachi. Bombay owes its enormous growth in recent times to the opening of the Suez Canal.

The total population of India was nearly 353 millions in 1931. The vast majority live in villages or scattered farms: hence, the only two cities with more than a million inhabitants are those in which English influence is strongest, viz. Calcutta (pop. 1,485,000 in 1931) and Bombay (pop. 1,161,000). And only thirty-three others exceed 100,000. Calcutta, which was long the English capital of India, is on the Hooghly, a distributary of the Ganges, and is the natural outlet of the valley. Its large industrial suburb of Howrah lies on the opposite bank of the river. Madras is the focus of the Carnatic and much of the Deccan, but suffers from want of a good harbour. Its population in 1931 was 647,000. Hyderabad, the capital of its state and the chief town on the Deccan, is largely native. Delhi, the seat of the viceregal government, is the fifth town in size, with nearly half a million people. It owes its suitability as capital to its central position on the Ganges-Indus watershed. Of similar size is Lahore, the capital of the Punjab. The large official towns on the lowlands have 'hill stations' connected with them, to which the European officials retire to escape the unhealthy rainy season. Simla and Darjeeling, the hill stations of Delhi and Calcutta respectively, are small places on the slopes of the Himalayas, and Ootacamund, the hill station of Madras, in the Nilgiri Hills is no bigger; but the Bombay station of Poona has a quarter of a million inhabitants and manufactures cotton goods. In the Indo-Gangetic plain there are a number of towns of similar size: Lucknow and Cawnpore, famous for sieges during the Mutiny;

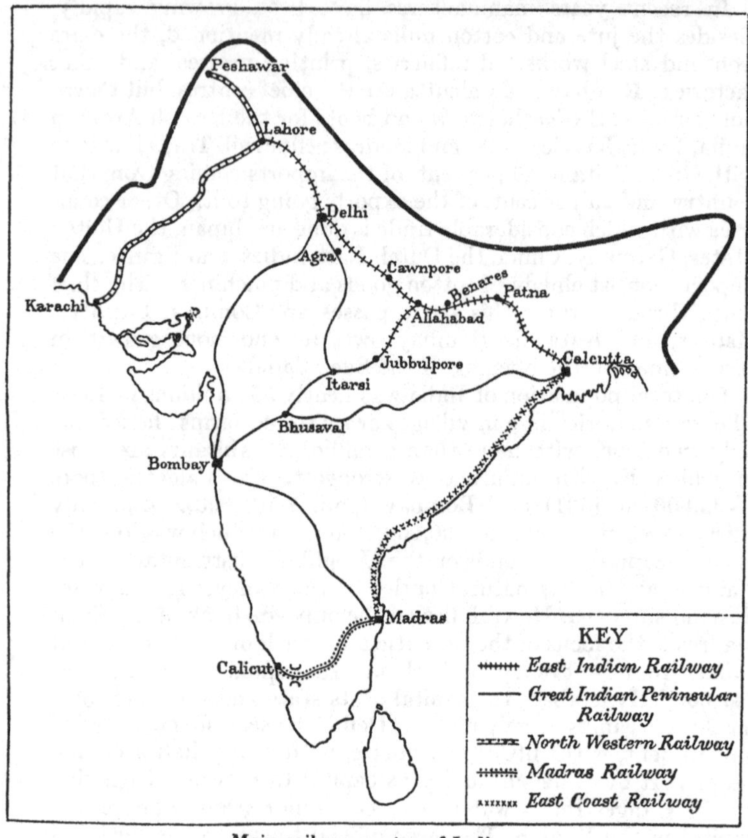

Main railway routes of India

Agra, the city of beautiful architecture and especially of the Taj Mahal; Amritsar, the sacred city of the Sikhs; and Benares, the sacred city of the Hindus.

The country is now completely enmeshed by a railway system. The East Indian Railway follows the Ganges valley from Calcutta

(Howrah) to Delhi and passes on to Lahore. Its lines touch at all the large towns in the valley. The Great Indian Peninsular Railway connects Bombay with Delhi, Calcutta, and Madras, and is the most extensive and important of all the sections of the system. The map on page 150 shows the principal lines. The total mileage in 1932 was nearly 43,000 miles, of which close on 32,000

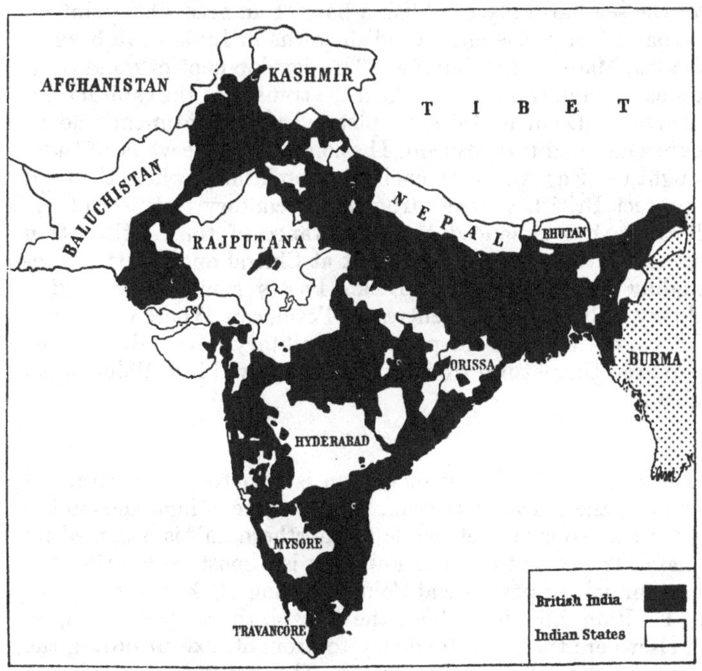

British India and the Native States

miles were Imperial State lines. Airways are being established. The main route of the Imperial Airways service from England to India and Australia passes from Karachi to Calcutta. Branches run from Karachi to Delhi on the one hand and to Ahmedabad, Bombay. and Madras on the other.

At the time of the first entry of the English into India the country was in a state of political chaos following the decay of strong central government from the capital of the Mogul emperors at Delhi, and provincial governors had seized the opportunity to set up independent principalities whose positions and boundaries depended more on the power of the usurper to hold than on any geographical conditions. English intervention suddenly 'froze' all political relations, but changed the outlook of the central government from the land to the sea. Invaders from over the sea naturally establish a base at an accessible point on the coast; hence, the chief English towns in India have been at Calcutta, Madras, and Bombay. The development of rapid communication and transport enabled the transfer of the capital from Calcutta to Delhi in 1912, to please Indian sentiment and to secure a more central position. The conquest of the whole of India brought the English to the encircling mountains, where the need to protect British subjects from the predatory instincts of the hill tribes has enhanced the importance of the northwestern passes. Fortresses have been built at Chitral and Quetta, commanding the Khyber and Bolan Passes respectively, and a military railway system centres on Peshawar. Where the mountains are not a bar to the passage of military forces, British rule has passed the natural frontier and overflowed into Baluchistan and Burma.

Ceylon

The pear-shaped island of Ceylon is structurally a detached portion of the southern Deccan. It has a core of highlands in the south rising to 8000 feet, while the northern half is occupied by a plain. The climate of the lowlands is almost equatorial, the mean annual temperature at Colombo being 81° F. and the range 3·5° F. Rain falls throughout the year on the southwest slopes, but elsewhere there is a tendency to drought, except during the season of the retreating monsoon. The natural vegetation is forest, though cultivation has replaced or modified the natural state.

The island is largely agricultural and produces both equatorial and tropical products. The former include rubber and spices, especially cinnamon; while the latter consist of copra, rice, tea, and cocoa. Several minerals are worked, but only graphite is important. Industry has begun in the form of refineries of agricultural products, the chief of which deal with coconut oil, rubber, and tea. Trade is mostly with the British Isles, the principal outlet being Colombo (pop. 284,000 in 1931). This port is of more than local importance, since it is the focus of the main ocean routes of the Indian Ocean. The railway system of the island centres on Colombo, the political capital, and joins up with the system of southern India by means of a ferry at Adam's Bridge.

Burma, Indo-China, South China

India is separated from its eastern neighbours by the Arakan Yoma and Patkoi Mountains, which, though not high, are yet difficult to cross owing to their steep gradients and the density of their forest covering. The path followed by this fold has been determined by the resistant tableland of Yunnan, which lies between the Bay of Bengal and the south China seas. Spurs run southward from it to form the backbone of the Malay Peninsula and the hills of Annam. Between the framework of these highlands are several considerable river basins, each forming a political division. Thus, the Irrawaddy basin, together with the narrower valleys of the Sittang and Salween, comprises the British province of Burma; the lower Mekong valley forms the French possessions of Cambodia and Cochin China; while the Song-koi lowlands are the French protectorate of Tongking. The Menam basin is occupied by Siam, a native buffer state set up by the Anglo-French Agreement of 1904 to prevent the clash of British and French interests. Farther east, the basin of the Si Kiang and its feeders forms the Chinese provinces of Kwangsi and Kwangtung.

The climate is influenced by the monsoons, and summer is therefore the season of rains. But Annam, which occupies the east coast of Indo-China, and lies in the rain shadow of the hills which form its western boundary, gets its maximum from the winter monsoon. During the change of the monsoon, violent storms known as typhoons trouble the China seas and the east coasts of this region. The temperature is hot in winter and very hot in summer, the annual range averaging 17° F. at Hué. Generally speaking, the range increases from south to north, being 12° F. at Saigon and 20° F. at Canton. Burma has a three-season year, as India has, with its temperature maximum in April.

The lower parts of the Yunnan tableland which fall within the region form extensive grassland areas in Laos and the Shan States, but are elsewhere densely forested. The teak is the predominant tree and forms one of the chief exports of the region.

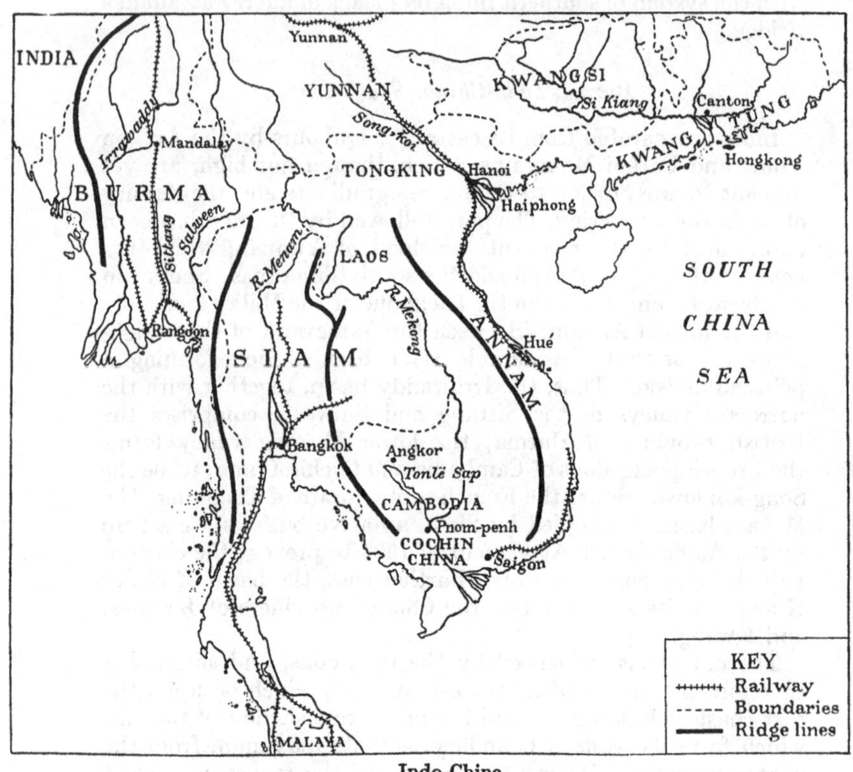

Indo-China

Showing political boundaries and places mentioned in text

Upper Siam exchanges much of its timber for Burmese cotton cloth. The river plains are thickly peopled and intensively cultivated; in fact, in south China and Tongking horticulture replaces agriculture. Rice is of course the chief crop and is grown in sufficiently large quantities to allow of important exports of the grain from Burma and Cochin China to India, China, Japan, and other countries. Besides rice, a large variety of tropical crops are grown. Cotton is cultivated wherever the soil is not too moist; tea plantations cover the hill slopes in Annam and south China, while sugar-cane shares the damp lowlands with rice. Maize is an export of Tongking, and is gradually spreading in south China. Silk, too, is extremely important, since the factories in France are relying more and more on Indo-China for their raw material, and south China has for ages shared the general Chinese love of sericulture.

The floods caused by the summer rains are carefully controlled in the delta lands of Tongking and Kwangtung, where the hard-working natives get every ounce they can from the soil. Large irrigation works are being established on the lower Menam for control of the river. In Cambodia the Mekong overflows into the lake basin of Tonlë Sap, swelling this inland water to four times its size in the dry season. This steadies the volume of the river below Pnom-penh, the capital of Cambodia, and also, when the floods subside, affords a large area of silt-covered ground which can be cropped in the dry months. This advantage has in the past given enormous wealth to the surrounding country and enabled the rise of a great kingdom, the ruins of whose capital at Angkor are one of the wonders of the East. The lake also yields fish, as do also the river and the coastal waters. Hence, Cochin China and Cambodia export fish and by-products such as fish oil and isinglass.

Minerals are found widely, the petroleum of Burma and the tin of the same country and of Siam being the chief kinds exported. But tungsten, silver, jade, and rubies are also worked in Burma, and various metals are found almost everywhere in the highlands.

Each of the great river plains has a port on or near its delta. Rangoon is placed so as to serve both the Irrawaddy and the Sittang, and indeed the railway to Upper Burma, which now reaches Myitkyina, runs up the valley of the latter. This valley is the former line of the Irrawaddy, the river having been captured by a tributary of the Chindwin just below Mandalay. The main stream is navigable as far as Bhamo. Rangoon (pop. 400,000 in 1931) is half-European in appearance and is not nearly as interesting as Mandalay, the capital (pop. 148,000 in 1931), which has picturesque houses and pagodas built in a peculiar half-Indian, half-Chinese style. The port of the Menam is Bangkok, the capital of Siam and forty miles upstream. It is a curious mixture of glittering palaces and wretched hovels, and evil-smelling canals take the place of streets in some parts of it. The trade of Siam mostly passes through Hongkong and Singapore, to which latter Bangkok is now connected by railway.

The navigation of the mouths of the Mekong is so difficult that the port of Saigon is placed on a small stream unconnected with the great river. It is the only town of any size in French Indo-China, outside Tongking, for Vientiane in Laos, Pnom-penh in Cambodia, and Hué in Annam are relatively small places. Formerly, the port of the Song-koi was Hanoi, the capital of French Indo-China, but the needs of modern ships have caused the growth of Haiphong just off the eastern edge of the delta. The railway from Haiphong to Hanoi continues up the river into Yunnan.

The most important town by far in the region is Canton, a city of 861,000 inhabitants (1926). It has large manufactures of cotton and silk, rice mills, and sugar refineries. Its backland includes much of the south of China besides the basin of the Si Kiang, and the town is a great centre for overseas trade. Off the mouth of the Si Kiang is the British island of Hongkong, whose capital at Victoria is an entrepôt for European trade with China.

Subtropical Monsoon Lands. This region differs from the last in having a distinctly cool or even cold season, while it is influenced by the monsoon winds of the great Asiatic system. It includes most of China, all of Japan south of Lat. 37° N., the peninsula of Chosen, and the island of Taiwan (Formosa). It will be convenient to treat the first three parts separately, combining Taiwan with China.

China

The relief of China is very complicated. Its western portion consists of lofty mountains and plateaus, and these send spur-like folded ranges eastwards to constitute the frame of the country. All these highlands are much eroded by fluvial erosion and broken by faulting. Between the eastward-running ranges lie extensive areas of lowland, connected with one of the two great rivers, the Hwang Ho and the Yangtse Kiang, which have added large alluvial expanses to the plains. Taiwan is a portion of the ridge of the great fold which runs from the Philippines to Japan, through the little Luchu Islands.

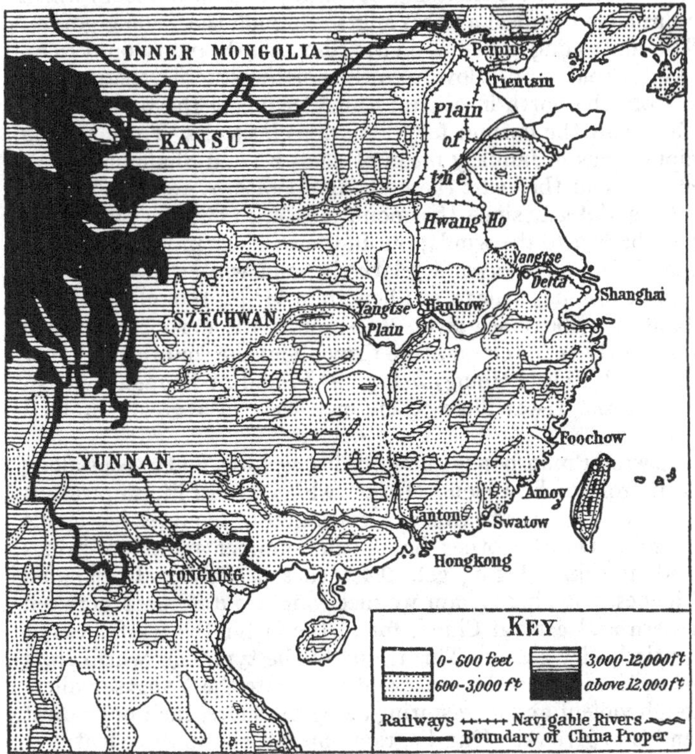

Physical features, routes, and chief towns of China Proper

Note the great north-south route which is now mostly followed by a railway, the east-coast route along the Yangtse, their crossing point at Hankow, and their outlets at Shanghai, Tientsin, and Canton

The western highlands consist of the Khingan Range in the north, the Alps of Szechwan in the middle, and the plateau of Yunnan in the south. Between the Khingan Range and the Alps of Szechwan the Tsingling Shan runs eastwards, reappearing after a break as the Hwai Shan. In the south the Nanling Shan extends under various local names from Yunnan to the mouth of the Yangtse. A branch of this range bends northward towards the Tsingling Shan, thus enclosing the basin of Szechwan from which the drainage escapes in the Yangtse through the Ichang gorge.

The controlling influence on the climate is the monsoon system. In winter cold winds blow from the northwest in northern China and from the north in central China, reducing the temperature to far below the normal for the various latitudes. Thus, Peking (Peiping) has a January mean of 23·5° F., and Shanghai one of 37·6° F. Near the coast these low temperatures are found as far south as Fukien, since the mountains are not high enough to form a barrier to the winds; but in the west the protection of the encircling uplands gives a relatively mild winter to Szechwan, where the January mean at Chengtu is 44° F., though this town is 1500 feet above sea-level. In summer the onshore monsoon gives a fairly even temperature throughout the country, the July mean at Peking (Peiping) being 79° F. and at Shanghai 80° F. North of the Hwai Shan, therefore, there is a greater annual range.

The winter monsoon is quite dry in the north and sweeps along with it from Gobi a cloud of dust which makes life unpleasant at that season. It is to this, however, that China owes her fertile *loess* soil. On the other hand, the summer monsoon brings abundant rain. Peking gets 9·4 inches in July and Shanghai 7·5 inches. But here again we meet another difference between northern and central China, for in the latter the winter is not an entirely dry season. This is due to the typhoons which from time to time sweep the coast and penetrate the Yangtse valley. Though violent and devastating along the shore, their only effect inland is to give rain. The diagram opposite shows these features. Yunnan, which is remarkable for the cloudlessness and healthiness of its climate, has a regular monsoon system of rainfall.

The portion of China within the region may be marked off into six natural subregions: (1) the southeast, (2) the uplands of the west and southwest, (3) Szechwan, (4) the lower basin of the Yangtse, (5) the plain of the Hwang Ho, and (6) the Shantung peninsula. Numbers 2, 3, and 4 comprise central China, numbers 5 and 6 northern China. The division depends partly on climate, but more on relief.

Southeastern China consists of a number of small valleys backed by irregular uplands. It has a hot, rainy summer and a mild, but not dry, winter. The island of Taiwan, which is included in this subregion though it is politically attached to Japan, owes its high rainfall to its lofty mountain backbone and its insular character. It is densely forested in the interior. The mainland faces the sea, where the coast is broken by many inlets and fringed with islands. The lowlands are intensively cultivated with rice and sugar-cane, while the terraced hill slopes produce

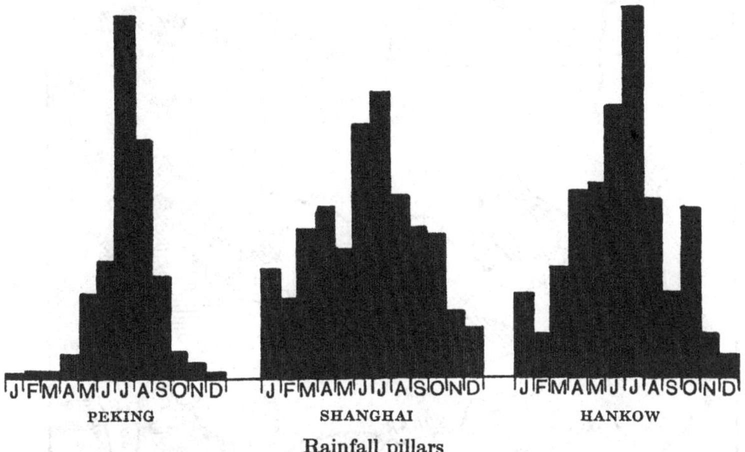

Rainfall pillars

1 inch of height in the pillars = 5 inches of rainfall. Note: (1) Peking has a typical monsoon rainfall; (2) Shanghai gets rain from typhoons; (3) Hankow is an intermediate type

tea and mulberry bushes for sericulture. This is in fact the main tea-growing area of China. It includes the Chinese province of Fukien and part of Chekiang. The largest towns are on the coast, whose broken nature encourages not only fishing, but also piracy, which even modern warships have been unable to suppress.

The mountains of the west are wild and infertile. Forested on their eastern and southern slopes, they are bare and merge into desert on the north and west. Yunnan is an upland savana country in which pastoral occupations are still predominant. Tea is cultivated on the southeastern slopes, and French influence from Tongking has caused some of its not very abundant mineral wealth to be exploited. Its chief town is Yunnan (='south of the clouds', i.e. of Szechwan), which is connected by rail with Hanoi.

Szechwan is often called the 'Red Basin' owing to the colour

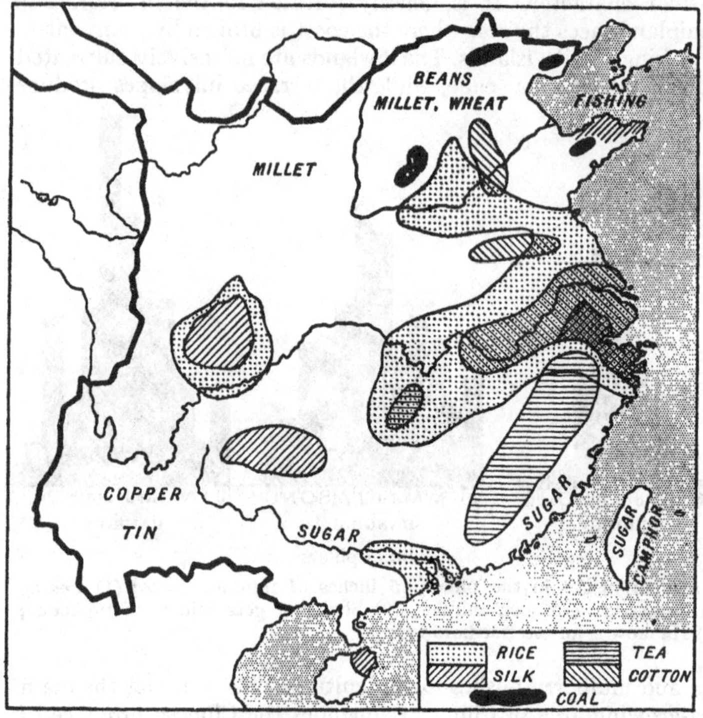

Simplified economic map of China, showing the chief areas of production of some of the most important products

of its soil. Its fertility together with the mildness of its climate makes it a productive area. Rice is the chief crop, but large quantities of tea and silk are produced on the terraced slopes. Chungking is the focal town, for all the feeders of the Yangtse in the basin flow towards its site. The surrounding ring of highlands is not unbroken, and there is moderately good communica-

Plate IX

(*Canadian Pacific Railway*)

'WILLOW-PATTERN' TEAHOUSE, SHANGHAI

The quaint, upturned eaves, the many windows, the foundation of piles, and the curious bridge should be noticed. The building on the left shows a strange mixture of Eastern and Western architectural designs

Plate X

(*Canadian Pacific Railway*)

PAGODA AT OMURO TEMPLE, KYOTO

The architecture and its setting of trees is typical

tion towards the north, west, and south. Towards the east the Yangtse forms the main route, but the gorge above Ichang is a distinct hindrance to navigation, for while boats can easily pass downstream, they can only ascend by being hauled upstream by large parties of men walking along the banks. Nowadays, powerfully engined river steamers are replacing human haulage.

At Ichang the river enters its lower course, flowing first through a wide plain and then through its extensive delta. On the latter are a number of large towns, among them Nanking, the present capital of China, Hangchow, a former capital, and Shanghai, a great port with a large European population. The plain includes the two large basins of Lakes Tungting and Poyang, and is considerably widened at intervals by subsidiary valleys, notably those of the Han and the Hsiang. Both these streams have cut back through the watersheds and in consequence have provided the main route from north to south. This partly explains the importance of the threefold town of Hankow-Wuchang-Hanyang which stands at the confluence of the Han with the Yangtse. The town is also the focus of the lower valley and can be reached by large river steamers. The subregion is one of the most fertile parts of China and is the only area which produces more rice than necessary to supply local demands. Tea is also much cultivated on the southern slopes, and mulberries are grown for making paper and for feeding silk-worms. The delta occupies parts of the provinces of Anhwei, Kiangsu, and Chekiang, while the rest of the lower basin is divided between Hupeh, Hunan, and Kiangsi.

Except in the delta area the lower basin of the Yangtse is separated by the low Hwai Shan from the plain of the Hwang Ho. The *loess* which forms the soil in the western parts of the plain is fertile, but needs irrigation, since it is very porous. It is easily cut by running water and the tread of man and beast, and since its angle of rest is very high, the streams and roads in this district pass through shallow gorges. The greater part of the plain is alluvial, and over its dead level surface the mighty river wanders between natural embankments which are usually reinforced artificially. When the banks break, ruinous floods often ensue, and at times the whole course of the river is changed. Between 1094 and 1853 the river flowed into the Yellow Sea, but in the latter year it suddenly shifted its course to the other side of the Shantung peninsula. Owing to the damage to property and the loss of life caused by these changes, the river is often spoken of as 'China's sorrow'.

As in the lower Yangtse valley, the fertility of the soil and the hot, wet summer cause the density of population to be extremely great. Failure of the monsoon to bring its normal rainfall has the same effects of dearth here as in India. Rice is the chief crop, but millet and soya beans are also grown in large quantities, and

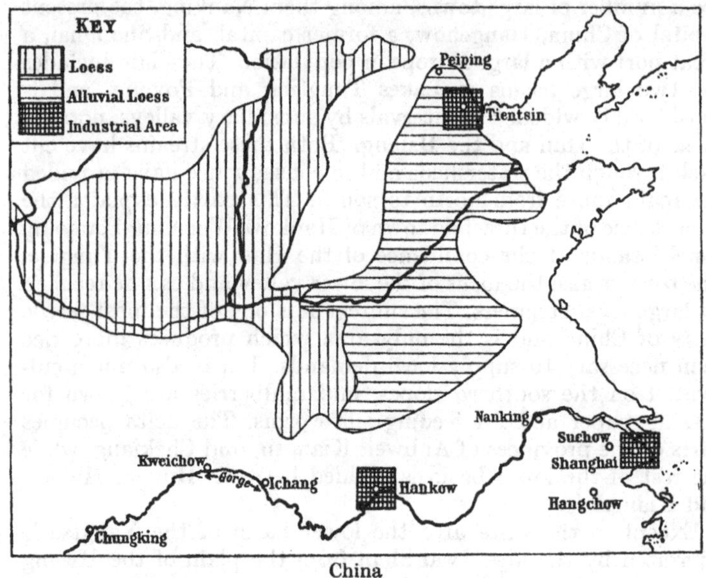

China

The loess area and the Yangtse towns

maize is increasingly popular. The river provides fish, which are caught with the aid of tame cormorants. The purely agricultural nature of the people causes the population to be distributed in farmsteads or in villages rather than to be concentrated in towns. Peking (now named Peiping) in the extreme north is by far the most important town and until 1912 was the capital of China. Its port of Tientsin is too far up the Pei Ho to be reached by large modern ships. The Grand Canal, now disused, connects Tientsin with the Yangtse delta and was formerly important.

The plain is bordered on the west by the escarpment of Taihan Shan, whose rocks of carboniferous limestone provide what has been estimated to be the richest coalfield in the world, anthracite having been worked since the time of Marco Polo at least. Lungan is, and has been for centuries, a centre of metal-working, since iron deposits occur nearby. Difficulties of transport hinder the exploitation of these minerals.

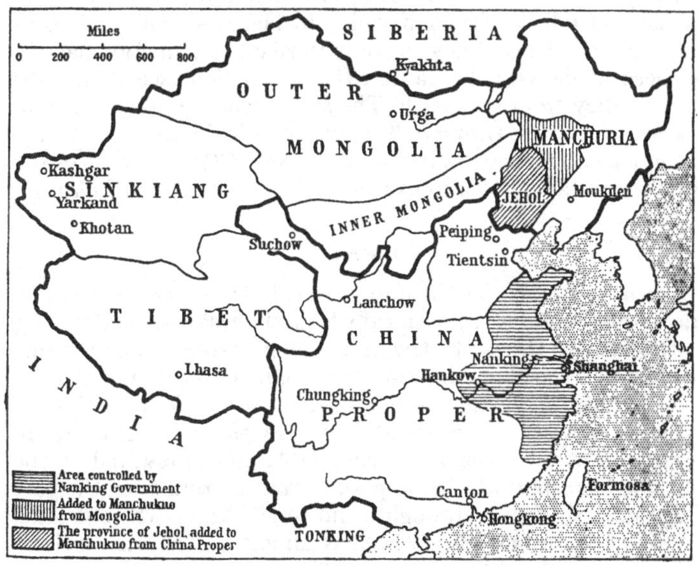

The Chinese Empire of former times

The outlying provinces have now broken away from China Proper

Shantung is a peninsula with two forested areas of upland. Its rocks contain a good deal of mineral wealth, especially gold, but the resources have scarcely been exploited so far. The narrow coast strip and the central band of lowland are cultivated with all the industry of the Chinese peasant, rice, maize, and tea being the most important crops. But the most typical product is the coarse silk made from the cocoons of 'wild' caterpillars which feed on the native oak forests.

The most important routes in China run north and south or

east and west. Of the latter, the valley of the Yangtse is the most important, while the Wei valley provides the first stretch of the great caravan route to the Lop Basin and Transcaspia. The Lunghai Railway now runs from the port of Haichow through Kaifeng to Cheng, and extensions have been planned into Turkistan. Two routes exist from north to south, both of which are now followed for most of their course by a railway. The first is the Kinghan Railway which starts at Peking, joins the Lunghai Railway at Cheng, and follows the Han valley to Hankow, whence it continues up the Hsing valley as far as Pinsiang. A junction has not yet been made with the Canton Railway which ascends the Pei Kiang valley from the south. The second route also starts from Peking, but goes through Tientsin and Tsinan to Nanking, Shanghai, and Hangchow (see map on page 157). In the north the Chinese railway system joins that of Manchuria (Manchukuo).

Chosen

Korea, or Chosen (as the Japanese call it), is a curving peninsula with mountains rising sharply from the east coast and sloping gently away to the west. Owing to the porosity of its limestone rocks, it offers good grazing country in spite of its abundant rainfall. For this reason, among others, it has been taken over by the Japanese, whose native islands have no pasture lands. Seoul, the capital, is the focus of a considerable area of lowland on the west and has a port at Chemulpho. A railway runs from the port of Fusan in the south through Seoul to Moukden, where it joins the Manchurian system. Chosen is almost entirely an agricultural country, the chief crops being rice and various other grains, beans, tobacco, and cotton. Fruit and silk are also produced. Stock-raising is important and includes the breeding of cattle, horses, pigs, asses, and goats. There is a good deal of fishing and some mining.

South Japan

Japan forms one of the festoons of islands which border the east coast of Asia. A second and smaller festoon is seen farther south in the Ryukyu Islands, which are politically joined to the larger group. The Japanese Islands consist of two lines of fold mountains with an intermediate valley, but, as earth movement has in the past been violent here and indeed is still in progress, the regularity of the folds is broken in various places, allowing the sea to divide the chains into numerous islands. The largest of these are Honshu, Hokkaido, Kyushu, and Shikoku. A comparison in size between these and the British Isles is interesting:

	sq. mi.			sq. mi.
Honshu	88,911	Britain	England	50,874
Hokkaido	34,276		Scotland	30,405
Kyushu	16,236		Wales	7,466
Shikoku	7,248			
				88,745
	146,671	Ireland		32,600
				121,345

Between Honshu and Shikoku the central valley has been drowned, to form the Inland Sea, whose shores have cradled the Japanese nation. The relief of the islands is extremely mountainous, and several peaks rise above 10,000 feet. The lowland areas are confined to a very narrow coast strip and two small plains, one around Tokyo and the other at the northeastern end of the Inland Sea. Between these plains is a highland area built up of volcanic rock and topped by a number of volcanoes.

There are more volcanoes in Japan than in any other country in the world. Over two hundred of these mountains exist, including Fuji-san ('the peerless'), Bandai-san, and Asama-yama. Fuji has not erupted since 1708. Its symmetry of outline has greatly impressed itself on the Japanese mind as a model of beauty. The other two cones have erupted with devastating effect within recent times. Hot springs occur everywhere and are used as hot baths in winter. Earthquakes are of frequent occurrence, the average number of shocks each year being 516 in the district of Osaka on the Inland Sea. Great damage often results from these earthquakes, as in 1923, when 150,000 people were killed and 170,000 houses destroyed in Tokyo. Earthquakes are sometimes followed by huge seismic waves in the sea, which sweep the coast, destroying villages and shipping. In 1896, one of these waves swept over 250 miles of coast and killed more than 30,000 people. The effect of these frequent dangers from natural causes has been to make the Japanese character brave and stoical.

Except on the west of Honshu, where the shores are low and sandy, the coastline is rocky and pierced by many inlets. This

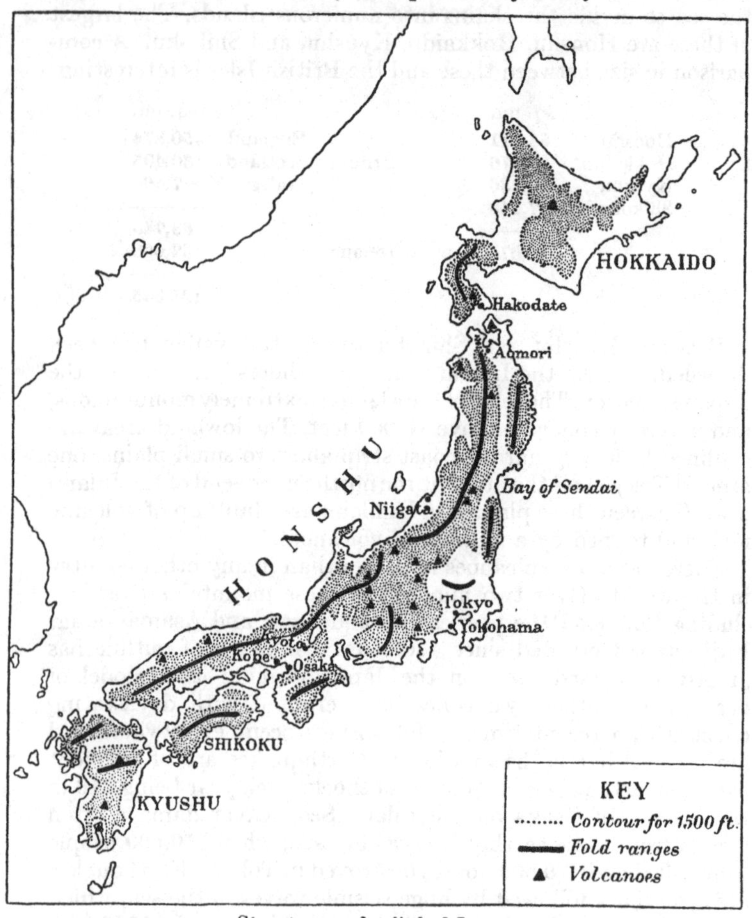

Structure and relief of Japan

has encouraged fishing and seafaring, and to-day over a million men make their living as fishermen, while large numbers are sailors in the mercantile marine or the navy. In 1930 the value of the fishing was estimated at £16¼ million, while the tonnage of the merchant shipping amounted to nearly 5 millions.

Since Japan stretches through 15° of latitude, its climate is far cooler in the north than in the south. In fact, only Kyushu, Shikoku, and the portion of Honshu south of latitude 37° N. come within the subtropical monsoon region, and the more northern parts will be dealt with in the next section. The main features of the climate of the south are a great range of temperature and a high summer rainfall. The summers are hot, the July mean being 78° F. at Kagoshima and 69° F. at Tokyo. This season is damp and enervating, since the monsoon brings much moisture from the Pacific and the highlands cause heavy precipitation. The rains begin in March and last until September, with maxima in June and September. In winter the temperature is mild, the January mean being 45° F. at Kagoshima and 37° F. at Tokyo. In this season the monsoon wind from the mainland picks up moisture as it crosses the Sea of Japan; hence, it gives a heavy snowfall to the highlands on the west. By a kind of Foehn effect, however, it causes clear skies and fine weather on the east. This normal state of things is broken at frequent intervals by the arrival of low-pressure disturbances which originate off the Philippines and travel along the Kuro Siwo to the Japanese coast. Sometimes typhoons are caused, but the usual effect is a heavy downpour of rain. Hence, no winter month is dry, Tokyo having 2 inches (in January) and Kagoshima 3·3 inches (in February) in the driest month.

The vegetation of south Japan is remarkable for its variety, since in it temperate pines grow with tropical bamboos, wheat with rice. The whole area is naturally one of forest. In Kyushu tropical plants predominate below the 3000-foot contour, and an evergreen flora persists, though up to a decreasing height, as far north as latitude 37° N. Above this plant belt is one of temperate deciduous trees; but in central Honshu the mountains rise above this and reach the belt of temperate conifers. The highest peaks attain the snow line. The evergreen belt of subtropical forest contains trees like the cinnamon, camphor, magnolia, and camellia; and its cultivated plants include rice, tea, cotton, and sugar-cane. The deciduous belt has the beech, maple, willow, and similar trees, and among cultivated plants the lacquer tree, the paper mulberry, the silk-worm mulberry, and, at lower

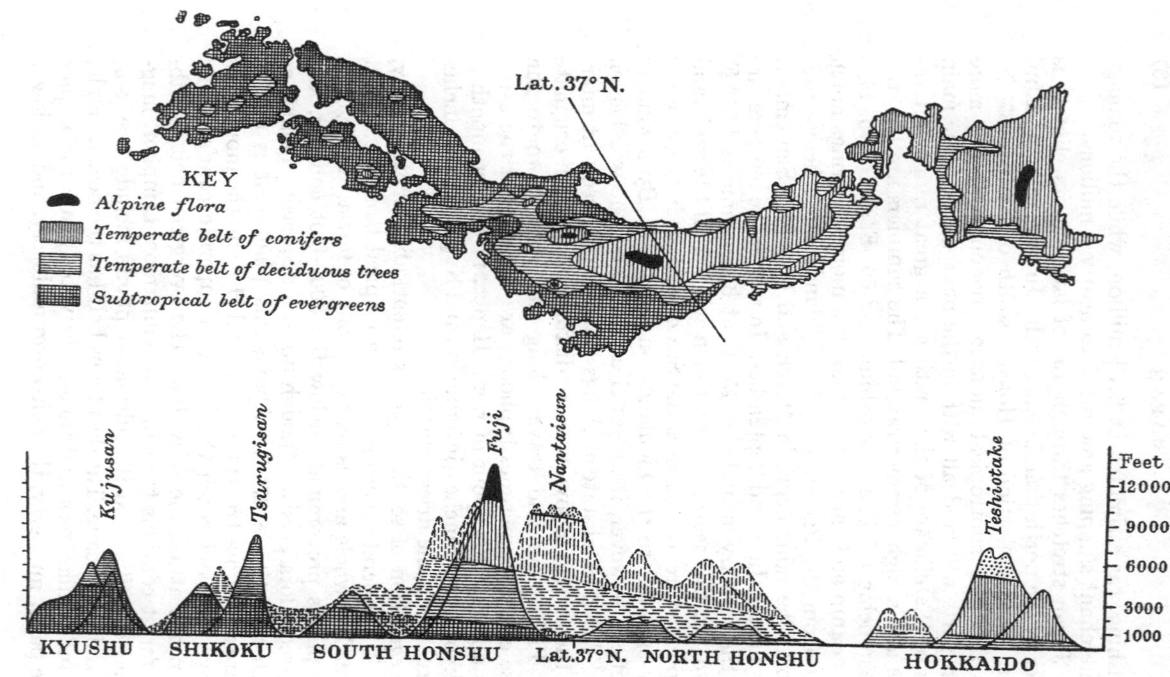

Distribution of vegetation in Japan

Above: map showing the superficial distribution of vegetation types. Below: diagrammatic profile showing the vertical arrangement of the vegetation types and the gradual descent of the temperate flora to the sea-level in the north

levels farther north, millet, maize, barley, and beans. Agriculture is of the intensive, horticultural type, as in China, and is carried on not only on the lowlands, but also on the hill slopes by means of terracing. The use of manure, irrigation, and rotation of crops is thoroughly understood and has been employed for centuries, but the methods and implements are still primitive.

Cereals form the chief crops, rice occupying more ground than all the other grains together. Wheat and rye take second place, and barley third place. But the quantity of cereals grown does not suffice for local needs and large shipments are imported each year to make up the deficiency. Among economic crops silk is by far the most important, the value of the export of raw silk amounting in 1932 to over £38 million, while that of silk tissues was £11 million. Next in importance is cotton, the value of the exports of this being £31 million. Tea, camphor, and sugar are also important.

Industry is increasing apace. Silk and cotton factories are being built year after year and are challenging the European products in world markets. Already the whole cotton crop is used for home manufacture, and the proportion of raw silk exported is being quickly lowered. Besides these main articles, Japanese industry produces pottery, toys, paper, and lamps. It is in these frail goods that the workpeople excel, for in cloth manufacture they are still capable of turning out only the coarser products. Hence, large quantities of the better class of cotton goods are imported.

Mining must be reckoned as important, gold, silver, copper, coal, iron, and mineral oil being found and worked in considerable quantities. In 1932 coal was among the chief exports.

The greater part of the commerce of Japan is with the United States, the total imports and exports in 1932 being £92 million. China comes next with £41 million, and India third with £24 million. Great Britain, Germany, and the Dutch East Indies also have important trade relations with Japan, the first-named supplying her with woollen and cotton goods.

Internal communications are largely carried on by water, the coastal traffic consisting of sailing ships of the junk type rather than of the modern steamboats which are gradually being introduced. Overland the roads are still narrow, crooked, broken in gradient, and of poorly built surface. Railways have developed rapidly, the mileage being 13,500, as compared with 20,500 in Great Britain and 14,000 in Italy. From Tokyo main lines radiate north, west, and south, following the coast strip or finding the

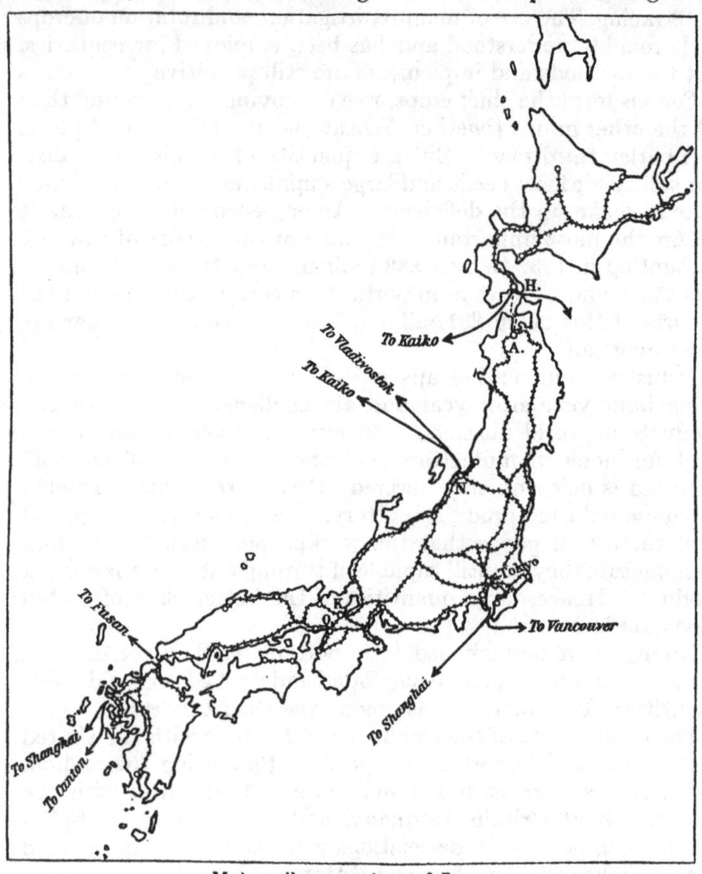

Main railway system of Japan

gaps through the mountains. Two lines run north to Aomori, while the western lines make for Niigata. The southern line curves round with the coast and reaches the Osaka plain, where there is a secondary focus with lines from the town of Osaka to Kobe and the ancient Japanese capital at Kyoto. Thence the main line skirts the Inland Sea, crosses the Strait of Shimonoseki, and ends at the port of Nagasaki. Air travel is also increasing, regular services now running between Nagasaki and Shanghai, and between Hiroshima and Port Arthur.

The insular position of Japan has enabled her in the past to maintain her political freedom against invasion from China and Chosen, but at the same time to absorb the best of Chinese civilisation. She has also escaped the deadening effect of tradition, and her change from 1868 onwards from a medieval and barbaric condition to a state organised on Western lines has been one of the wonders of the modern world. She has felt and is still feeling keenly the wrench involved by exchanging one civilisation for another. The chief trait in Japanese character is a love of nature, which may be seen in Japanese pictures and poetry. Flower-viewing is a national recreation. Another trait is patriotism, a feature which has made possible the rise of Japan to her present position among nations.

The population of south Japan is largely rural, yet there are a number of big towns, seven of which have each over a quarter of a million inhabitants. Tokyo (pop. 5,000,000 in 1930) is the focus of its plain, has easy communication in all directions, a central position, and facilities for reaching the sea at its port of Yokohama (pop. 620,000 in 1930). It is the modern capital of Japan. Osaka (pop. 2,450,000 in 1930) is the chief industrial centre, a port, and the focus of its plain. Kobe and Kyoto, each with close on 800,000 inhabitants, are not far from Osaka and share its industrial development. Nagoya (pop. 900,000 in 1930) is on a small plain of its own and is also industrial.

Temperate Monsoon Lands. The Maritime Province of Russian Siberia, the islands of Sakhalin and Hokkaido, and Honshu north of Lat. 37° N. have a monsoonal régime of winds, but are marked by far lower temperatures than the rest of Monsoon Asia. It is a hilly region, with little lowland except along the valley of the Ussuri River. A break between the hills of Chosen and the Sikhota Mountains gives room for the port of Vladivostok, whose usefulness is limited by the necessity for keeping its harbour ice-free by means of ice-breakers during two or three months of the year. This great cold—the port is in the latitude of the north of Spain—is due partly to the winter winds from the 'cold pole' of Siberia and partly to the influence of cold ocean currents issuing from the Okhotsk and Bering Seas. The January mean in Vladivostok is 5° F., but at Sapporo in Hokkaido, which has the benefit of sea influence, it is 21° F. Summer is hot, the July mean being 70° F. at both places mentioned.

The aborigines of the region have been nearly ousted by the Japanese and Russians. But Russia is far away, and the Maritime Province has not been greatly developed. Some wheat is grown in the Ussuri valley, but the main occupation of the people is cattle-rearing. The uplands are too heavily forested with pines and firs to be of much use, though they yield some gold. Between 1904 and 1945 Russians shared Sakhalin with the Japanese; but neither nation was able to make use of the island's dense coniferous forests. Even Hokkaido, which still contains survivors of the Ainu people who once occupied all Japan, is far from being fully developed as yet. It has good fisheries off its north coasts, where cod, herring, and whales form the greater part of the catch. The climate is too cold for rice, and beans and potatoes are the chief crops. Hakodate and Sapporo are the chief towns, each having about 200,000 inhabitants. A railway connects them and has branches to other smaller places.

A ferry connects Hakodate with Aomori, the terminus of the railway from Tokyo. This northern portion of Honshu is distinctly temperate in climate and vegetation, and in it, as in Hokkaido, beans and potatoes replace rice as the popular food crops, though wheat is grown now in large quantities. Niigata on the west coast has regular steamship communication with Vladivostok. North Japan is far less densely peopled than the portion farther south.

II. The Rest of Asia

The Subtropical Southwest. This consists of the peninsula of Arabia, an old crustal block which forms a worn tableland with its surface generally tilted down towards the east. As in the Deccan, the western edge stands up as a mountain ridge. Its shores are washed by the Red Sea whose waters fill the depression formed by the northern portion of the Great Rift Valley of Africa. The line of sinking is continued, though in smaller proportions, through the Gulf of Aqaba, the Dead Sea, and the Jordan valley up to and beyond the Sea of Galilee.

The position of Arabia within the high-pressure belt of the

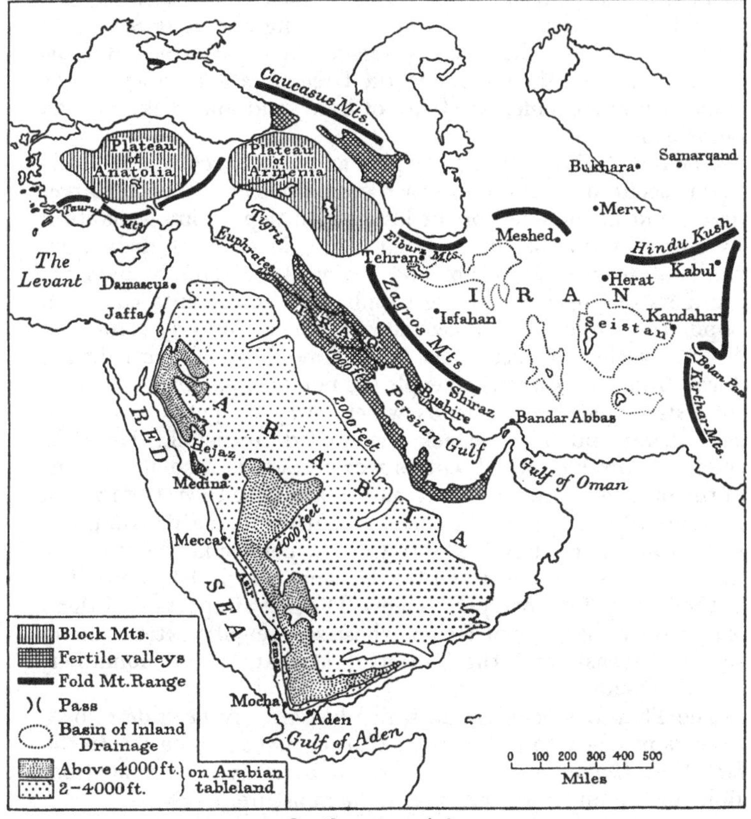

Southwestern Asia

Showing important structural and physical features

Tropic of Cancer and in the midst of the three great land masses of Asia, Africa, and Europe makes the peninsula into a barren region. Only in the extreme southwest in Yemen, the old Arabia Felix of the Romans, and in the east in Oman, does sufficient rain fall to support more than a sparse, xerophilous vegetation. In the Hadhramaut, or southeast coast strip, the hollows between the broken, rocky hills are sand-filled, with extensive thickets of acacia and tamarisk. In the centre the Nejd has a similar, though far sparser, vegetation, and along the Red Sea coast palms grow and cultivation is possible in small areas. But the Nefud Desert in the north and the Great Arabian Desert of the south are unredeemed stretches of bare sand and rock, with an occasional oasis.

The population is therefore small and widely scattered. Little towns occur along the west coast and on one or two of the larger oases, and in these a more or less settled population feeds itself by cultivating millet in irrigated fields. Mecca and Medina are the headquarters of Islam and are visited annually by large numbers of pilgrims, on the supply of whose wants the town populations live. The Hejaz railway, which formerly connected Medina with Damascus, has been allowed to fall into disuse. Away from the towns the people are nomads, passing from oasis to oasis in patriarchal bands which carry on petty wars with each other and often attack the settled townsfolk. The only economic product which enters world trade is a small amount of the famous mocha coffee, so named after the town from which it is exported. Aden, in the extreme southwest, is a British naval station and port of call. The rest of the peninsula was ruled by the Turks up to the war of 1914–18, since which it has been left in the hands of its own petty chiefs. The most important of these is naturally the King of Hejaz, whose sovereignty extends along the west coast, and the Sultan of Muscat, whose dominions include Oman.

The Thar Desert of India, which has already been described, belongs properly to this region. It owes its barren nature to the fact that the monsoon winds which blow over it come from Baluchistan and therefore bear little moisture.

The Central Toplands. From the Persian Gulf northeastwards to Manchuria central Asia is occupied by a series of toplands separated by high mountain ranges. The largest of these is Iran, which is divided politically between Persia, Afghanistan, and Baluchistan. It is enclosed by the ranges of the Elburz and Hindu Kush on the north, the Suleiman and Kirthar on the east, and the Zagros on the west. Its rough surface averages between three and four thousand feet above sea-level, but rises in places to highlands above 8000 feet and in others falls into enclosed depressions which form areas of inland drainage. The largest of these are Seistan and the Lut Desert.

The winters are cold, while the summers are hot. Furthermore, owing to the rareness of the air, the daily range is large, and changes of temperature are apt to be sudden and great in all seasons. There is little precipitation, except on the mountains, whose tops are snow-clad and which therefore provide melt-water for irrigation. Hence, a number of towns like Tehran, Isfahan, and Shiraz have grown up in favourable spots where millet can be cultivated. The mountain slopes have sufficient grass to feed flocks of sheep and goats, from whose wool and hair has sprung the one ancient industry, carpet-making. In modern times the south-west has been exploited for oil by the Anglo-Persian Oil Company (see map on page 180).

The poverty of the country has given rise to professional brigandage which has sometimes been carried on on a large scale. The objects of attack have been the settled agricultural people of the country itself, the wealthier folk of the lowlands of India and Iraq, and the traders using the old caravan routes which traverse the topland. The chief of these run from Tehran, the capital of Persia, through Isfahan, Kerman, and Shiraz to the ports of Bushire and Bandar Abbas; from Tehran through Meshed and Herat to Kabul or Kandahar and so to India. The British have curbed the marauding propensities of the natives to some extent by annexing Baluchistan, but India is still subject to raids from Afghanistan.

The next topland is that of Tibet, which is enclosed between the Himalayas and the Altyn Tagh. The eastern end, which is not bounded by a mountain barrier, abuts on to the broken highlands of western China. The topland is divided into two parts, southern and northern Tibet, by the Karakorum and Kunlun Mountains. The surface averages over 12,000 feet above sea-level and is dry and swept by bitterly cold winds. The Chang, or central area lying astride of the Kunlun, is mostly of bare rock and has a severe climate. North and south of this there is an abundance of grass, and flocks and herds are kept. The great valleys of the southeast, particularly that of the Tsangpo, are sufficiently sheltered and fertile to produce millet. It is here that lies Lhasa, the capital of Tibet.

The yak plays a great part in the domestic economy of the people, who use its shaggy coat for clothing, its milk for food, and its dung for fuel. Religion, in a form of Buddhism, enters very largely into the life of the people, for the chief priest, or Dalai Lama, is the ruler of the country and monasticism is

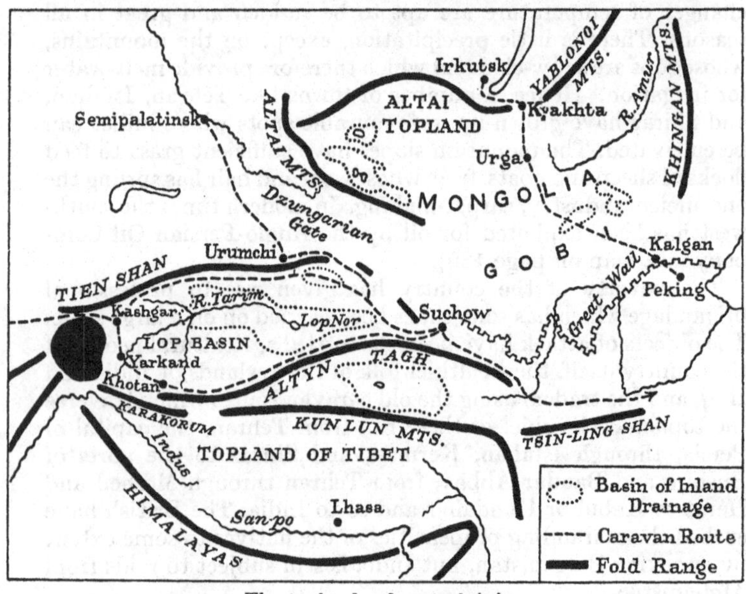

The toplands of central Asia

a popular occupation. The 'devil dances' performed by the monks and the use of the prayer wheel are among the striking customs of the land. Tibet was formerly a Chinese province, but is now independent. Its chief connexion with the outside world occurs in the purchase of tea from China.

The Lop Basin is enclosed between the Altyn Tagh and the Tien Shan. Its surface, which is largely unexplored, is about five thousand feet above sea-level, though the extraordinary, pit-like Turfan valley in the northeast actually falls below sea-level by some fifty feet. Distance from the sea and the enclosing mountains cut off all precipitation from the basin, which is therefore extremely barren. The melt-water from the surrounding mountains moistens a strip along the foot of the ranges, but the streams are soon swallowed up by the thirsty soil. The Tarim alone struggles on as far as the salt swamp of Lop Nor. Hence, the central area is occupied by a great sand desert, the Takla Makan. The watered strip at the foot of the hills is fertile and where irrigated produces millet, melons, and other fruit, and cotton. Towns, of which the chief are Yarkand, Kashgar, and Khotan, occur at intervals and are connected by the great caravan route from China to the west. Higher up the slopes of the mountains live a nomadic population of herdsmen.

East of the Lop Basin lies the vast country known as Mongolia. It consists of three parts: Inner and Outer Mongolia, and Gobi. Inner Mongolia comprises the western slopes of the Khingan Mountains and a fringe of grassy country at their foot. A belt of semi-desert divides it from Outer Mongolia, which lies to the south of the Yablonoi and its western continuations. Much of the country is poor steppe, but it improves towards the mountains, whose slopes are forested. Grazing land of moderate quality is found in the Dzungarian Gate, a wide gap between the Altai and Tien Shan. Gobi is a vast desert of irregular surface. Little systematic knowledge has as yet been collected of Mongolia, but travellers' reports attribute to it a very cold winter and a hot summer. High winds often cause a thick dust haze which makes travelling difficult. The scanty rainfall occurs in summer. Its poor steppe lands are inhabited by nomadic tribes who live on horseback and feed partly on the flesh and milk provided by their herds and partly on the fruits of brigandage. In days gone by, hordes of these wild horsemen used from time to time to invade China. This led to the building of the famous Great Wall which stretches from the Gulf of Chihli nearly to Suchow, a distance of 1200 miles. Suchow is the starting place of the caravan routes from China to the west, while from Kalgan in Inner Mongolia ran another route through Urga and Kyakhta to Irkutsk in Siberia. As Chinese tea was sent to Russia by this way it was known as the 'tea route'.

The Mediterranean Region. This includes Asia Minor, Transcaucasia, Syria, Palestine, and the valley of the Euphrates-

Tigris. The differentiating factor is the transitional position of the region, which brings it within the influence of the westerly winds in winter, but under tropical desert conditions in summer. The main features of the climate are a hot, dry summer, windless and cloudless, when everything becomes grey with dust and the vegetation is parched; a mild, pleasant winter with rain enough to revive plant life and to enable crops to be grown; and an absence at all times of violent weather. The rainfall decreases from the Mediterranean coast eastwards and is due partly to relief, as in Palestine, and partly to the effect of low-pressure disturbances, as in Iraq. The following table gives some means at typical places:

	Temperature (° F.)		Rainfall (in.)		
Place	January	July	January	July	Year
Jerusalem	45	73	6	0	25
Batum	43	74	10·2	6·0	93·3
Baghdad	47	94	1	0	7

The vegetation is peculiar. It consists of shrubs in favourable places, degenerating to scrub in poorer conditions. Much of the soil is bare, especially in summer; but the hill tops are often clad in forests of cedar and of the species of oak and pine characteristic of this region. The drought conditions of summer are resisted by various xerophilous adaptations, the most striking of which are the reduction of transpiration by means of a coating of wax on the leaves, as in the olive, or by a glossy surface, as in the laurel; water-storage in a bulb; and æstivation, as in the hyacinth, daffodil, etc. Many species of flowers and soft fruit (peach, grape, apricot) are either indigenous or have long been acclimatised.

The peninsula of Asia Minor consists of the tableland of Anatolia, but the name also applies to an extension farther east which is known as the plateau of Armenia (see map on page 173). The Taurus Mountains form the southern boundary of the former, while the Ararat ranges separate the latter from Iran. The Black Sea and Levantine coasts are unbroken, but on the Ægean there is an alternation of promontories and deep-set inlets with a fringe of small islands. The inlets are continued by valleys of great fertility, the most important being that of Smyrna. The narrow Black Sea coast strip widens out behind Scutari. The chief products are tobacco, raisins, figs, olive oil, and cotton. Manufactures are relatively unimportant, being limited to some cotton cloth and fruit-preserving in Smyrna, cement works in Ankara, and silk-making at Brusa, whither the industry was moved wholesale from the island of Kos by the Turkish authorities some centuries ago.

Transcaucasia includes the valley stretching from sea to sea between the plateau of Armenia and the Caucasus. It is divided between the three little states of Armenia, Azerbaijan, and Georgia, all of which are Soviet Socialist Republics in union with Russia. The soil is fertile, and good crops of cotton, grapes, tobacco, and silk are produced. The chief economic product is mineral oil, which is raised near Baku on the Caspian and taken by a pipe-line through Tiflis, the capital of the three federated states, to the port of Batum on the Black Sea (see map on page 180). The Ararat ranges cause a good deal of relief rain, most of which finds its way into the Tigris and Euphrates. These two rivers formerly entered the sea somewhere near Baghdad, but they have overlaid their valley with alluvium and have half filled the Persian Gulf with sediment. When irrigated and drained, the flood plain is very productive, and crops of cotton, wheat, and barley are grown. But owing to Turkish misgovernment the irrigation system has been ruined and has not yet recovered. Oil is being exploited near Kirkuk and in other districts.

Syria and Palestine rise in steps from the Mediterranean to the desert tableland behind. The country is of limestone and has many karstic features; but the valleys are fertile and the rainfall sufficient for crops. The coast plain, which is of sandstone, is the most fertile part of the country and produces wheat, barley, millet, olives, lentils, and oranges. Farther inland is the hill country, where cattle, goats, and donkeys are pastured, while the valleys produce wine and oil. The Jordan valley grows grain by irrigation. It falls below sea-level before it reaches the Dead Sea, whose shores are salt-encrusted and barren. The principal towns are Jerusalem, Damascus, the route-junction of Aleppo, and the port of Jaffa. These are all connected by rail with the Turkish system.

The region has one of the longest histories in the world, for it is the scene of Old Testament story, of the Trojan War, and of the Crusades. The remains of ancient fortresses built by the Venetians and Genoese are still to be seen in Rhodes, Kos, and other islands in the Ægean. The people are a confused mixture of Turks, Arabs, and Greeks, as well as many other older races now submerged. The misrule of the Turks has checked progress, and the region is therefore of greater interest for its past than for its importance in the present.

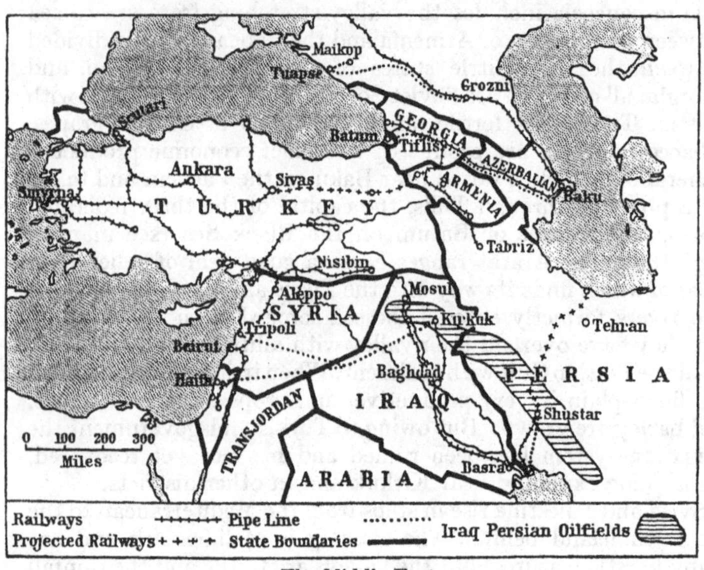

The Middle East
Modern boundaries, and developments in railways and oilfields

Subtropical Interior Lowlands. The northwest of Asia consists of the largest expanse of plains in the world. The southern portion, which is distinguished from the northern by its scanty rainfall, is so low that not a little of it is actually below the level of the sea surface. The shallow Caspian and Aral Seas cover the lowest ground. Formerly, the area under water was much greater than now, and the Caspian and Aral were one and were joined to the Black Sea (see map on page 6). Recent desiccation has caused the seas to shrink and has formed a desert, the Qara Qum, south of the Aral. The Caspian is maintained by the Volga, which rises outside the region and contributes so large a proportion of the water of the sea that the northern part is merely brackish; while the Aral depends entirely on the melt-water brought to it by the Oxus from the Pamirs. It is thought that when desiccation first began, hordes of nomads were forced to migrate to Iraq, Asia Minor, and Russia, where they conquered new homes for themselves. Such a theory would account for the historic invasions of those countries in early times.

The climate is cold in winter and very hot in summer. The January mean is 3° F. at Irgis in the north of the region, 23° F. at Krasnovodsk in the southwest, and 30° F. at Tashkent in the east. In July the mean exceeds 78° F. at all these places. The rainfall is exceedingly scanty and does not exceed 8 inches a year, except on the eastern mountain slopes, and as a rule the summer is dry. The region is therefore one of poor steppe, with patches of desert here and there. North of the Aral there is sufficient grass for pasturage and the support of nomadic herdsmen, but farther south in the Qizil Qum, or area between the Sir Darya and Oxus, vegetation becomes very sparse. These two rivers are used for irrigation by a settled population which grows rice, cotton, and mulberries for silk production. The chief town here is Khiva, once the centre of an independent state. Along the foot of the highlands on the east and south, streams fed by melt-water moisten a narrow strip of land and allow of cultivation. Wheat, fruit, hemp, and cotton are grown, together with mulberries for silk-worms. It is in this area that we find the towns of Tashkent, Khoqand, Samarqand, and Bukhara, well known of old and situated on the caravan route to China. Samarqand is a romantic old city with some 85,000 inhabitants and was once the capital of the empire of the famous Genghis Khan, a nomadic conqueror whose deeds were known as far west as England.

The region is now divided into three soviet republics allied with Russia. On the southeast is the little state of Tajikistan, which is of little importance. North of it is Usbekistan, which includes the fertile hill-foot district and the Oxus valley. Its chief town, Tashkent, is the capital of the Transcaspian republics. The third division is Turkmenistan, which contains most of the desert country. The Transcaspian Railway runs in a wide curve from Samara in Russia through Tashkent, Samarqand, Bukhara, and Merv to Krasnovodsk, a little port on the Caspian. The Soviet Government of Russia is endeavouring to exploit the region for tropical and subtropical products, such as cotton and hemp, and irrigation schemes have been established at various points.

New political divisions in Trans-Caspian Russia

The Temperate Interior. Between the Ural Mountains and the Yenisei River northwestern Asia offers a great expanse of temperate grassland and rolling plains. Northwards and eastwards this open country merges into the belt of taiga, while in the south, as we have seen, it is separated by a low divide from the steppes of Turkistan. The region forms the basin of the Ob, though its eastern portion drains into the Yenisei. Its climate is extreme, cold winters alternating with hot summers. Tobolsk, which may be taken as a fair representative of the whole region, has a January mean of $-2°$ F. and a July mean of 65° F., i.e. a range of 67° F. This means that outdoor work is suspended from November to March inclusive, the thermometer showing mean records of well below freezing point for those months. The snow cover is slight, since precipitation is small, owing to the distance from the sea. The mean rainfall in January at Tobolsk is 0·5 inch. But in summer convection rains give a July mean of 3·5 inches. Evaporation is not active, so that the total rainfall of 19 inches is abundant, and the drainage is sufficient to swell the Ob into a large river.

Formerly, the plains were the home of nomadic herdsmen, but Russian immigrants have displaced them and settled down to an agricultural life. The soil is fertile and yields good crops of wheat. In 1931 the area under this grain was about 6 million acres. The cost of transport has, however, checked wheat-growing and has led to the development of dairy-farming. Butter is the chief trade product. The valley of the Irtish, a feeder of the Ob, is the richest area, and its focal town, Omsk (pop. 160,000), is the second largest town in the region. Semipalatinsk on the upper Irtish is a grazing centre and a terminus of the caravan route through the Dzungarian Gate to China. Barnaul and Tomsk on the upper Ob are also grazing centres. Another feeder, the Tobol, has its focus at Tobolsk near the junction with the main stream.

The region forms the greater part of western Siberia, one of the territories of the Union of Russian Socialist Soviet Republics. Two railways enter it through the Urals, one from Leningrad via Sverdlovsk, the other from Moscow via Samara, converging at Omsk and continuing eastwards as the Trans-Siberian Railway as far as Vladivostok. Before the building of these lines at the end of the last century, transport was mainly by water along the Ob and its feeders, which are conveniently arranged with an east-and-west direction and have low watersheds. This feature of the river system had great influence on the immigration of Russian settlers. Unfortunately, since the mouths of the rivers are frozen for months each year, and in any case the outlet is into the Arctic Sea, the main streams are useless for commerce, though the Ob empties into a spacious gulf.

On the eastern side of the continent there is another, though far smaller, region of similar relief and climate. It lies south of the Amur between the Khingan and Sikhota Mountains and is partly in Manchuria and partly in Mongolia. The northern portion is drained by the Sungari, the southern by the Liao Ho. Like western Siberia, it has extremely cold winters and hot summers. The January mean at Harbin is −2° F. and the July mean is

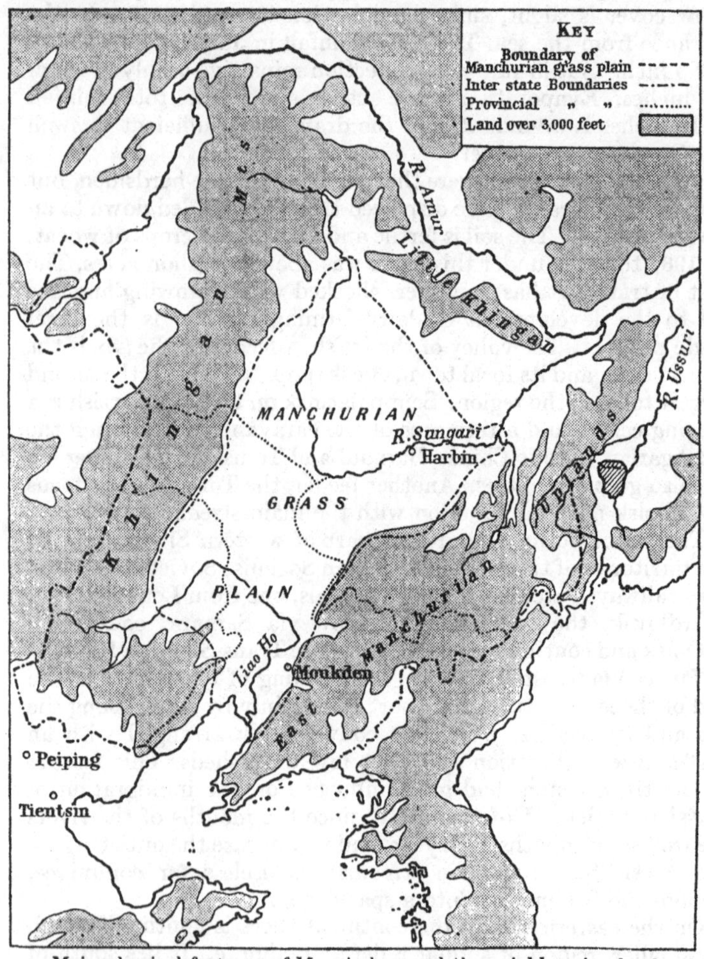

Main physical features of Manchukuo and the neighbouring lands

72° F., a range of 74° F. The winds are of the monsoon type, and the rainfall maximum is in summer. The régime shows its affinity to the monsoon type by having a drier winter and a wetter summer than western Siberia.

Formerly, it was the home of nomadic horsemen who, some hundreds of years ago, conquered China. In recent years Chinese farmers have been emigrating into the country and cultivating its fertile soil. Wheat, soya beans, and millet are the chief crops. At one time the Russians tried to take possession of the country, but were checked after a war with Japan in 1904–5. The Japanese then established a puppet state under the name of Manchukuo; but their defeat in 1945 ended its existence.

In the political no less than in the economic development of the region the railways are playing an important part. The Trans-Siberian Railway runs along the northern side of the Amur valley, making a long *détour* to Vladivostok. To shorten the route, the Chinese Eastern Railway was built. At Harbin the new line forked, one branch going on to Vladivostok, the other turning south to Moukden, Port Arthur, and Tientsin. Since the Japanese controlled this railway, the Chinese built another line from Peking through Manchuria and Mongolia to join the Trans-Siberian Railway. The map on page 186 shows these lines.

The Cold Lands of the Northeast. From the River Yenisei to the Pacific the country consists of broken uplands whose eastern portions are known as the Yablonoi and Stanovoi Mountains. The general slope of the land is northwards down to the sea, and all the valleys open towards the Arctic—two facts which have a bearing on the climate. The temperature is one of extremes, the winters being as cold as any on earth, while the summers are hot. Thus, Yakutsk has a January mean of $-46°$ F. and a July mean of 66° F., a range of 112° F. At Verkhoyansk these extremes are even greater. Farther east, sea influence begins to operate, and Okhotsk on the shores of the sea of that name has a January mean of $-11°$ F. and a July mean of 55° F. The length of the winters is also abnormal, the thermometer rising above freezing point only in the five months from May to September. The rainfall is slight, the annual mean total at Yakutsk being 14 inches, with a maximum in summer. In the south along the slopes of the mountain boundary of Outer

Mongolia there is greater precipitation, but at no station does it exceed 15 inches. In a region of such low evaporation as this the

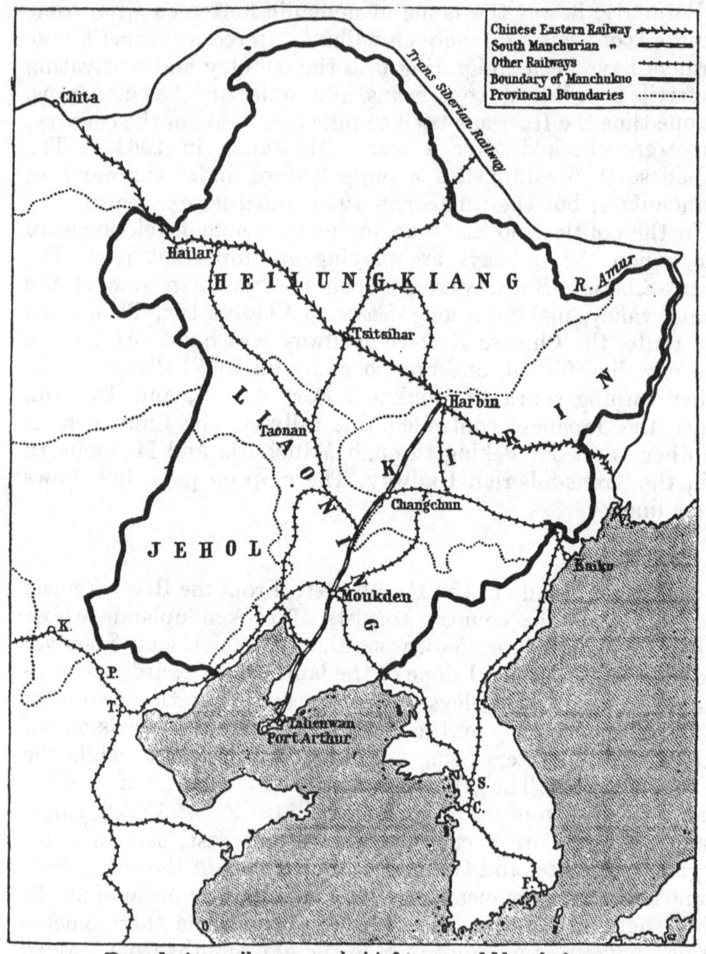

Boundaries, railways, and chief towns of Manchukuo

rainfall is sufficient for forest growth, and here is found the taiga of Siberia. This is a mixed forest of deciduous trees and conifers in the south, but northwards the deciduous species gradually disappear. The highest ground rises above the tree line, which near Yakutsk lies at about 1200 feet.

The natives, who are Yakuts and Samoyedes, live mainly by hunting. Under Russian influence some effort is being made to develop the resources of the country and to produce for export minerals, timber, and furs. On the sea coast fishing takes the place of hunting, and in the south there is agriculture, in which wheat is the chief crop, and pasturage. The minerals have the best future, for besides vast resources of coal (especially in the Kuznetsk basin) there are also silver, gold, lead, and copper. Development is hindered by transport difficulties, the rivers being open for only a short season and the railways being almost limited to the main Trans-Siberian line. This runs from western Siberia through Krasnoyarsk and Irkutsk to Chita, whence its original line follows the north side of the Amur valley to Khabarovsk. A regular air service now transports gold and furs from Yakutsk to Irkutsk. Another service plies between Irkutsk and Moscow through Omsk. The whole of the region lies within the territories of the Union of Russian Socialist Soviet Republics.

The Tundra Region. North of the taiga the temperature, which never rises above a mean of 50° F. in any month, precludes the growth of trees. The ground itself is frozen for the greater part of the year and thaws only on the surface. The lower ground is swampy, especially near the courses of the great rivers which inundate their valleys in summer. Vegetation consists of mosses, lichens, dwarf trees, and, in the short summer, of annuals which complete their life cycle in a few weeks. Owing to these natural conditions the region is of small importance. In fact, it is only occupied by a sparse population of hunters and fishermen who eke out a squalid and hand-to-mouth existence. The only trade product of value is the fossil ivory which is mined in the New Siberian Islands. The former existence here of mammoth elephants, to whom the ivory tusks belonged, suggests that in an age gone by the climate was far warmer than it is to-day.

HUMAN GEOGRAPHY

Exploration. As the home of the three chief civilisations, habitable Asia has in a sense been known from earliest times. India was known to the Greeks, whose armies under Alexander the Great penetrated as far east as the Indus. Even China had been heard of in a vague way. But the upheaval of European social life which accompanied the fall of the Roman Empire and the rise to importance of Muslim peoples checked the growing acquaintance of East and West and even caused the slight knowledge of the East to be forgotten in Europe, whither the centre of Western civilisation had now moved. It was not until the Middle Ages that relations began to be re-established through the exchange of courtesies between the Crusaders and their foes, through trade, and through adventurous travellers whose curiosity led them far afield. By 1250 Transcaspia had been visited by monks or traders, some of whom wrote accounts of their journeys.

In 1271, a Venetian named Marco Polo achieved the greatest voyage of all. Reaching Bandar Abbas on the Persian Gulf, he made his way over the topland of Iran and across the Pamirs to Khotan and thence to Peking. There he attracted the attention of the Emperor and rose to be an ambassador. In this character he journeyed far and wide in China and noticed various features, such as the use of coal for fuel, in which Chinese civilisation surpassed that of the West. After some fifteen years he returned home, journeying by ship through the Straits of Malacca to the Persian Gulf. He wrote an account of his adventures, but unfortunately he had so many strange things to tell that the whole book was regarded as a fictitious narrative.

Two hundred years later, the obstacles put in the way of trade by the Ottoman Turks caused the Portuguese to seek an ocean route to the East, and in 1497 Vasco da Gama rounded the Cape of Good Hope and coasted northwards to Zanzibar. There he found Arab seamen who were accustomed to use the periodic monsoon winds for voyages to India and back. In fact, these Arabs had penetrated as far as Java. With the help of Arab pilots da Gama reached Calicut, and a few years later other Portuguese navigators sailed to Malaya. The new route brought English, Dutch, and French ships to southeastern Asia, and gradually the coast became known as far north as Japan. It was not until the eighteenth century, however, that Cook and La Pérouse explored the northeastern waters.

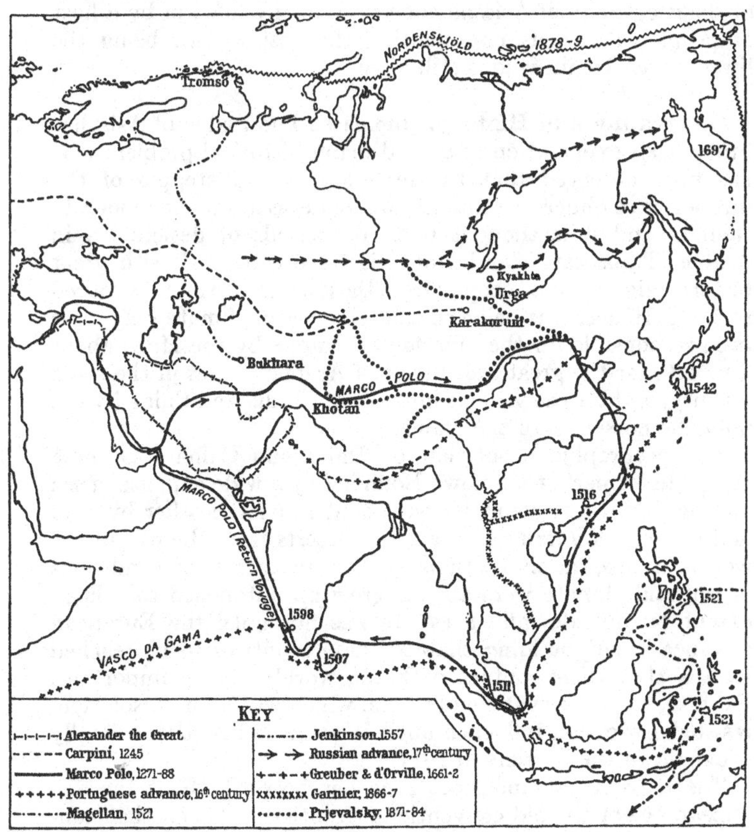

The exploration of Asia

Meanwhile, the Russians had been sending expeditions across Siberia, and in 1728 Bering reached his strait and crossed to Alaska. The Russian advance was on the whole achieved gradually by many explorers, but between 1871 and 1880 Prjevalsky made three journeys through central Asia which enabled the chief features of the country to be put on the map. In 1879, too, Nordenskjöld made his famous voyage in the *Vega* from Tromsö to the Pacific along the Siberian coast. Since that date the details of the geography of Asia have been gradually filled in by a host of people of various nationalities, the best known being the Swede Sven Hedin who is still alive.

Geography and History. Increased knowledge of Asia has brought an explanation of a good many historical problems. It has been observed that the semi-deserts and steppes of the interior have undergone alternations of desiccation and moderate rainfall, and it is thought that the periods of desiccation in Arabia, Transcaspia, and Mongolia have coincided with great historic migrations of peoples from those areas into more favoured regions. Hence, it is believed that changes of climate may have been responsible for the invasion of Europe by the Huns about A.D. 400, for the great movement of Arabian tribes in the sixth century, and for the various inroads into India and China by the nomadic horsemen of the plains.

The geographical isolation of India and China was more complete formerly than now. Bounded by a wide expanse of sea on one side, these countries were shut in on the other by wild and broken mountains, forests, and deserts from the rest of the world. Hence, the civilisations which were fostered by conditions of soil and climate were able to grow up untouched to a large extent by outside influences. It was not until the European advance in shipbuilding enabled these countries to be reached by sea that their contact with the world became important. Even so, the conservatism of China, which was born of isolation, resisted European influence until Western industrialism finally broke down the barriers in 1911.

The Russian advance into Asia has brought about the disappearance of the old conventional frontiers between that continent and Europe. The Ural Mountains are in truth no physical barrier, since the ascent to them is gradual and they are pierced by the wide Ekaterinburg Gate. They presented no obstacle to the building of railways and indeed are more imposing in appearance on the map than on the ground. Between their southern end and the Caspian Sea there is no trace of a boundary, and so the Asiatic tribes have overflowed into Russia. The conventional line of the Ural River exists in books only, the Russian provincial boundaries taking no account of it. In the same way, even the rapid current of the Bosporus and Dardanelles has not sufficed to separate the peoples on the opposite shores, and history contains a series of attempts on one side or the other to overflow into what is after all a region similar to their own in relief and climate.

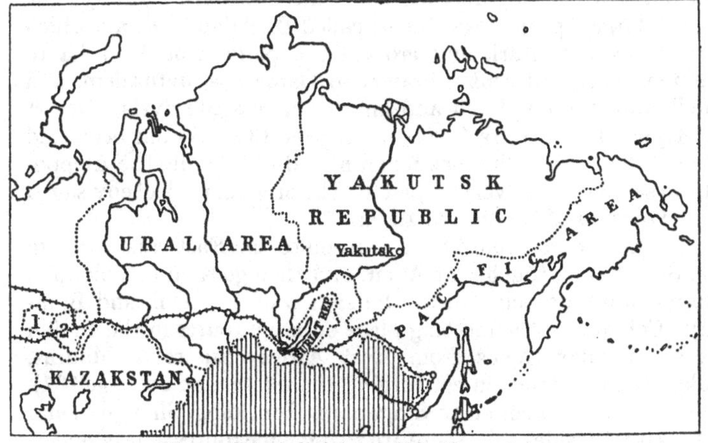

New political divisions of Northern Asia

The non-Russian areas are shaded. 1 = the Tatar Republic; 2 = the Bashkir Republic

Communications. Asia contains some of the most ancient overland routes in the world. From all corners of Asia Minor, Iraq, and Iran old caravan routes converge on Meshed, whence starts the main road to China. After reaching Merv, it makes for Bukhara, Samarqand, and Yarkand, where it forks. The more popular route follows the line of hill-foot towns on the south side of the Lop Basin, finally crossing a stretch of semi-desert before reaching Suchow. The other branch skirts the north side of the basin and makes for the same goal. Another route starts from Semipalatinsk, passes through the Dzungarian Gate, and then forks, making for Suchow in one direction and in the other crossing Gobi to reach Kalgan. The tea route from Kalgan to Kyakhta has already been mentioned. Other less important caravan paths led from Khotan across the Karakorums into Kashmir, and from Szechwan into Tibet.

Nowadays, railways are being constructed, but mostly on a regional basis. The Trans-Siberian line from Moscow or Leningrad through Omsk is the most ambitious. From Omsk it passes through Krasnoyarsk to Irkutsk and Chita, where the northern arm goes to Khabarovsk and Vladivostok and the southern to Harbin, after joining up with the Chinese Eastern Railway. The second largest project was the so-called Baghdad Railway, which started from Scutari, and crossed the plateau of Anatolia to Aleppo, whence it sent a branch to Damascus, Jerusalem, and Medina on the one hand and another across the Syrian Desert to Baghdad and Basra (see map on page 180). As has been said above, the Medina line has fallen into disuse, while the Aleppo-Mosul section was never completed. The original project envisaged the extension of the line to India.

The great ocean route from Europe to the East runs through the Suez Canal, touches at Aden, and then goes on to Colombo, after sending branches to the Persian Gulf, Karachi, and Bombay. Colombo is the meeting place of ocean routes in the Indian Ocean, regular tracks from East Africa, Australia, Madras, Calcutta, and Rangoon converging there (see map on page 137). The main route goes on to Singapore, Hongkong, Shanghai, and Nagasaki. Besides this, there are two Pacific routes, one connecting Shanghai and Hongkong with Australia, the other crossing from Yokohama either on the great circle to Vancouver or else via Guam and Honolulu to San Francisco.

During the last ten years a number of airways have been established. The chief line is that of the British Overseas Airways Corporation, which runs from London to Lydda in Palestine and thence through Basra to Karachi. From this port secondary lines branch out to Delhi and to Bombay and Madras, while the main route crosses the peninsula to Calcutta, whence it continues through Rangoon to Singapore. Here it will connect with the Australian airway from Darwin, thus affording a through route between England and Australia. Another important route goes from Moscow to Irkutsk, but this has already been referred to above, as have also the numerous short, local services that are being increasingly established, especially in Japan, China, the East Indies, and India.

Plate XI

(*Canadian Pacific Railway*)

A FELUCCA ON THE NILE

An ancient type of river-craft

Plate XII

(*Canadian Pacific Railway*)

THE PYRAMIDS AND THE SPHINX

One of the Seven Wonders of the World. The constructions are now much eroded by wind-blown sand

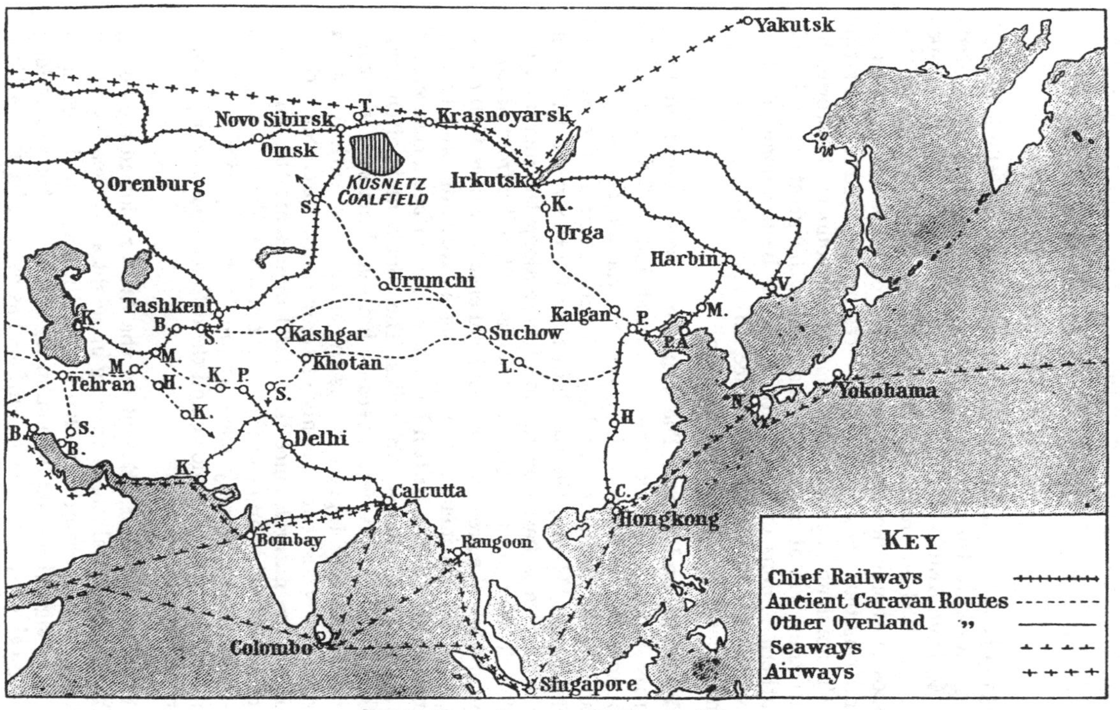

Main lines of communication in Asia

PART III

AFRICA

AFRICA is a part of the Old World and is actually joined to Asia by the isthmus of Suez. Although it is separated from Arabia by the Red Sea and from the south of Europe by the Mediterranean, yet the geographical affinitics of the opposite shores make the sea a bond of union rather than a sundering barrier. But circumstances have caused a mere strip of the north and northeastern coasts to be connected historically with the civilisation of Eurasia, while the great central and southern regions of the continent have been among the last portions of the Earth's surface to yield to the exploring tendencies of the western European.

Position and Size. The Equator passes almost midway across Africa, whose extreme ends reach Lat. 37° N. and 35° S. The meridian of Greenwich crosses Algeria and the Gold Coast, placing a large slice of the bulge of the Sahara in the western hemisphere. Hence, the standard time for the whole Union of South Africa is based on the local time along the meridian for 30° E. The greatest length of Africa is 5000 miles from north to south, while the greatest breadth from east to west is 4000 miles. In area the continent is twice as big as Australia and twenty times as large as the British Isles, having a surface measurement of about 11 million square miles. The continent is often compared with South America because both land masses are cut by the Equator and taper towards the south, and with Australia because they are both mainly of similar structure. Such comparisons are natural, since these land masses are the only ones with considerable areas in the southern hemisphere; but in fact the likenesses are only such as are to be expected between similar geographical regions, and the differences are conspicuous and important.

Build and Relief. Except for the Atlas region in the north, Africa is a vast tableland varying in height above sea-level from 6000 feet in the south to 1200 feet in the Sahara. Large areas show no sign of having been submerged beneath the sea for long geological ages, and the land surface has assumed a comparative evenness unbroken by lofty mountain ranges or deep depressions. Such a land mass is sometimes called a crustal block. Similar formations are noticed in the tableland of western Australia, the Deccan, and the Highlands of Brazil and Guiana, areas which for this and other reasons are thought to have once formed part of the ancient continent of Gondwana Land.

The evenness of the surface of Africa is broken in the north, as we have seen, by the fold ranges of the Atlas, and in the Sahara and Rhodesia there are lines of hills whose original formation is obscured by ages of weathering. The boulder-strewn, granite Matoppo Hills in Rhodesia and the fantastically wind-carved

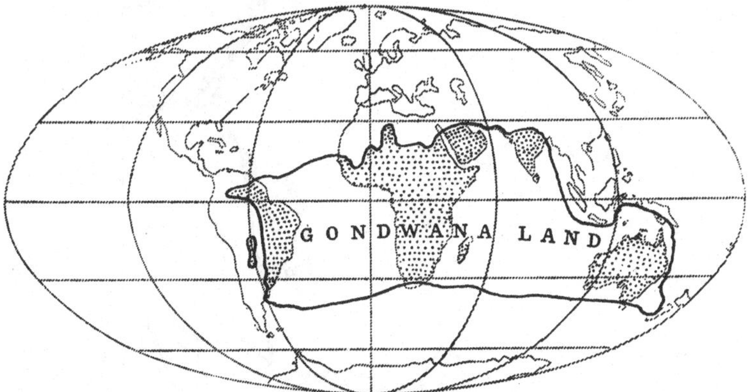

Map of Gondwana Land

heights of Tibesti are well-known examples. The eastern side of the continent has suffered from the earth movements of Asia, and a great rift valley which begins in Syria and forms the Red Sea continues through Abyssinia and Lake Nyasa to Beira. Its course is marked on the map by a line of lakes, and on the ground itself the parallel, inward-facing escarpments are clearly cut. A branch of this great rift valley contains Lake Tanganyika and other bodies of water west of Lake Victoria. The edges of the rift have in parts been disturbed by volcanic activity, large areas having been built up of volcanic rock in the Mufómbiro Range, Ruwenzori, and above all in Abyssinia. Two lofty extinct cones, Kilimanjaro and Kenya, which rise to about 19,000 feet, raise their snow-capped peaks almost on the Equator.

The most noticeable feature of the relief of Africa is perhaps the absence of large, fertile plains such as form the centres of population in all the other continents. The Nile, the Congo, and the Niger have all lowered their basins to a considerable extent, but they have not yet reached the stage in the cycle of erosion when they will be free of rapids and falls in their lower courses.

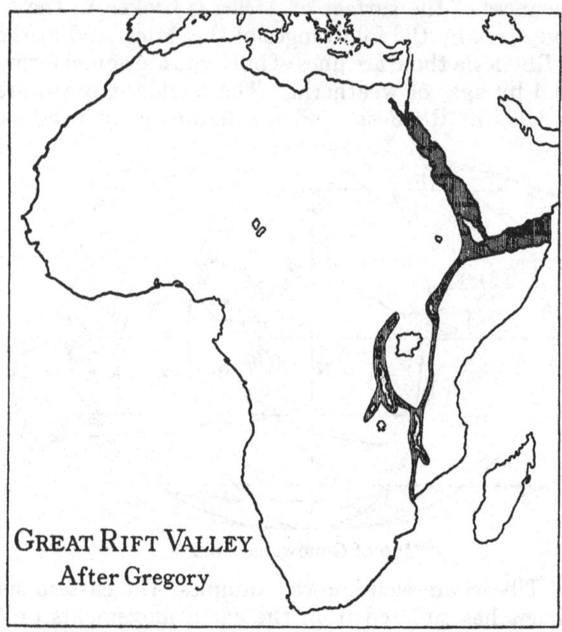

The Rift Valley is shaded. Note the line of lakes which lie on its floor

The Nile is interrupted six times between Khartoum and Aswan by rapids or cataracts, the Congo has twenty miles of unnavigable rapids between Matadi and Leopoldville, and the Niger is broken by rapids at Busa, where Mungo Park met his fate. These breaks in the gradients of the rivers are caused by the necessity to descend the plateau edge. Naturally, the usefulness of the rivers has been greatly impaired, and indeed these interruptions, by increasing the inaccessibility of the interior, were one of the chief causes of the long delay in the opening up of Africa.

The height of the plateau edge and its distance from the sea varies a good deal. In Natal the edge is tilted up to simulate a mountain range to which the name Drakensberg has been given. On the other hand, along parts of the north and northwestern coasts the drop is small. Between the plateau edge and the sea there is everywhere a coast strip. In South Africa there is a tendency for the descent from the plateau to be made by terraces, a tendency which is best seen in the Cape Province, where the two terraces are known respectively as the Great and the Little Karroo. The coast strip is broadest in Nigeria, Kenya, and Nyasaland.

Climate. Africa lies so equally on either side of the Equator, with no part extending beyond the subtropics, that its climatic belts are duplicated in the two hemispheres. The set of maps on the next page shows the distribution of temperature in January, April, July, and October. In July the area north of the Equator is subject to a high midday sun and runs a mean sea-level temperature of over 80° F., the shaded space running one of over 90° F. Around the coast in the north and northwest the isotherms clearly indicate that the sea is cooler than the land. In the southern hemisphere at this season the general trend of the isotherms is east and west, though the influence of the cold Benguella Current makes the west coast about 8° F. cooler than the east. In October the midday sun has moved south, and the area with a temperature of 90° F. and over has conformed with the movement and split, one part lingering north of the line, the other lying south of it. The influence of the sea is more pronounced in both hemispheres.

Then, in January the area with a temperature of 90° F. and over is well to the south, though, owing to the tapering of the land, it is smaller in size than in July. The influence of the Benguella Current reaches its maximum, but the difference between land and sea temperatures is not great in the northern hemisphere, and the isotherms run fairly east and west. The April distribution is similar to that of October.

The areas of highest temperature on the maps at which we have been looking become areas of lowest pressure on the maps on page 199. These areas of lowest pressure move north and south with the sun. In January there is a large centre of low pressure over South Africa, and winds blow into it from all around. Here

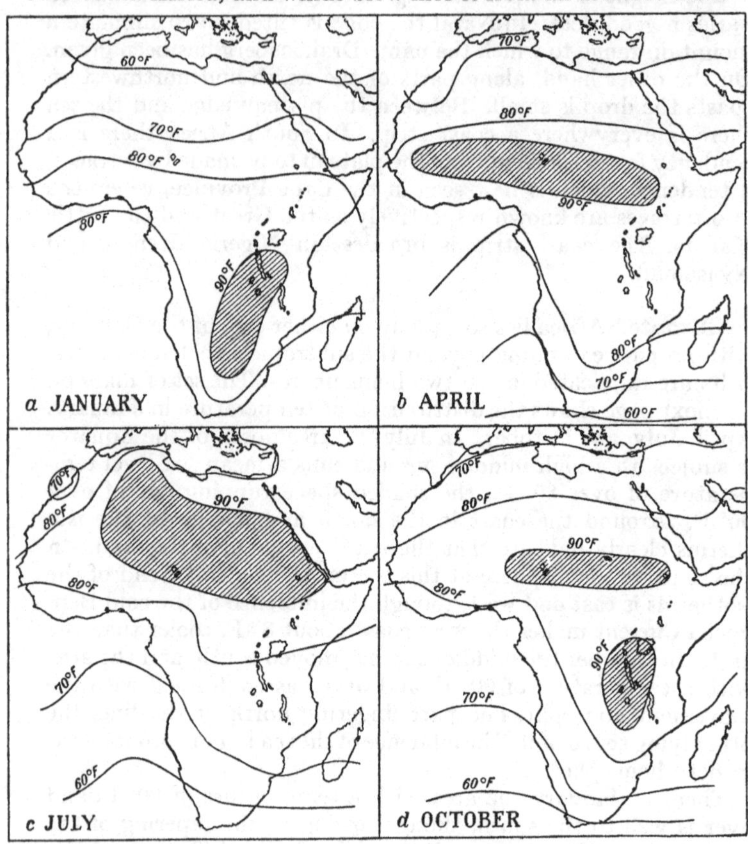

Maps showing distribution of temperature in Africa

it should be noticed that while the extreme south of the continent lies in the Horse Latitude belt of calms, the extreme north has entered the belt of Westerlies. On the horn of East Africa winds are northeasterly, while on the Guinea coast there is blowing the pleasant, dry wind from the Sahara, which is known as the Harmattan. As April comes the low pressure moves north to the Equator and a transitional set of conditions are seen. July finds a deep low-pressure centre in the north, whose effect is

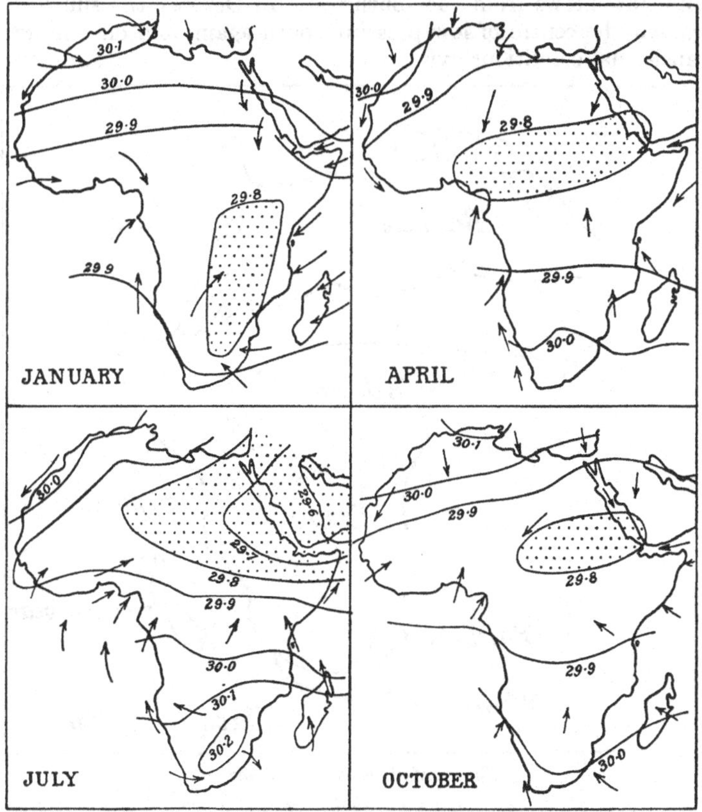

Distribution of pressure and winds. Note the correspondence between these maps and those on the opposite page

increased through the great breadth of the continent in this latitude and through the influence of the neighbouring land mass of Asia. Winds move towards this centre, and it should be noticed that the Horse Latitude belt of high pressure and calms is now well developed over the extreme north, while the south has just reached the belt of the Westerlies with its southwestern tip. Monsoonal conditions prevail in the Gulf of Guinea and on the horn of East Africa, in both of which regions the wind now blows from the southwest. In October the sun has brought the centre of low pressure south again, and once more transitional conditions exist.

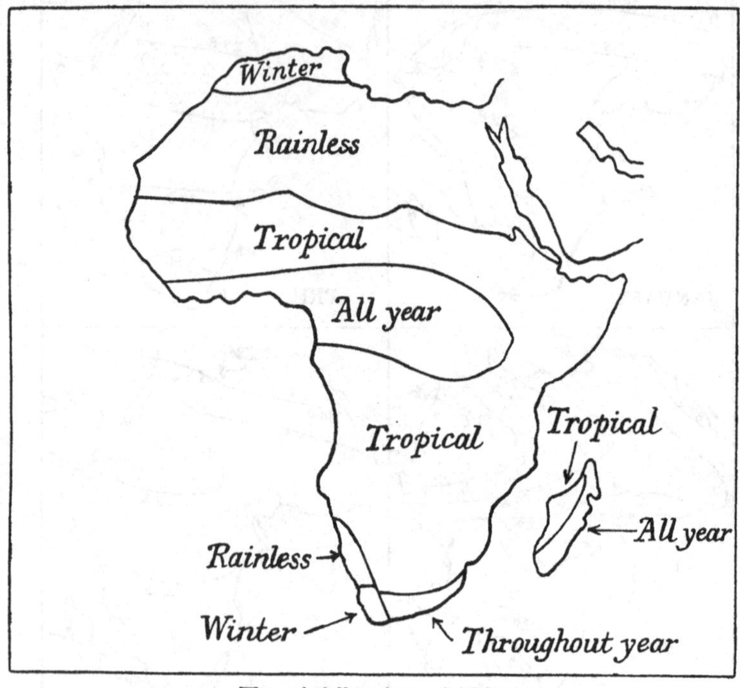

The rainfall regions of Africa

The general uniformity of surface of Africa gives the rainfall system a regularity not found elsewhere. The chief factor in the rainfall is convection, and accordingly the period of maximum precipitation follows the sun. A fairly large area around the Equator has rains throughout the year, with its maximum when the sun is overhead at midday. North and south of this area is a vast extent of country with a summer rainfall, the period when the altitude of the midday sun is least being a season of drought. This belt of tropical régime is succeeded b one of perpetual drought and uncertain showers. In the nort the great breadth

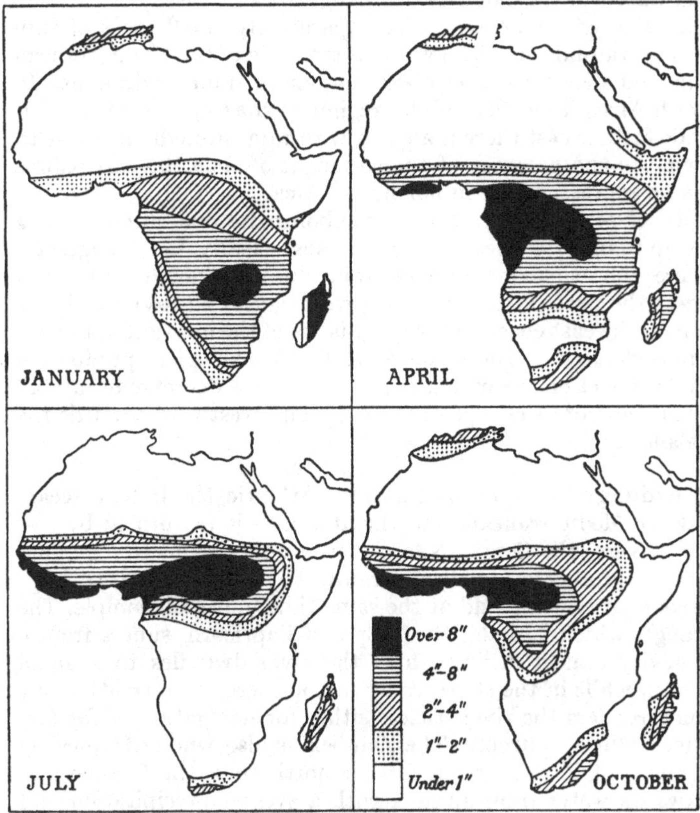

Distribution of rainfall

of the continent has caused the area of drought to be vast and to form the Sahara Desert, while in the south the influence of the sea has redeemed the east coast from drought and restricted the Kalahari Desert to a narrow strip of the west coast. It will be noticed that both on the west of the Sahara and of the Kalahari desert conditions reach right down to the sea, partly because the prevailing winds do not blow onshore and partly because such winds as do blow from sea to land have been cooled by the cold waters of the Canaries and Benguella Currents. On the poleward side of the deserts are comparatively small areas of subtropical rainfall. In the north this is wholly of the Mediterranean type and shows a winter maximum and a summer drought. In South Africa there is a similar régime in the Cape Town district, while farther east there is an area with rain throughout the year. Owing to the presence of the continent of Asia this last régime has no representative in northern Africa.

Relief plays a part, though a subordinate one, in controlling the rainfall. Its effects are seen most clearly in Madagascar, where the backbone of mountains running north and south across the path of the prevailing wind causes a heavy precipitation on the east coast and the highlands of the interior, but forms a rain shadow on the west coast. The Drakensberg produces a similar effect on the mainland, where a rapid decrease in rainfall is noticed from the edge of the tableland westward towards the Kalahari.

Drainage. Africa drains into the Atlantic, the Indian Ocean, and the Mediterranean, and this drainage is performed by five large rivers: the Congo, Nile, Niger, Zambezi, and Orange, and a large number of smaller streams. The régime of these large rivers is interesting and at the same time unusually simple. The Orange, which is under the Tropic of Capricorn, suffers from a long, dry season during which the river dwindles to a small stream, while in the short, rainy season there is a torrential flow which renders the river useless either for navigation or for any other ordinary purpose. The Zambezi is also wholly tropical in régime, but it rises much farther north than the Orange and draws its water from an area with a greater precipitation and two seasons of rain. Hence, its volume is always considerable, though there is a summer maximum. The river has cut its way back through the plateau edge as far as the Victoria Falls, one of the world's most spectacular waterfalls. Below the falls is a gorge whose sides have not yet been reduced from the perpendicular by erosion. From the falls to the river mouth the gradient is so steep that the current is too swift for navigation upstream. A hundred miles from the sea the river receives the overflow of Lake Nyasa through the River Shiré. It empties into the Indian Ocean through a delta, whose distributaries

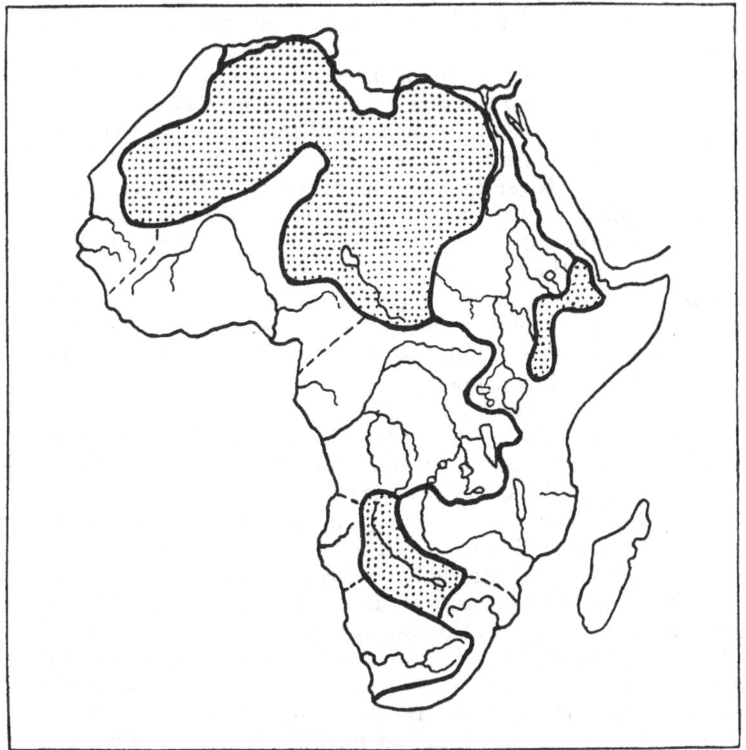

Drainage of Africa

The stippled areas have inland drainage

(the chief of which is the Chinde) are obstructed at their mouths by bars.

The Congo has a vast circular basin with many tributaries, the largest of which is the Ubanghi. The main stream rises in Lake Tanganyika and flows in a great curve to the Atlantic. Since a great deal of the basin lies in the belt of equatorial rains, the volume of the river is fairly regular, though the Kasai and other tropical feeders from the south give it a distinct summer maximum in January and February. Its course is broken by the Stanley Falls and again, more seriously, by rapids at the plateau edge between Matadi and Leopoldville. It enters the sea through a broad estuary.

The Niger rises on the rim of the great tableland and flows in a great curve, leaving the plateau by rapids at Busa in Nigeria. It receives one large tributary, the Benuë, a stream with an equatorial régime. At its source, and again for the last 500 miles of its course, the Niger draws its supply of water from the region of rain throughout the year; but for most of its length it flows through country with low rainfall and a tropical system. In fact, at the most northerly part of the great bend it actually crosses the Sahara. Hence, in this long section of its course it gets little water, but loses much by evaporation and seepage. It is thought that at one time the Sudan was wetter than now and that a far larger Lake Chad overflowed westwards to swell the volume of the Niger. The river enters the sea through a large delta.

The Nile is of course the principal river of Africa. It begins as the outflow of Lake Victoria, which it leaves by tumbling over the Ripon Falls, and it reaches the sea after a journey of 3473 miles. Every type of hot climate, including that of the desert, is represented in the basin of the river. This basin includes an equatorial corner which has a heavy rainfall and contains the huge Lake Victoria as well as the far smaller Lakes Albert and Edward. As far as Gondókoro the river has an equatorial régime, and it will be seen from the diagram on page 205 that the graph of the volume of the river at this point resembles the typical graph of equatorial rainfall (page 211), the differences being largely due to the fact that much of this part of the basin is in the southern hemisphere. In the Sudan the basin widens out enormously and the gradient of the river bed becomes far gentler. Here the rainfall has a tropical system, and during the clear skies of the long day season evaporation is enormous. Hence, in spite of the reinforcement brought from the west by the Bahr el Arab, the river nearly ends here in a salt swamp. The flow is hampered by masses of papyrus and other floating vegetation which, under the name of Sudd (=Arabic 'block'), also impedes navigation and gives the remarkable green colour to the water of the river as it passes through Egypt. The influx of the Sobat enables the river to reach Khartoum. This feeder has a tropical régime, which it imparts to the main stream after their junction at the town of Sobat, as the diagram below clearly shows. At Khartoum the river receives another large feeder, the Blue Nile, which rises in Lake Tana and has a very marked tropical régime. A third Abyssinian tributary, the Atbara, joins the main stream later. It is these three feeders that cause the summer floods in the Nile which have been of such great im-

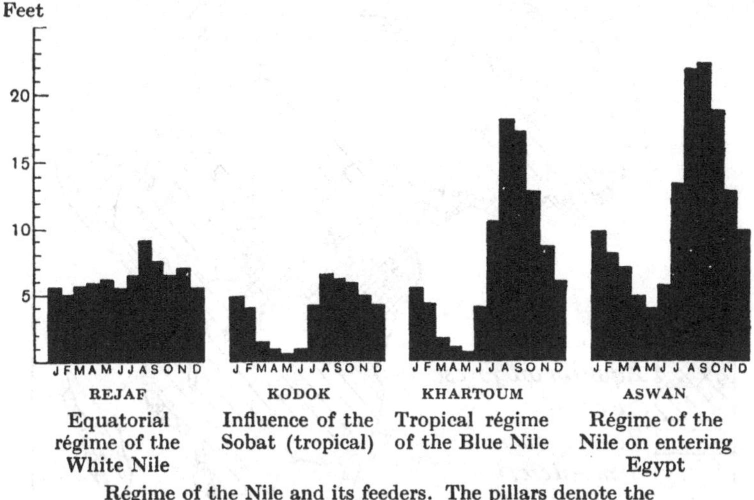

Régime of the Nile and its feeders. The pillars denote the variations in depth of the stream

portance to Egypt. From the junction of the Atbara with the Nile the basin contracts into merely a drainage channel cut by the river to carry off the rainfall from regions farther south, and no further supplies of water are received. Hence, the Nile, like the Murray, the Indus, and the Euphrates-Tigris, is a great river which diminishes in size in its lower course. Its famous triangular delta, through its likeness in shape to the Greek letter Δ, gave its name to the topographical formation.

Vegetation. The vegetation of Africa is determined to a very great degree by the rainfall, as will be seen by comparing the map below with that of the rainfall systems on page 211. The equatorial belt of rain throughout the year is one of dense rain forest, whose storeyed, creeper-bound tree-life man has found so difficult to exploit. Within this area the Ruwenzori mass rises to 16,000 feet and in unique conditions produces a podocarpus

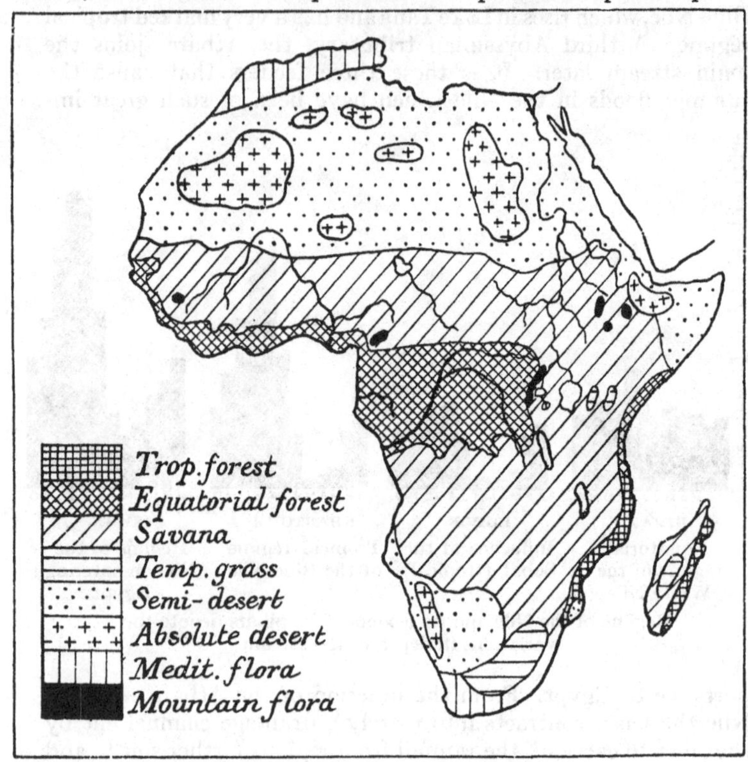

Distribution of vegetation in Africa

forest, in which groundsels grow as large as trees, mosses take the place of lianas and festoon the woody plants with weird meshes like those of a spider's web, while the ground is buried knee-deep in squelchy fungus. A forest not unlike that of the equatorial belt lines the east coast strip from the Juba River in Somaliland to the Ruvuma and covers the eastern slopes of the mountain backbone of Madagascar. On the high tableland of Kenya and Tanganyika Territory the lower rainfall and the cooler temperature do not allow of the growth of forest, but cover the land with the short grass of temperate climates. On river banks and other favoured places there are long, narrow stretches of gallery-forest of the open tropical type, while the Rift Valley degenerates in places into scrubland.

On either side of the equatorial belt lies a broad band of tropical vegetation, in which open forest occupies moister ground and savana holds the rest. Away from the equator drought conditions gradually affect plant life and after a time cause the savana to merge into semi-desert. In the south, where the tapering continent allows sea influence to moderate the tendency of the land to climatic extremes, this tropical band is broader than it is in the north. The height of the South African *veld* and of the slopes of the Abyssinian topland introduces other areas of temperate grass. At the Tropics lack of rain produces semi-deserts which contain cores of utter, barren desert. In the south relief rains and the influence of the Southeast Trades prevent desert conditions from developing on the east coast, but where their influence fails on the west there is a large area of desert. This is the Kalahari. Its area of absolute desert is confined to a comparatively narrow coast strip, but there is a broad fringe of semi-desert. In the north the Sahara stretches from the Red Sea to the Atlantic with only the insignificant break caused by the ten-mile-wide Nile valley. Large areas of it are uncompromising sand or rock desert, but at wide intervals there are oases in which the rise of a spring nourishes groves of palms and other vegetation. Finally, the Atlas region and its 'Mediterranean' counterpart in the south have that special vegetation of which the olive, the grape, and the cork oak are the typical trees, though they are not indigenous in South Africa.

Animals. Africa is the continent *par excellence* of wild animals and 'big game'. Its fauna may be put into four classes: the grasslands, the forest, the desert, and the Atlas region. The grassland class includes the 'big game' and contains the greatest number of well-known species. The animals are either herbivorous or carnivorous. To the former class belong many different kinds of hoofed or horned animals, including many species of antelopes and zebras, the giraffe, the buffalo, and the rhinoceros. In the rivers lives the hippopotamus. The carnivores, including the lion, leopard, panther, jackal, and hyena, feed upon the grass-eaters. For this reason they are provided with powerful jaws and teeth, and they have a rapid burst of speed over short distances. In order to escape the notice of their prey, they are tawny in colour and variegated with spots and stripes. The grass-eaters are very swift and mostly timid and live in herds. Like their pursuers, they are protectively coloured so as to blend with their surroundings. Since they must continually migrate to keep pace with the fresh pastures that spring up under the rains that follow the sun, they have great powers of endurance and can go long distances at a fast pace.

The forest animals are mostly arboreal, but include some varieties that live in the thinner woodland at the edges of the dense equatorial or the open tropical forest. Chief of these is the elephant, an animal which in Africa has been killed for its ivory tusks rather than domesticated as it is in India and Burma. In the denser forest the fauna is arboreal and consists chiefly of monkeys, birds, and insects. In parts of the Guinea coast and west of the great lakes in central Africa the gorilla is found. It will be remembered that the earliest explorers mistook these anthropoid apes for men. The desert fauna consists of animals, like the horse and camel, which have great endurance and can go long distances without food or water. Both these animals have been domesticated and are no longer found in their wild state in Africa. In the semi-desert parts of the Sudan the gazelle is the largest wild beast, and in the Kalahari there is the wart hog. In both regions there are large numbers of baboons. Swarms of spiders and lizards live under the rocks and in the sand of the deserts. The Mediterranean fauna has long since been domesticated. Its characteristic animals are the goat, the ass, and the sheep.

Natural Regions. The preceding paragraphs will have enabled us to summarise the various geographical conditions as they are distributed over the continent and to understand how and why Africa is divided into nine main natural regions by various combinations of the conditions. Owing to the uniformity of the relief, climate is the determining factor in this continent.

The regions are as follows: (1) equatorial lowlands; (2) equatorial highlands; (3) subequatorial belts; (4) tropical deserts of the west coast; (5) the Lower Nile Valley; (6) tropical lands of the east

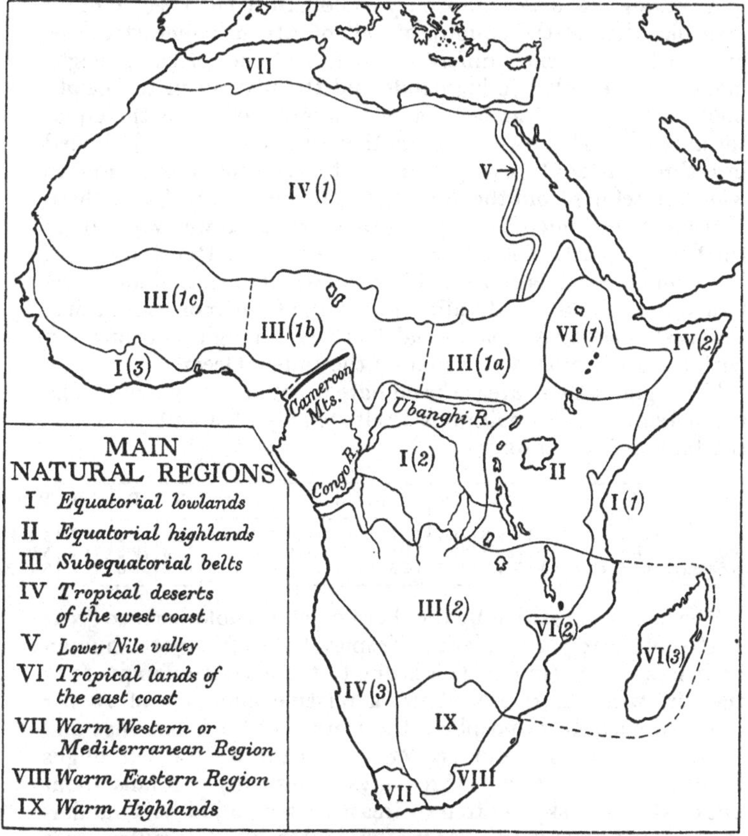

coast; (7) warm western or Mediterranean Region; (8) warm eastern or Chinese Region; (9) warm highlands. We shall now discuss each of these in detail in the next section.

THE REGIONS

1. The Equatorial Lowlands

Anything like dead uniformity could hardly be expected of an area like that of the equatorial belt of Africa, which stretches over districts of such different relief; and the effect of height marks off the volcanic highlands and the plateau in the neighbourhood of the great lakes as a separate variety of the equatorial region. The lowland areas themselves may be subdivided into three parts: (1) a strip of the east coast some hundred miles wide, stretching from the Juba to the Ruvuma; (2) the northern portion of the Congo basin with (3) an extension westward along the Guinea coast. The difference between the latter two is that the Guinea coast is more subject to sea influence than is the Congo basin, and the dividing line is the Cameroon Mountains. The east coast strip has special features through its connexion with the great monsoon system of the Indian Ocean.

Throughout these areas the climate is hot and very damp. The mean temperature maintains itself steadily at about 78° F., as the table below shows:

	J	F	M	A	M	J	J	A	S	O	N	D	Year	Range
Duala	79·7	80·4	79·5	79·3	78·7	77·1	75·0	75·0	75·9	76·3	78·3	79·1	77·9	5·4° F.
Libreville	79·7	80·1	80·6	80·4	79·9	77·4	75·6	76·3	77·7	78·1	78·3	79·2	78·6	5·0° F.
Dar es Salaam	81·5	81·3	80·4	77·9	76·1	74·3	73·4	73·6	74·5	76·6	79·2	80·8	77·5	8·1° F.

The heat as measured by the thermometer is not nearly so great as that of the regions under the Tropics, but as felt by the human body it is overwhelming. This is due to the extreme dampness of the air, which normally shows a relative humidity of 79 per cent. In such an atmosphere the leather of book bindings and of boots and shoes becomes covered with a green growth of fungus if not touched for two or three days. Every day rain falls about dusk, while the sky is often overcast even at midday. Thunderstorms are frequent and of terrifying intensity. The sultry air is not only most unpleasant, but saps the energy of the inhabitants and weakens the health of the white man so much that he not only always feels tired, but is also liable to catch diseases. The European often grows insensibly into the habit of artificially stimulating himself with alcohol and thus lays himself open still

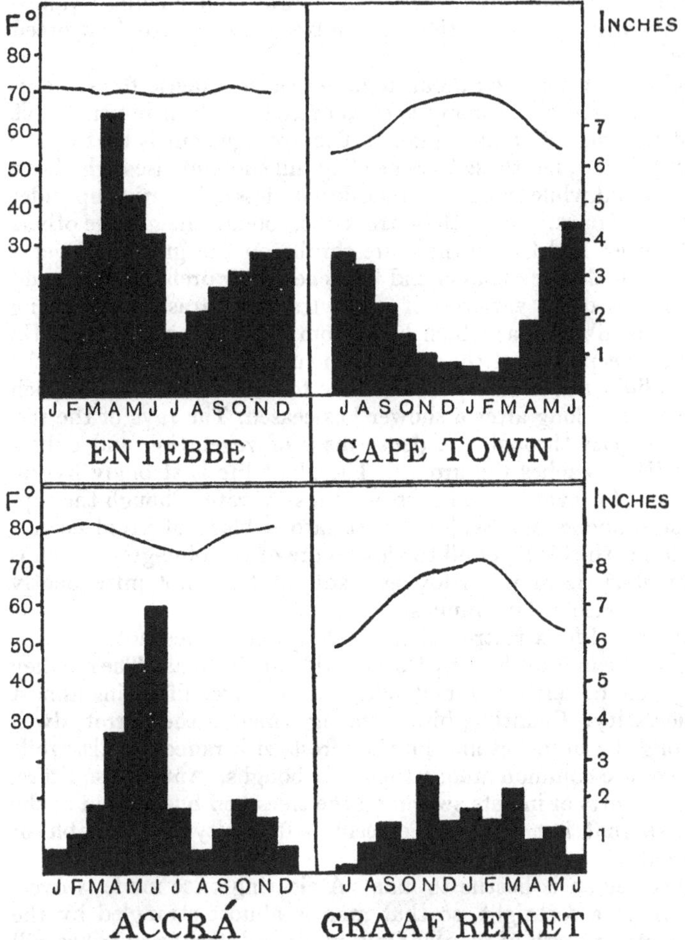

Temperature and rainfall diagrams of some African types of climate

Entebbe is in the equatorial highlands, Accrá in the equatorial lowlands, Cape Town in the southern 'Mediterranean' region, and Graaf Reinet on the warm plateau of South Africa

further to disease. Near the coasts there are evil-smelling swamps from which rise during the evening treacherous mists that breed fever.

The great heat and damp together cause a dense forest vegetation consisting of many species of trees and an impenetrable undergrowth of smaller plants. The arrangement is that of the storey-forest, i.e. those trees needing full sunlight rise high above the ground, while those which can do with less light spring up under the taller ones. Below these trees of secondary height are others still lower, and finally there are shrubs on the ground. Everywhere on the tree trunks and branches grow orchids, ferns, and numerous other varieties of epiphytes and parasites, including the rope-like lianas which hang from the boughs to the earth. Other creepers bind the whole into such a dense mass that the sun's light can hardly penetrate the foliage and raindrops reach the ground long after a shower has ceased. The rays of the sun cannot pass through the dense mass of vegetation and only a dim light reaches the ground. The plant life is strongly hygrophilous, for evaporation goes on at a slow rate. Though the topmost branches not seldom burst into a blaze of vivid colour, underneath this layer all the leaves are of the blue-grey which is characteristic of water-loving plants and are not infrequently covered with a dark fungus.

Animal life is restricted to small species, since not even the elephant can force his way through the tangled mass. The monkey is the characteristic animal, adaptation to tree-life being almost a necessity. Countless birds, among which is the parrot, dwell among the branches and fill the forest with raucous calls, while snakes are common among the lower boughs. Ants, mosquitoes, and all sorts of insects swarm on the trees and bushes and in the air, so that but little space seems left empty of vegetable or animal life.

The region is hostile to man. A clearing once made is overgrown in a fortnight, so that man is almost strangled by the energetic growth of plants. Only weak and backward tribes will attempt to live here, and then only in favourable places, like the river banks, where nature helps to make a clearing. Most of the people are of poor physique, if not actually pygmies. Their mental powers are small and they have little conception of social life. Their mode of life is collecting, their food consisting of fruit, insects, birds, and small animals. Their shelters are little better than piles of bush, they wear little or no clothes, and their only belongings are primitive weapons made of wood and rough ornaments of nuts, shells, or teeth.

In modern times, the white man has tried to exploit the region and, after many years of failure, he seems about to succeed. His powerful tools and instruments can remove the forest, and his knowledge enables him to replace the natural woods with useful plants like the rubber or the cocoa tree. So far he has been unable to make a permanent home in the region because of the diseases which assail him, but the study of tropical medicine has advanced much and, largely by killing off the mosquitoes, has made it possible for him to avoid actual illness. But until he is acclimatised to the extent of developing the negro's natural power of living at a slower rate and of radiating animal heat more quickly, he is unlikely to be able to dwell permanently in the region.

1. *The Congo Region.* This is a vast horseshoe-shaped area stretching from the Angola tableland in the south to the Cameroons in the north, and from the Mufómbiro highlands on the east to the sea on the west. It includes all but the southernmost part of the Congo basin, together with the basins of the Ogowe and other minor streams between the Congo River and the Cameroon Mountains. The Congo and its feeders, the biggest of which are the Ubanghi and the Kasai, all flow in an unusual anti-clockwise curve, beginning with a northeasterly and ending with an easterly or southeasterly direction. They concentrate on a focal point within the basin in a manner which suggests that there formerly existed a great area of inland drainage of whose central sea Lakes Tumba and Leopold II are survivals. The main stream has cut a passage to the ocean through the plateau edge so recently that twenty miles of rapids interrupt navigation from Stanley Pool to Matadi.

The central belt of the region lies in the Doldrums and gets the extreme type of equatorial climate with its convection rains and oppressive heat, but north and south of this belt, where the land rises to the watershed, the summer is noticeably rainier than the rest of the year. The mean height of the basin is 1000 feet, but round the rim and along the ridges dividing some of the tributaries it rises to 1500 feet and in the extreme south and east to over 3000 feet. Broadly speaking, all the area below 1000 feet is covered with equatorial forest, but on the higher ground the undergrowth thins out, and towards the boundaries of the region the woodland gives way gradually to savana. In these parts where the forest is less dense the elephant once lived in great numbers, and for many years ivory derived from their tusks was

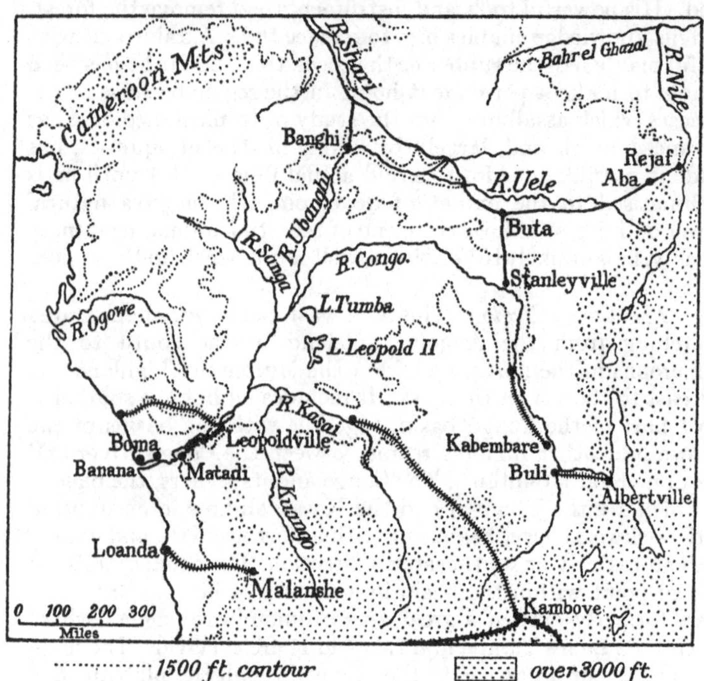

The Congo region. Relief, rivers, towns and railways

a chief product of the region; but reckless destruction has vastly diminished the number of animals and consequently the amount of ivory available.

Naturally, the chief products are forest-derived. Rubber was once plentiful, but wasteful methods of collection have caused the industry to die away before the competition of plantation rubber in Malaya. Efforts are now being made to put production on a more scientific basis, the native trees being supplanted by the *Hevea brasiliensis* which is imported from South America. The oil palm yields nuts from whose covering and kernel is obtained a vegetable oil used for making scent, hair-oil, soap, margarine, etc. Copal, a kind of resin, is also produced for use in the manufacture of varnish.

The Congo and its feeders form natural routes along which economic products are carried to the sea, and the basin offers a perfect example of a simple backland in a 'new' country. A railway, parts of which are in operation, has been planned to run from Rhodesia northwards along the Congo to Stanleyville and thence eastwards to join the British system in Uganda. If and when it is finished, it will form part of the once famous ideal of the Cape to Cairo railway. Boma, the outlet port of the region, is the only town of any size. It has an outport at Banana. Matadi and Leopoldville are 'portage towns', being placed at the ends of the railway built to go round the rapids of the Congo.

2. *The Guinea Coast.* This is an uneven strip running from the Cameroon Mountains to Cape Verde. It varies in breadth from 300 miles at the mouth of the Niger to under 50 in parts of the Gold Coast and it gradually narrows off in the northwest, to widen out again in the valley of the Gambia. Along the coast runs a fringe of unhealthy mangrove swamps, the home of swarms of fever-carrying mosquitoes. Behind this the ground rises—very gradually at first, but after a time it slopes steeply up towards the edge of the tableland, where the region ends. At the mouths of the Niger and Volta the area of the marshy coastal fringe is increased by the new land formed by sediment. The equatorial climate is modified by sea influences aided by the monsoon. During the greater part of the year the high-pressure centre north of the Equator causes an inrush of wind at right angles to the coast all along the region, and relief rains add to the effect of convection. A rainfall maximum occurs in July when the Saharan high-pressure centre is at its greatest intensity. In January, when the continental high-pressure centre is in the south, the wind blows from off the Sahara, bringing a dry heat, mingled however with dust. This is the Harmattan, so welcome to the human dwellers in the region. It naturally brings but little rain. The subregion is therefore not truly equatorial.

There is the usual dense forest, but larger areas have been cleared and replanted with useful trees than in the Congo area. Rice, which has been introduced from India and thrives in the wetter parts, has become one of the chief foods of the natives. In the drier districts manioc and maize, both American in origin, are also grown to some extent. The forest products are cabinet woods, such as ebony and mahogany, as well as palm oil and kernels. The ivory yielded by the tusks of the elephants which inhabit the forests here forms an important product. Indeed, it has given its name to a part of the coast. Little gold is now derived from the Gold Coast, but cocoa plantations established by British companies have been very successful and have deprived the northern states of South America of much of the world trade in this product. The natives chew a kind of nut known as the kola for much the same reasons as Europeans smoke; hence, the kola is also cultivated.

The original people were negroes of a backward type, who believed in magic and worshipped fetishes. But they have improved greatly by intermixture with the better tribes which have pressed down to the coast from the north. Many of these were Arabs who brought with them some of the ruder arts such as the weaving of coarse cloth, the making of crude pottery, and somewhat better methods of tilth. The influence of European settlers has also raised their standard of civilisation, especially in Nigeria and the Gold Coast. Except some small islands in the Gulf of Guinea and the free republic of Liberia, the region is governed by the British and French, who have built towns at suitable points along the coast as collecting stations for the products of the interior. Lagos, which is on an island just off the northeastern corner of the delta of the Niger, has a large hinterland beyond the coast strip, and Sekondi and Accrá are stations of the Gold Coast. Owing to modern needs an artificial harbour was built in 1930 at Takoradi. Kumassi is a native town of some size in the Ashanti district of the Gold Coast.

3. *The East Coast Strip.* This consists of a low-lying, unhealthy margin very like that of the Guinea Coast, though narrower. There is a mangrove fringe backed by a stretch of broken country which finally rises to the African tableland. The Southeast Trade, strengthened by monsoonal conditions in July, is the prevailing

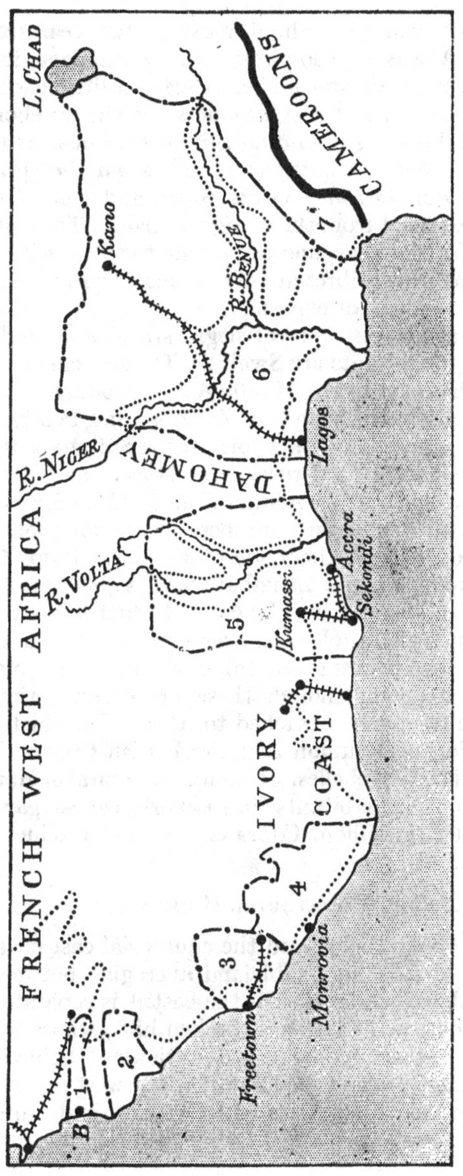
L. CHAD
CAMEROONS
Kano
R. BENUE
7
6
Lagos
DAHOMEY
R. NIGER
FRENCH WEST AFRICA
R. VOLTA
Accra
Sekondi
Kumassi
5
IVORY COAST
4
Monrovia
3
Freetown
2
1
B

West Africa

wind and brings rain throughout most of the year from the Indian Ocean. A season of low minimum in January decreases the density of the forest and makes clearing a little easier than it is in West Africa. The forest products are chiefly ebony and copal. The oil palm is not found on this coast; but near the sea, where the wind contains particles of salt from the spray, the coconut palm flourishes and yields copra and coir. Bark for tanning purposes is got from the mangrove fringe. The cultivated crops consist of rice, maize, and sugarcane together with sisal, a kind of succulent whose fibres are used in making rope. Zanzibar exports large quantities of cloves.

The aboriginal natives were of negro stock, and to-day the majority of the inhabitants are Swahilis. The settlement of Arab traders along the coast in the Middle Ages introduced a Semitic element, especially in the islands of Zanzibar and Pemba, which with the adjoining mainland once formed the Sultanate of Zanzibar. The islands are now a British protectorate and have been separated politically from the mainland area, which forms Kenya Protectorate. In recent years numbers of Indian coolies have been allowed to settle in the region. Mombasa (pop. 57,000), Dar es Salaam (38,000), and Zanzibar town (45,000) are the outlets for the region, but the backlands of the first two stretch far inland and include the whole of the next region to be described.

The Indian Ocean contains a number of islets with a maritime equatorial climate, and, though these are oceanic, they may conveniently be treated as attached to Africa. The chief are the French possession of Réunion and the British Crown Colonies of Mauritius and the Seychelles. St Louis, the capital of Mauritius, is a town of 55,000 inhabitants and exports cane-sugar to the annual value of £1½ million. Copra is the chief product of the Seychelles.

2. The Equatorial Highlands

Between the Congo region and the equatorial east coast strip stands a broad area of high tableland averaging between 3000 and 4000 feet above sea-level. On the east it is separated from the coast by a narrow belt of low, broken hills, whose poor soil causes them to be scantily clad in vegetation and behind which rises the steep edge of the tableland. On the west one region is bordered by the volcanic masses of Ruwenzori and Mufómbiro which form a ridge on the western side of the Tanganyika branch of the Great Rift Valley. Southwards its boundary is indefinite, since the high tableland here runs uninterrupted for hundreds

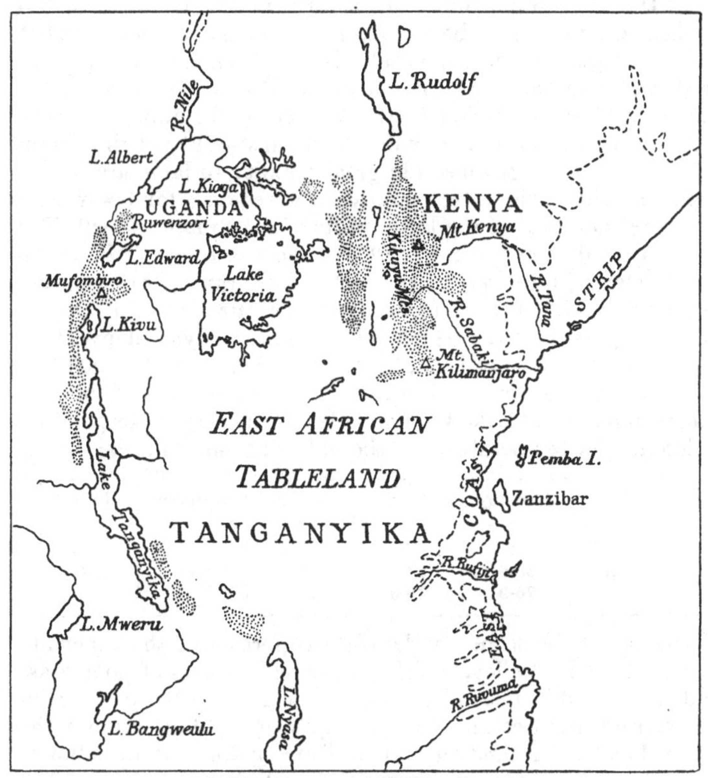

The Equatorial Highlands

The broken line along the coast is the 600 foot contour. The highest ground is stippled

of miles, ending in the High Veld of South Africa. Northwards the limits are climatic also, but they are defined in the northwest by the descent of the land to the lower tableland of the Sudan and in the northeast by the rise of the surface to the even higher region of Abyssinia.

The Great Rift Valley and its branches make a mighty trough through the region, and the lakes which occupy parts of the valley bottom offer an excellent means of communication. This is all the more important since the rivers are too small or too broken in gradient to be navigable. It should be noticed that Lake Victoria, which incidentally is one of the largest bodies of fresh water in the world, is not in the Rift Valley, but merely fills a shallow depression in the surface of the tableland. The outpourings of volcanic rock due to the instability of the region in ages past have given soil of great fertility to large areas.

The height of the tableland gives the region a far lower temperature than that of the lower parts of the equatorial belt. The mean annual temperature is 70° F., and the range 7° F. As in all tablelands, however, there is a big daily range, from midday temperatures of 80° F. to morning temperatures of 40° F. In the northern portion of the region, which forms Kenya Colony, there are great variations in level, for the tableland rises to 8000 feet in the Kikuyu district and falls to 2000 feet in parts of the Rift Valley and around Lake Victoria. Hence, there are corresponding differences in temperature, as the following table shows:

	January	April	July	October	Range
	° F.	° F.	° F.	° F.	° F.
Nairobi	63·8	63·9	58·5	64·8	6·3
Kisumu	76·9	73·3	71·8	75·4	5·1

The rainfall is scant on the eastern portion of the tableland and in the Rift Valley, which are therefore areas of poor grass and scrub; but it rises to an annual mean of 40 to 50 inches in the Nairobi district and on the slopes of the hills around the Great Lakes. In Uganda and in Tanganyika it is less, but is generally enough for the savana which clothes these areas. The Kikuyu district is clad in forest or, in places of lower rainfall, with temperate grass. The margins of the Lakes are forest-clad.

The country is the ideal home of grass-eating animals, and vast herds of many kinds of antelopes, zebras, giraffes, and buffaloes live here, accompanied by the lion and other big cats which feed on the grass-eaters. Near the rivers live rhinoceroses and in the water itself are hippopotamuses. As the grass tends to dry up in the season of drought, the animals wander from place to place in search of fresh pastures. The herds are now much smaller than formerly, since the natives have acquired more efficient weapons and since big game hunters and settlers have recklessly killed off the animals. Efforts are being made to check the slaughter by imposing an expensive licence on big game hunting and by legally restricting the number of animals which may be killed; but, as the country is gradually being opened up for civilised settlement, the big game is bound to disappear sooner or later from the open country. To save some of the animals for future study, large areas of the least productive land have been reserved exclusively as homes for them.

The vast majority of natives are Bantus, but in the north of the region there are infusions of Hamitic and Sudanese tribes, while on the northeast are similar traces of Semitic people. Among the chief tribes are the Kikuyu and the Masai. The natives of the tableland are well built and warlike and are superior physically and culturally to the lowlanders. When uninfluenced by European settlement, they are chiefly nomadic herdsmen, as might be expected in a grassland region; but most of them engage in agriculture to a certain extent, producing crops of millet. Their dwellings are of wood and are thatched with the long savana grass. Their personal possessions were formerly such as could be made from wood, hide, or bone; but certain arts had reached them from the north and east, and they were able to forge sword blades and spear heads and to make rough pottery and coarse cloth. Now, of course, they are able to obtain goods of European manufacture and the use of such articles is rapidly spreading. They were organised into tribes, each under a chief who ruled according to traditional custom. They speak various languages mostly of the Bantu family, but Swahili is understood and spoken nearly everywhere.

The chief native crop now is maize, which with beef and millet forms the main food of the people. Under European influence other crops are grown on the plantation system for export. Of these the most important is cotton, which flourishes in the drier parts, especially in the north. Coffee is growing in importance and has become the second greatest export. Sisal, tea, wheat, and wattle (for tanning) are the other main crops. Of the pastoral products hides form the principal export, but dairying and wool are increasing in importance. The forests provide hardwoods, of which camphor, olive, and pencil cedar are the chief, while the bamboo is being used for pulp. The minerals are not yet fully exploited, though there is some export of gold, tin, and marble.

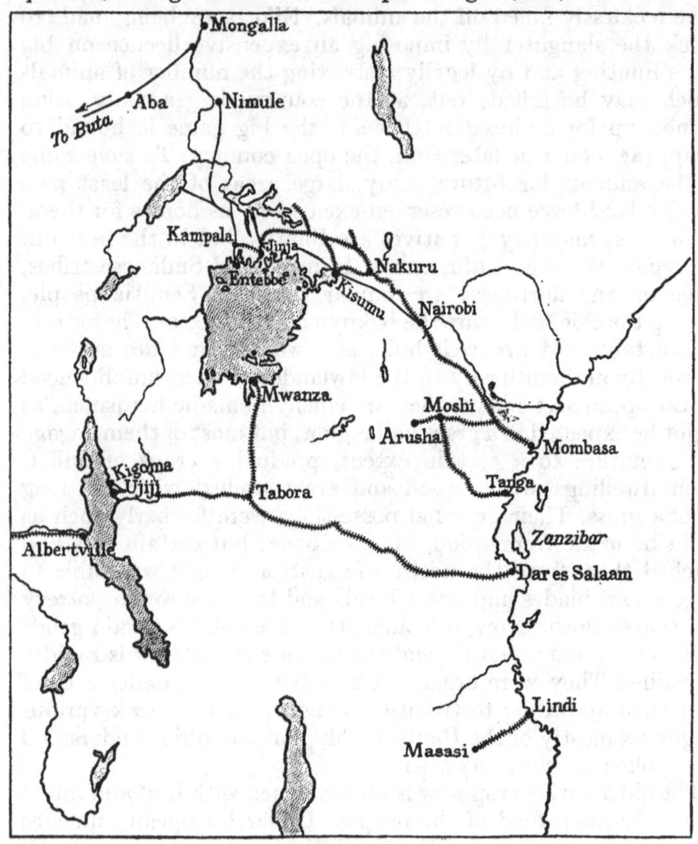

The Equatorial Highlands. Railways and main roads

In most countries in which overseas European settlement has taken place, the headquarters of the settlers have naturally been established at the seaports. This is true in the Congo and on the Guinea Coast, for example. But in this region, once European settlement had reached the more suitable climate of the tableland, the seat of government and centre of life was transferred to the inland town of Nairobi. This town, which has a population of 85,000, including 7000 Europeans, has one of the most pleasant climates in the world and is centrally placed in the region. Its district is so suitable to European settlement that it contains no fewer than 3000 British farmers, the whole of Kenya having nearly 17,000 Europeans.

Transport is by steamer, road, and rail. A main railway line runs from Mombasa to Kampala and another from Dar es Salaam to Kigoma, with a branch from the large native town of Tabora to Mwanza. A number of smaller branches and minor lines complete the system (see map). Roads are growing in importance through the increasing use of cars and lorries. The chief highway runs from Nairobi to Mongalla, but there is a fairly close network in the neighbourhood of Nairobi. Wireless stations at Nairobi, Kampala, and Kigoma connect the region with the world outside. On the Lakes steamers ply between the various small ports and carry goods to and from the railhead. A regular service runs from Lake Albert to Nimule, whence a motor road connects with Rejaf at the head of the navigation of the Nile. The regular air mail between London and Cape Town touches at various points in the region and gives connexion with both those cities.

The region includes the greater part of the British colony of Kenya, the protectorate of Uganda, and the mandated territory of Tanganyika.

3. The Subequatorial Belts

North and south of the equatorial regions lies a belt of country in which the climate and therefore the vegetation and conditions of human life are quite different. The northern belt is known by the general name of the Sudan. It runs right across the broadest part of the continent, but is nowhere very wide. The southern belt, which contains parts of Nyasaland, Rhodesia, Belgian Congo, and Angola, is narrower from east to west than its northern counterpart, but is wider from north to south. The main distinguishing feature of the two belts is the tropical rainfall system in which there is a marked dry season in the cooler months and a smaller total precipitation. Also, there is a greater difference in temperature between the coolest and warmest months. The vegetation is less dense on account of the smaller rainfall, and, though damp valleys and other favourable places contain lines of forest, they are without the equatorial tangle of undergrowth. Ordinarily, the vegetation is savana, though in drier parts and in general progressively towards the Tropics, the grass gives way to scrub.

1. The Sudan. This belt falls into three well-marked physical divisions: Eastern Sudan, the Chad basin, and Western Sudan. These all form part of a low tableland, with similar, though not quite the same, climate all over. In the north where the region touches the equatorial region there is dense forest which gradually thins out towards the north, and soon there is savana, followed by scrub, which adjoins the desert of the next region. Animals are not numerous. In the wooded parts there are herds of elephants and in the more open country gazelles and wild asses. The people as a whole belong to the Hamitic and Sudanese branches of the negro race, though they are largely mixed with Arabs and other peoples from the north and east. In the rainier parts they are agriculturalists, but in the grass and scrub lands they are as a rule herdsmen, tilling the ground to eke out their food supply.

Eastern Sudan is largely a basin centring round the inflow of the Bahr el Ghazal into the White Nile, but it has an extension down the Nile valley as far as Berber. Its climate is hot, with a noticeably cooler season in January. Thus, Khartoum with a January mean of 70·3° F. and a June mean of 91·4° F. has an annual range of 21·1° F. Owing to the clearness of the air there is a great daily range. The rainfall, which occurs mainly in summer, is scanty, but fortunately the White Nile and other smaller streams bring an abundant water supply from the equatorial region. From Gondókoro to Kodok the fall in height is only 250 feet, and, as in this stretch a large number of streams converge on the main river, a wide belt of marsh occurs in which the Nile all but loses itself. The flooding of the country is assisted by the large masses of floating vegetation known as 'sudd'. In autumn the Sobat, the Blue Nile, and the Atbara bring from the highlands of Abyssinia the summer rainfall of that region and add to the supply of water available for irrigation in the Sudan.

The people of this division are a mixture of Sudanese negroes

(*Mondiale*)

KENYA

Natives at Ngao on the River Tana

Plate XIV

(*South African Railways*)

NATAL

In the foreground are fields of sugarcane, behind which can be seen the town of Verulam

with Semites who have penetrated from the east. Along the river banks they are agricultural, but become more and more pastoral towards the desert and on the savana. The staple food crop is millet. Under European influence cotton cultivation is greatly increasing, especially in the district of Gezira, a triangular area between the main Nile and the Blue Nile, where the great dam at Sennar on the latter river provides water for a widespread

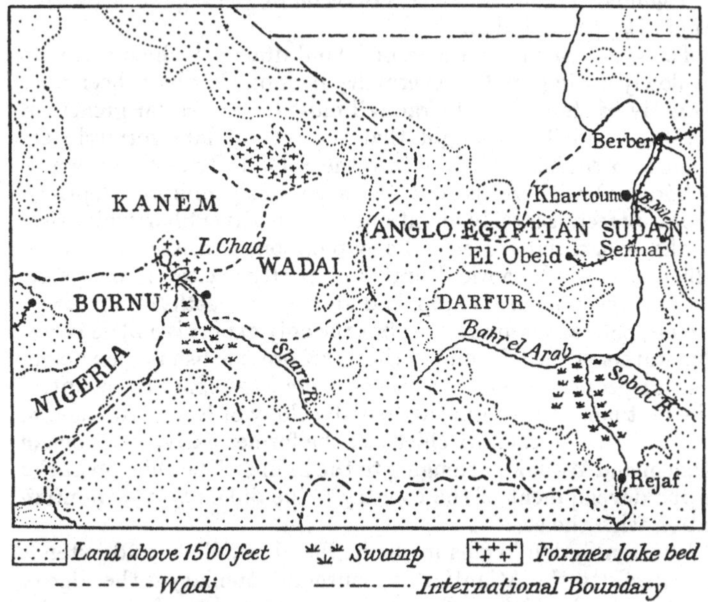

Eastern and central basins of the Sudan

irrigation scheme. Another important cotton district lies around Kassala at the foot of the Abyssinian highlands. Second in importance to cotton is gum-arabic, most of the world's supply coming from this division of the region. Cattle and sheep breeding is capable of great development, the quantity of hides and skins exported being at present small.

The interruption of the Nile by cataracts causes the cotton and other exports to go through Port Sudan and Suakin on the Red Sea. The main railway line runs from Khartoum down the Nile to Berber, where it forks, one branch going to Halfa, the other to the Red Sea ports. Branch lines go to Kareima, Sennar, and El Obeid. Another main line starts from Port Sudan and goes through Kassala to Sennar. Regular steamer services ply along the Nile from Khartoum, where the cataracts begin, to Rejaf at the head of the navigation of the river. Motor roads lead from Mongalla southwards to Aba in Belgian Congo and to Kampala in Uganda. The whole of this division of the region lies within Anglo-Egyptian Sudan.

The Chad basin is an area of inland drainage which seems to be slowly drying up, for several dry beds of what have been once large rivers show that the rainfall must have been far greater at one time and it is thought that the present lake formerly extended northeastwards well into the Sahara. Lake Chad, which is bordered with reeds and has a curiously uniform depth of about six feet, is largely maintained by the River Shari which rises in the equatorial region. The population, consisting of tribes which are partly agricultural and partly pastoral, is sparse. Millet is the chief grain crop, and numbers of cattle, sheep, asses, camels, and horses are raised; but the only export is a little ivory. The waters of the lake and of the Shari provide fish for the neighbouring people. So far little has been done to exploit the country. The best land lies within the British colony of Northern Nigeria, the rest—which consists chiefly of the former native divisions of Kanem and Wadai—belonging to the French. The chief obstacles to the development of the country is the difficulty of communication with the sea.

Western Sudan begins in the highland of Sókoto and extends westwards to the Atlantic. Its southern boundary is the edge of the tableland which ends the equatorial region of the Guinea Coast. Northwards, this tableland falls away to the Niger and Senegal, which more or less form the boundary with the desert. Round the lower course of the Senegal are broad stretches of lowland covered with tropical open forest. The Niger originally descended the edge of the tableland at Idda in Nigeria, but it has cut back its lower valley to Busa, where rapids interrupt its course. Just before entering the equatorial lowlands it is joined by the Benuë. Except where trenched by these rivers, the tableland has a monotonous surface.

The climate of this division of the region is marked by high midday temperatures and a high daily range. There are two maxima during the year, at the times when the midday sun is overhead. The annual range varies from about 16° F. on the coast to 24° F. at inland places. The dry Harmattan blows in the cool months, but at other times damp monsoon winds bring rain, especially in the middle of the year. Most of the rainfall is convectional and the showers are accompanied by thunderstorms. The amount of rain quickly decreases with distance from the coast, and naturally the luxuriance of the vegetation acts correspondingly. At the plateau edge and in damp valleys there is tropical forest, but towards the interior there is first savana and later scrub.

Except in Nigeria and to a less extent in Senegal, there has been little development so far. Cotton is the chief economic crop and is cultivated in Nigeria and in parts of French West Africa. The oil palm flourishes throughout the region, and its nut is exported. Ground nuts are largely grown in Senegal. At Udi in Nigeria copper, tin, and coal are mined, but the exploitation is hardly developed as yet. Except for the Senegal district, which has its own outlet at Port Louis, this division of the region forms the backland for ports already mentioned as lying on the equatorial coast strip. But there are several inland towns of some little importance. Thus, Kano in Nigeria is an interesting old walled native town, acting as capital of the province of Sókoto and as a starting place for caravans bound for the Eastern Sudan (see map on page 235). On the great bend of the Niger is another famous old town, Timbuktu, which derives its importance from the junction of ancient caravan routes. Most of this part of Africa belongs to either the French or the British. The British portion will probably become of immense importance to Great Britain later on, since it will tend to become the special area from which she will draw tropical products in the future.

2. South Central Africa. The southern subequatorial region occupies the high tableland to the south of the Congo basin and the equatorial highlands as far south as the Tropic of Capricorn. Including parts of Belgian Congo, Angola, and Tanganyika Territory and all of Nyasaland and the two Rhodesias, it stretches from the eastern edge of the continental tableland right across to the Atlantic, with a broad, tongue-like extension southwards in Bechuanaland between the barren region of Southwest Africa and the warm highlands of the Transvaal. It differs from its northern counterpart in being higher above sea-level and so cooler. Its surface is that of a rough peneplain studded with relict hills of ancient rock and pitted with enclosed hollows which form basins of inland drainage. The largest of the latter consists of the dry bed of the so-called Lake Ngami and the swamps of Makarikari. On the east the comparatively youthful rivers Zambezi and Limpopo have gouged out wide, deep valleys whose bottoms are hotter than the rest of the region and have a more luxuriant vegetation.

Over such an extensive region the temperature naturally varies a good deal. Other things being equal, there is from north to south a general decrease in the mean temperature and an increase in the range; but altitude frequently counteracts the effect of latitude. The following statistics are representative:

	January	April	July	October	Range
	° F.	° F.	° F.	° F.	° F.
Lauderdale	72·8	70·2	62·4	74·5	13·5
Livingstone	75·7	72·8	64·6	80·8	16·9
Bulawayo	71·0	65·8	56·6	71·0	15·0

The prevailing wind is the Southeast Trade, whose belt extends farther south in summer (December) and whose effects are more marked in the same season. It is this wind which provides the moisture for the rainfall. The eastern portion, therefore, gets abundant relief rains, but precipitation decreases progressively towards the west. Convection rains, which are responsible for most of the rainfall, occur during the period of the year when the midday sun is highest, and the system is typically tropical. Both the tendency to a decrease in rainfall towards the west and the incidence of a summer maximum are noticeable in the diagram on page 253. In general the vegetation is of the savana type, but in the moister east and in the two great river valleys there is a good deal of tropical open forest, while in the west the savana becomes poorer and poorer until it finally degenerates into scrub. Naturally, the big game of the equatorial highlands is found in this region also.

The natives are Bantus who have within the last two or three centuries displaced the former inhabitants, the little Hottentot, who in turn had previously dispossessed the Bushman. Survivors of these latter two races now inhabit the Kalahari. At one time a people of some degree of civilisation seem to have occupied the district in which still remain the ruins of the ancient city of Zimbabwe. Little is known of them, however. The Bantus are chiefly herdsmen, but they also plant crops of millet. Their dwellings are frail, beehive-shaped structures which in some

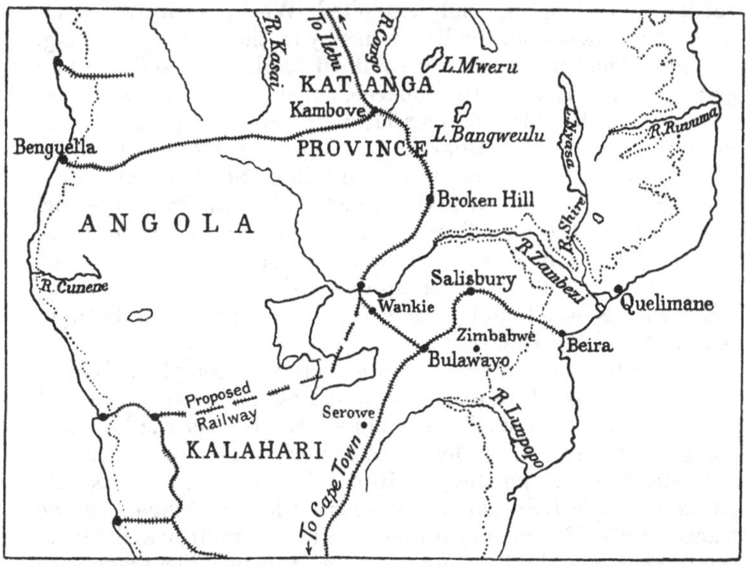

South Central Africa

areas consist of frames of twigs covered with grass and in others are made of mud. They live together in villages, under the rule of a headman, often enclosing the whole or a part of the ground occupied by the houses with a quickset hedge or wicker wall. They are comparatively few in numbers, owing to the bloody tribal wars which formerly decimated the population; but the village of Serowe in the Bechuanaland Protectorate contains 30,000 persons and is the largest native cluster in Africa south of the Equator.

Like the Sudan, this region has great possibilities for the future, though whether it can become a permanent home for Englishmen remains to be seen. At present there are about 60,000 British settlers and a somewhat smaller number of Portuguese and Belgians. Cattle-rearing is the chief occupation of the settlers as well as of the natives, and live animals, meat, hides, and bone manure are exported. The attacks of the tse-tse fly, whose bite is deadly to cattle, are a great drawback to the business. Some of the best grazing land is in Angola, but it has not yet been anything like fully exploited. Wheat is cultivated on parts of the savana in the Rhodesias by methods of dry-farming, though agriculture is largely confined to the east and the two great valleys. Here maize, tobacco, coffee, and some cotton are grown. But the chief commercial wealth of the region is found in various minerals. Gold is found in various parts of the Rhodesias, but especially at Broken Hill in Northern Rhodesia, while coal is mined at Wankie, near the Victoria Falls. The fuel is used in the Katanga Province of Belgian Congo, where there are rich copper mines near Kambove, whose exploitation has led to the building of a special railway to Benguella. Numerous other metals are found in less important quantities both in Katanga and the Rhodesias.

Communications in the region are growing complex. Nyasaland and the rest of the district around Lake Nyasa are connected with Quelimane by small steamers which ply over the lake and along the River Shirë, a feeder of the Zambezi. But this is the only simple backland, for the Rhodesias export their products either through Beira or Cape Town, while Katanga has the choice of the Benguella railway, the Congo route via Ilebu, or the river, rail, and lake route to Dar es Salaam. For trade with Europe the two western routes have the advantage of a shorter distance. Salisbury (pop. 29,000), Bulawayo (pop. 31,000), and Livingstone are the three chief towns. The first is connected directly by rail with Beira and has a line to Bulawayo; the two latter are on the main line of the so-called Cape to Cairo Railway which runs from Cape Town to the Congo border and on to Katanga.

4. The Tropical Deserts

In the latitude of the Tropics and within a belt ten degrees wide on either side of them there is found in every continent an area of desert occupying the west coast. In South Africa this area occupies parts of Southwest Africa and Bechuanaland and is relatively restricted in extent. But in the north, owing to the presence of Asia and Europe which practically continue the land mass of Africa, the area of desert stretches right across the continent from the Atlantic to the Red Sea. Nor does it end at this narrow cleft, for the Arabian peninsula is a continuation of the Sahara. Furthermore, a narrow strip along the coast of Eritrea and the whole peninsula known as the Horn of Africa must also be considered desert, partly because of the influence of the surrounding lands and partly because of the direction of the monsoon winds which prevail.

In days not long gone by the area regarded as desert was far larger than it is to-day. This is not merely due to an increase in geographical knowledge, but rather to man's vastly augmented powers of overcoming the niggardliness of nature in barren regions. Better organisation, the railway and the motor vehicle, systems of dry farming and irrigation—these and other advances in human power have led to the reclamation and use of much land which was formerly looked upon as useless and uninhabitable. And geographers and politicians are increasingly insisting upon the shrinkage of 'absolute desert'. Certainly, the old idea that the Sahara and the Kalahari are immense stretches of sand dunes and rock without a vestige of plant, animal, or human life is wrong; but we must be careful not to go to the other extreme and regard these areas as being even capable of abundant production.

The Sahara is bounded on the south by the Senegal and Niger and by an indefinite line following the parallel for 14° N. from the Niger to the Nile. Thence it stretches northwards to the sea, except in the Atlas region, where it ends at the mountains, and in the little peninsula of Barka. On the east the Nile valley is reclaimed by the water supply derived from the floods of the river. but between the narrow valley strip and the Red Sea lies a further belt of desert, the northern part of which is known as the Arabian and the southern as the Nubian Desert. The Sahara is a low tableland averaging 1200 feet above the sea. The monotony of its surface is broken by the sharp edges and cliffs produced on the higher ground by aeolian erosion. A central highland area known as Ahaggar rises in parts to 8000 feet and sends out

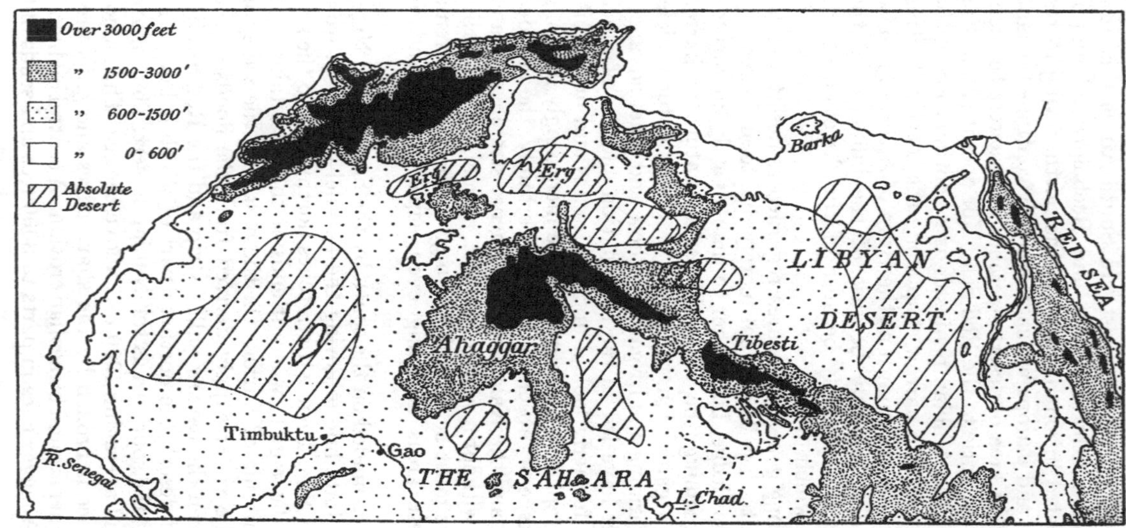

The Sahara

Note the relatively small area of absolute desert. The nodal positions of Timbuktu and Gao are clearly shown, the routes being forced to avoid the absolute desert on the one hand and the stony uplands on the other. **See map on page 235**

low spurs northwards, southwards, and south-eastwards to join the Atlas and the high tableland farther south. Of these spurs the Tarso Mountains in Tibesti are the highest, rising to over 6000 feet. Here and there are enclosed depressions which in the Libyan Desert fall below sea-level in more than one case. In the Western Sahara and in Libya there are vast stretches of monotonous country either ridged with sand dunes or covered with pebbles. East of the Nile the ground rises above the general level of the Sahara and is ridged with hills whose grain takes a northwest-southeast direction.

Since the Sahara lies in the Tropical belt of Horse Latitude, it has clear skies, which allows the sun to pour its rays pitilessly on the ground. Day temperatures therefore run up to 160° F. in many areas, while over a vast extent of country the mean annual temperature is more than 90° F., the greatest mean attained to by any large district in the world. The clear skies also cause, by facilitating radiation at night, an enormous daily range. Frost is not uncommon, e.g. Insalah averages nine nights a year with frost. Moreover, the annual range of temperature is great for the latitude, Insala showing one of 43° F. Except in the extreme south, the wind direction has a northerly component throughout the year. In July there is little air movement, but the southwest monsoon from the Gulf of Guinea penetrates beyond the coast strip at this season and gives rain as far north as Lat. 14° N. and even beyond this on the mountains of Ahaggar and Tibesti. At other times the wind is the Northeast Trade with the usual drying characteristics of trade winds. In January occasional rain is brought to the higher ground in the north by the low-pressure disturbances that sweep along the Mediterranean in winter. Rainfall is fitful everywhere, however, and does not exceed an annual mean of 5 inches.

When a shower occurs, the water runs off in wadis, or watercourses which are usually dry. These lead towards the sea, one of the great rivers, or one of the centres of inland drainage; but the drainage waters seldom or never reach their goal, being taken up by the dry soil long before. This causes a vast quantity of water to be stored up underground, and in many places it comes to the surface spontaneously or may be reached by sinking wells. At such points there are oases, where dwell the sparse and relatively few inhabitants of the desert. In actual fact, if absolute desert be taken to mean utter absence of plant life, only certain areas, which are shown on the map on page 232, answer to this description. Elsewhere there is a scant vegetation of wiry tuft grass, tamarisk, and acacia thorn. In the oases the date palm is characteristic, though probably all other cultivated plants—wheat, millet, gourds, onions—have been introduced. Camels, goats, donkeys, and horses are kept for domestic purposes. The inhabitants are Arab tribes which comprise people of different races, but mainly of Mediterranean type. They are nomadic for the most part, each tribe moving on a regular beat from oasis to oasis. Tribal wars are frequent, though the French in the west and south and the Italians in the Libyan Desert are gradually establishing peace.

From ancient times certain established caravan trade routes have run across the Sahara from the coast of the Mediterranean to various points on the great east-and-west route through the Sudan. The once famous city of Timbuktu derived its importance from the junction of some of these routes. The map on page 235 shows the chief ways across the desert. Until quite recently the crossing of the Sahara was a perilous affair, partly owing to the natural dangers of desert travel and partly owing to the attacks of the bedawin tribes. Within the last eight years the French Compagnie Transsaharienne has, however, established a regular motor service twice a week between Colomb Béchar in Algeria and Gao on the Niger, and airways will soon cross the desert from Algeria to Dakar and Lagos.

The extension of the Sahara along the coast of Eritrea and in Somaliland contains no considerable areas of absolute desert. Eritrea consists of a narrow coast strip backed by hills, while Somaliland is a tableland with an escarpment facing the Gulf of Aden and sloping gradually away to the Indian Ocean. It has a far smaller annual range of temperature than the Sahara, owing to sea influence. Its lack of rain is due to the path of the summer monsoon winds, which move parallel with the coast. The coast strip adjoining the Indian Ocean is barren, but the highlands get sufficient rain to support a vegetation of scrub and coarse grass on which the natives keep cattle. The people, who are Semitic in race, are nomadic herdsmen. The only town of any size is the little port of Jibuti, in French Somaliland, which serves as the outlet for Abyssinia and is connected by rail with Addis Ababa.

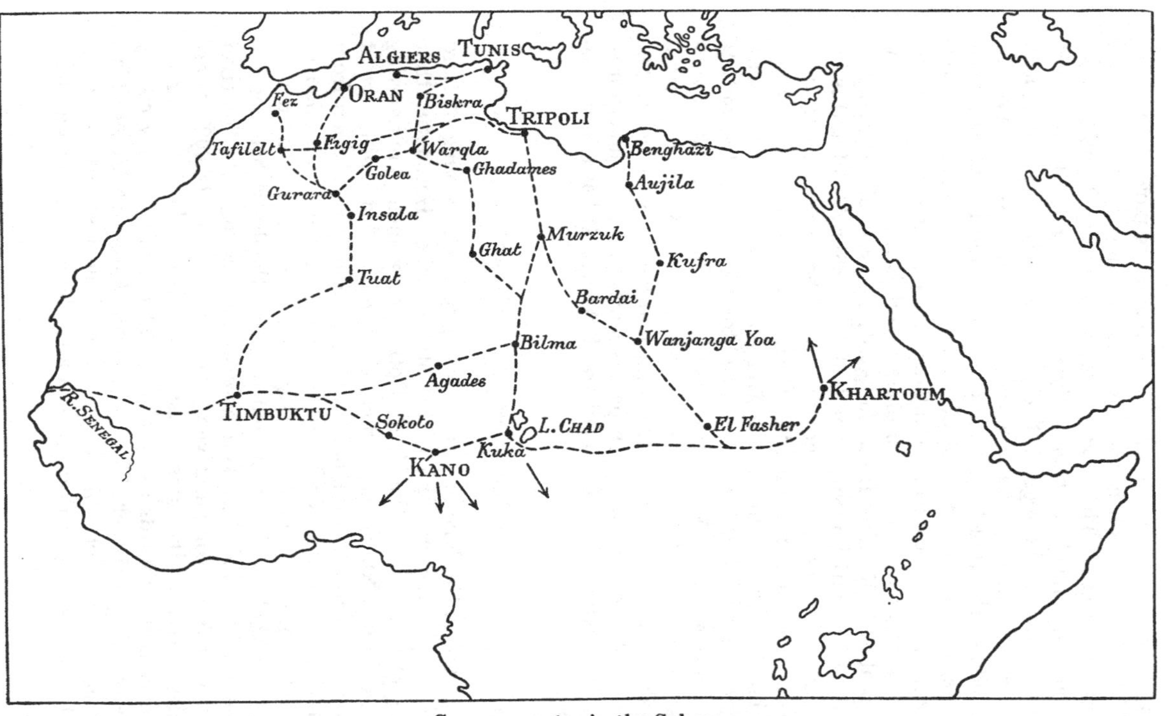

Caravan routes in the Sahara

The southern counterpart of the Sahara is part of the high tableland of the south and includes the narrowest of coastal strips of lowland. The Orange and a wadi named the Malopo, which must once have been a big river, have gouged out a deep valley through the region. The presence of the cold Benguella Current off the coast reduces the temperature, as does also the altitude of the tableland. The prevailing wind is the Southeast Trade, which has lost most of its moisture in crossing the continent. Hence, the mean rainfall does not exceed 5 inches. Nevertheless, the only area of absolute desert is a relatively narrow strip about 100 miles from the coast. The rest of the area, including the part known as the Kalahari, is scrubland and is inhabited by tribes of nomadic herdsmen. These are either Hottentots or, farther south, the little Bushmen, who have been driven off into this barren land by the stronger Bantu tribes. At present, exploitation of the region is confined to mining. The chief outlets are Swakopmund, Walvis Bay and Angra Pequeña, and there is a small inland settlement at Windhoek. A railway runs from Swakopmund to Grootfontein and another to Windhoek. The latter is the main line which passes south to de Aar junction to join the system of the Union of South Africa. It has a branch to Angra Pequeña.

5. The Lower Nile Valley

'Egypt is the Nile and the Nile Egypt.' This saying of Lord Rosebery's was an echo of the statement of Herodotus that 'Egypt is the gift of the Nile'. Both of these men were referring to the Lower Nile Valley, where the summer floods of the great river provide water for a narrow strip of land and reclaim it from the desert conditions which prevail on all sides. The modern kingdom of Egypt includes a large slice of the Sahara to the west and the whole of the Sinai Peninsula and the desert between the Nile and the Red Sea. But here we are dealing with the river valley only, a comparatively small area some 13,600 square miles in extent. It begins at Aswan, just below the first cataract, and runs northwards as a narrow, flat-bottomed valley cut by the river in hard rock. This gorge from Aswan as far as Cairo, which averages ten miles in width and is enclosed by low cliffs, is known as Upper Egypt. At Cairo the sediment-made delta spreads out its fan-shaped area as Lower Egypt. In the valley may be included the Faiyum, one of the depressions of the Libyan Desert which is near enough to be watered by a canal dug from the Nile.

The climate of the valley is Saharan, except at the seaward edge of the delta, where the Mediterranean type prevails. Elsewhere there is the same cloudless sky as in the desert, the same hot days and cool nights, the same almost total absence of rain. At Cairo, for instance, the mean annual rainfall is 1·3 inches. The prevailing winds are northerly, but, when low-pressure disturbances pass along the Mediterranean, they sometimes have the effect of bringing with them a parching wind from the south and clouds of dust which make this type of weather very unpleasant. Locally, this dust-laden wind is known as the Khamsin. Apart from this climatic conditions are stable and congenial. For four months in the year the temperature is as cool as that of an English spring, though during six months the mean exceeds 70° F. On account of its pleasant temperature during December, January, and February it is used as a winter resort by many English as well as other Europeans.

Nothing but the river prevents the valley from being as barren as the desert on each side of it, and the existence of the region as it is depends wholly on the Nile; indeed, the life and outlook of the people are bound up with the peculiarities of the river. From the earliest times of which there is record, the people of the valley have been agriculturalists and have used the flood waters of the river to irrigate the land. The floods occur in summer when the rains are falling in the Sudan and Abyssinia and reach the Lower Valley in August or September, bringing with them much fertile silt from the volcanic rocks of the latter country. Hence, the farmer enclosed his fields with low mud embankments, and let the silt-loaded water enter and flow slowly through so that during its passage it deposited a layer of sediment on the earth. In this way he gained not only the water needed for his crops, but also a fresh layer of good soil. Thus, the land needed no rest and no manure. Under English influence in recent years this system of 'basin' irrigation has been changed to the 'canal' system. A vast dam capable of holding 2423 million cubic metres of water —a capacity which was vastly increased in 1933—has been constructed at Aswan and barrages have been established at

Esna, Naq'Hammadi, Asyut, and Zifta. These erections store up flood water which when needed is led by pipes and canals to the fields. The use of pumps to raise the water permits the cultivation of land which is out of reach of the floods. It is a drawback to this system that much of the sediment is laid down within the dam or behind the barrages, thus depriving the fields of some of their fertility.

The peasants, or *fellahin*, form 62 per cent. of the population and are mostly smallholders. They grow food crops of beans, lentils, and millet for their own consumption. These, together with wheat, barley, and clover (for fodder), are winter crops. On the higher and drier ground autumn crops of maize and sugarcane are usual, while the delta produces rice and the chief export crop, cotton. The cultivation of this last crop has so increased that large imports of rice have become necessary. In 1931 the value of the exports of raw cotton amounted to £19,200,000. The date palm flourishes wherever sufficient water can be got. Water, indeed, is the main problem of the *fellah*, whose whole object is to secure an adequate supply for his land. The quantity due to each man is regulated by custom, but constant squabbles go on over water rights.

The valley is the home of one of the oldest peoples and earliest civilisations. Interesting remains of the architecture and art under the Pharaohs are found in the Pyramids of Gizeh and in the temples and royal tombs at Luxor. These attract numerous tourists and form the special branch of study of the Egyptologist. For over two thousand years, however, the land has been subject to foreign rule, and the original inhabitants now form the *fellahin*, whose standard of life is very low. The upper classes are the descendants of the Turkish conquerors. In 1927 the population of Lower Egypt was 6½ millions and that of Upper Egypt 5¾ millions, the whole totalling over 12 millions. The agricultural pursuits of the people do not favour the growth of large towns, and only two have more than 100,000 inhabitants, viz. Cairo (pop. 1,064,567 in 1927) and Alexandria (pop. 573,063 in 1927). Cairo, the largest city in Africa, is at the head of the Nile delta and centrally situated relatively to the valley. It is the capital of Egypt. Alexandria, on the western edge of the delta, is the chief port. Although the sweep of the tide is eastwards, the harbour suffers from constant silting. Apart from the towns,

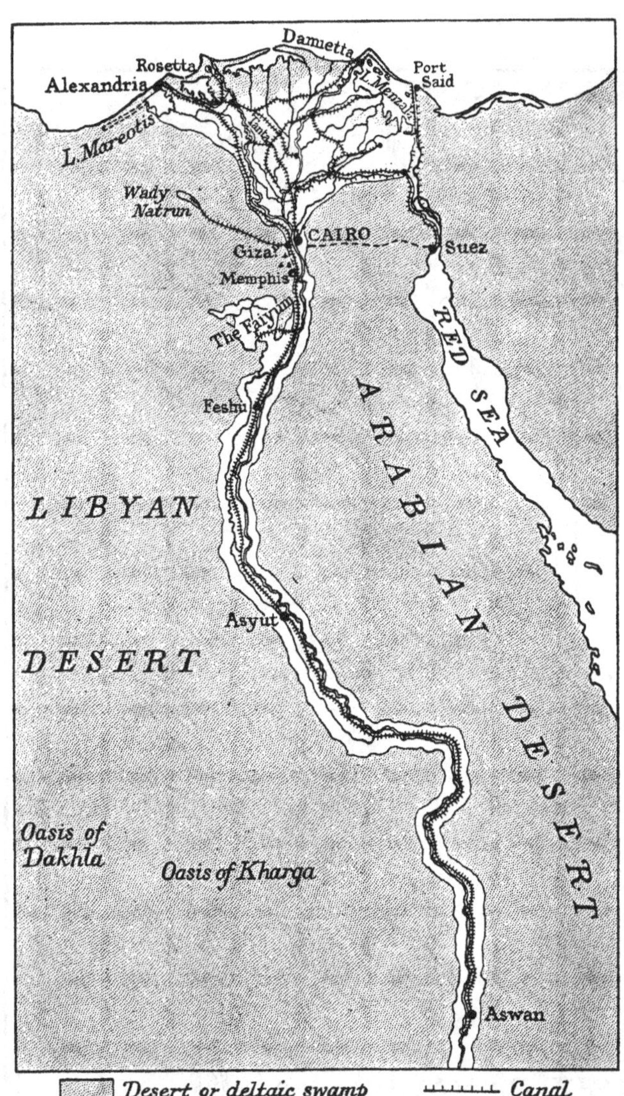

Desert or deltaic swamp *Canal*

Railway *Old Mecca pilgrims' caravan route.*

Note the canal which takes fresh water to Suez from the Nile.

The Lower Nile Valley

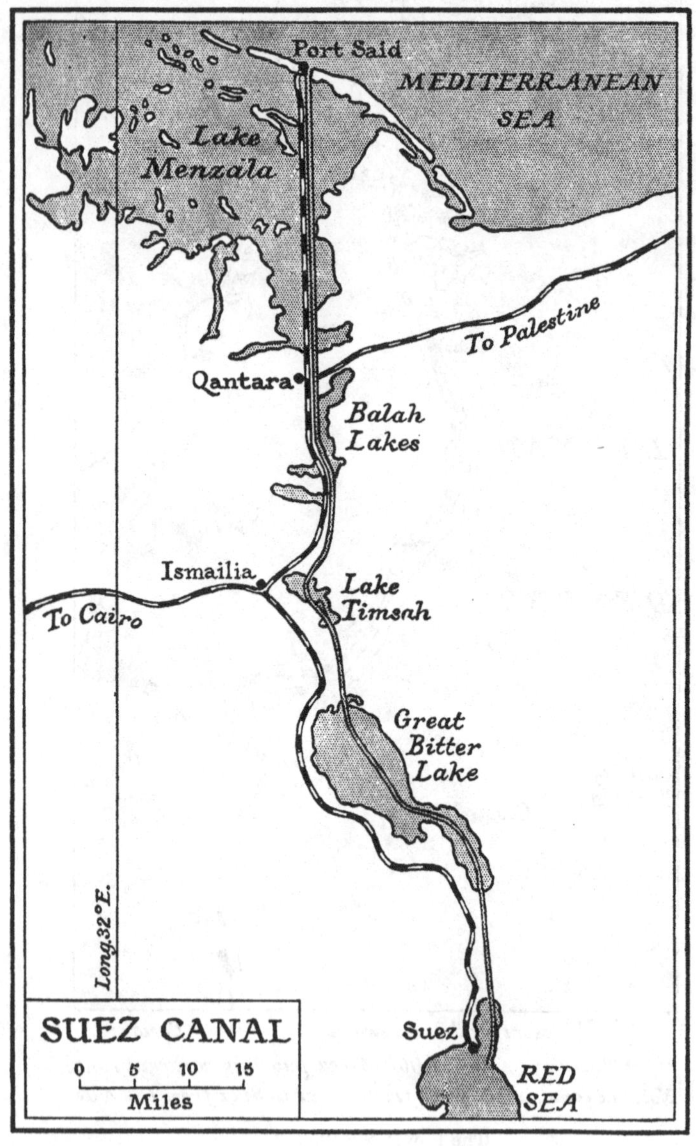
Port Said
MEDITERRANEAN
SEA
Lake
Menzala
To Palestine
Qantara
Balah
Lakes
Ismailia
Lake
Timsah
To Cairo
Great
Bitter
Lake
Long.32°E.
SUEZ CANAL
0
5
10
15
Miles
Suez
RED
SEA

the population is dense, since the fertility of the soil allows a few acres to support whole families. Taken as a whole, the valley contains 900 persons to the square mile, a density not reached in the most thickly peopled areas of the north of England.

The river is used for communication up to Aswan and also between the cataracts. One sees both the modern river steamer and the ancient Nile boat plying up and down stream. On the delta there is a close network of railways, and a main line ascends the valley as far as Aswan, with a narrow gauge branch to the Western Oases. Lines also run eastwards to Port Said and Suez and on to join the railway system of Palestine. Alexandria is one of the stopping places of the British air mail to India and South Africa.

6. The Tropical Lands of the East Coast

These areas, partly by position and partly by height above sea-level, have the common features of warmth, fairly heavy rainfall in summer, and an open forest or grassland vegetation. They are the southeast coast lowland strip from the Rovuma to just south of Delagoa Bay, the Abyssinian Highlands, and the island of Madagascar.

The first of these areas coincides almost exactly with Portuguese East Africa. It is a continuation of the equatorial east coast region, without any definite line of separation so far as relief and climate are concerned. The tropical strip differs from the equatorial, however, in having a typically tropical rainfall régime, with a well-marked maximum in summer and a distinct minimum in winter. During the equinoctial periods of the year the prevailing wind blows from the southeast and is the normal trade. At midsummer (January) the low-pressure centre then developed over the Transvaal and the Rhodesias causes a shift of the wind to the northeast. It is this wind, which is practically a monsoon, that brings the rainfall maximum. In winter (July) the winds are light and blow off the land: hence there is little rain at this time. The vegetation consists of tropical forest over the whole area, though between Beira and the Limpopo River there is much scrubland. Mangrove swamps fringe the coast.

The Portuguese have not done much to exploit the economic resources of the country, and so far the development of the region has depended on the use of the ports of Lourenço Marques, Beira, and Quelimane by the English and Dutch who have settled on the tableland in Nyasaland, the Rhodesias, and the Transvaal. Railways run from Lourenço Marques to Pretoria and from Beira to Salisbury, while the Zambezi and the Shirë connect Quelimane with Nyasaland and parts of Rhodesia. The Portuguese have planned a railway along the coast and, as soon as the Zambezi has been bridged, there will be continuous railway connexion between British Nyasaland and Beira.

The Abyssinian Highlands are a mass of volcanic rock rising to more than 12,000 feet, but cut into deeply by rivers and almost divided into two parts by the Great Rift Valley. The whole forms a very high plateau or topland and is the most extensive area of ground of similar height in Africa. Its climate varies greatly with height above sea-level: the tableland is cool and at times cold and bleak, while the valleys, which are often deep and steep-sided, are subject to tropical heat. The great Asiatic monsoon system just touches this region, so that the winds blow from the southeast in summer and from the northeast in winter. The summer winds come from over the Indian Ocean and give relief rain on the highlands of the south and east, while convection rains fall on the west side. The Rift Valley is here, as farther south, subject to drought and to poor vegetation, except in the neighbourhood of streams and of the lakes which occur on the floor of the valley. A strip of very dry country lies round the inland basin of Lake Rudolf.

The chief rivers are the Blue Nile, which rises in Lake Tana, the Sobat, the Atbara, and the Hawash. The last flows along the floor of the Rift Valley, but fails to cross the thirsty stretch of lowland which it must pass to reach the Gulf of Aden. Owing to the marked tropical system of rainfall, these rivers are in flood from June to October, and then it is that the Sobat, the Blue Nile, and the Atbara pour their vast quantities of water into the Nile to irrigate the ribbon valley of Egypt. The river valleys are often gorge-like, and have their sides and bottoms heavily wooded with open tropical forest which contains valuable trees including rubber. Higher up the slopes and on the tableland the trees disappear, leaving the country covered with grass. As in the equatorial highlands adjoining, the vegetation type is savana, but the grasses are shorter and less coarse than normal. The people are naturally herdsmen, keeping cattle, sheep, and goats, and, to a less extent, horses, mules, and donkeys, which serve as baggage animals for transport along the tracks that serve as roads. Hides and skins are exported in small quantities. There is, however, a fair amount of agriculture, especially on the lower ground, and the produce is sufficient for local needs, though without a surplus for export. Coffee is the only crop grown for the purposes of foreign trade. It is either of the Mocha variety or a wild plant peculiar to the region. The last grows in extensive forests on the southwestern slopes.

The region forms most of the independent native kingdom of Abyssinia, which has preserved its freedom through its isolation amid a ring of desert and other country difficult to cross. The people are, for the same reason, very suspicious of strangers. They prefer to dwell in villages, and consequently the only towns are Addis Ababa (pop. 65,000?) and Dire Dawa (pop. 30,000?). The former, which is the capital of Abyssinia, owes its importance to its position at the converging point of several valleys. It is connected by rail with the port of Jibuti on the Gulf of Aden, but the train service is very slow and only three trains run in each direction every week. Dire Dawa is also on this railway. The people are a mixed race in which Semitic and Hamitic stocks prevail. They are superior to the natives of central and South Africa, having a fairly high standard of civilisation among the upper classes.

Madagascar is a block of Gondwana Land which has been severed from Africa in the general foundering of the ancient continent. It is topped by a plateau from which the ground falls away sharply on the east and rather more gently on the west. Its climate is similar to that of the coast strip opposite, from which it is separated by 400 miles of sea. The eastern side gets heavy relief rains from the Southeast Trades, but the western slopes are far drier, while the southwestern corner is very dry owing to a rain shadow cast by the plateau. The eastern side is covered with tropical forest, but the southwest is savana land. The natives, who are known as Malagasies, are of Malay stock in the east and of negro origin in the west. The island is a French possession, but is little developed for exploitation. Cattle-grazing is carried on in the west, and tinned beef and hides are exported in small quantities. When cleared, the forest areas produce rubber and coffee on the uplands and rice, cotton, and sugarcane on the lowlands. The chief town and capital is Antananarivo (pop. 71,000), which is situated on the central plateau and connected by rail with its little port of Tamatave.

A number of islands, such as Réunion and the Comoro group, are politically subordinate to Madagascar. They are of little importance.

7. The Warm Western, or Mediterranean, Region

The Atlas system, together with a narrow strip on the north coast, and a small area around Cape Town are marked off very distinctly by their climate from the rest of Africa. These regions indeed are transitional between the tropical deserts and the temperate belts which lie farther polewards. The belt of high pressure which in summer lies along Lat. 30° N. and 30° S. causes dry, clear air, an absence of rain, cloudless skies, and few disturbances of weather—in fact, desert conditions. And, in fact, these regions are extensions of the tropical deserts in that season. In winter the high-pressure belt moves a few degrees towards the Equator, leaving the regions in the belt of Antitrades (Westerlies). These winds blow from the west in the northern hemisphere and from the northwest in the southern. Hence, they pass from the sea to the land giving winter rains and moderating the temperature.

The Atlas system consists of two parallel lines of fold mountains, with a topland of old rock pinched up between them by the forces of earth movement which produced the folds. On the west the topland descends gradually to the Atlantic, while on the Mediterranean coast there is a narrow strip of fertile lowland known as the Tell. The more northerly of the two fold ranges is the Tell Atlas, which is a simple fold running from the Straits of Gibraltar along the coast to Cape Blanco. The more southerly fold range is more complicated and consists of three main portions. The Saharan Atlas runs from Cape Bon to the neighbourhood of Colomb Béchar, though with a break at Biskra. Farther west the continuation is known as the Anti-Atlas and reaches the Atlantic at Ifni. A secondary continuation known as the High Atlas runs north of and parallel to the Anti-Atlas, sending a branch northeastwards to join the Tell Atlas near the town of Fez. This branch divides the intermontane topland into the Mesa and the Plateau of the Shotts. The rest of the north coast of Africa consists of lowland, except for the Nefusa hills which lie south of Tripoli and the Barka uplands.

The Atlas region comprises four climate subdivisions: (1) the Mesa, which is influenced by the cold Canaries Current and has an equable temperature and a certain amount of rain throughout the year; (2) the Tell, with a typical Mediterranean climate; (3) the Plateau of the Shotts which, sheltered from the moisture-bearing winds in winter, has a mean rainfall of only 12 inches a year; and (4) the Saharan slopes of the Atlas in which desert conditions are almost complete. The Mesa proper consists of a

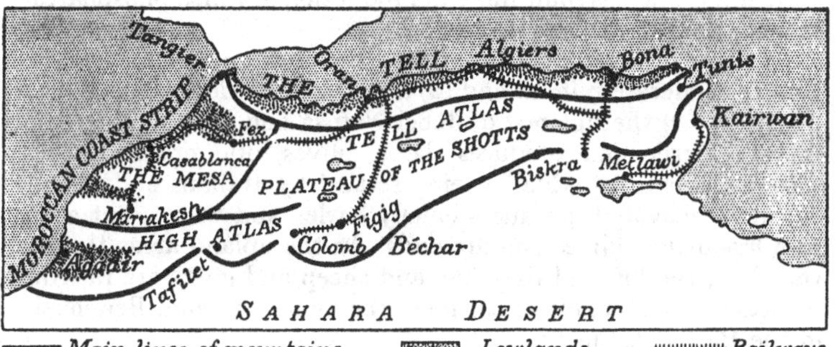

The Atlas region

small tableland whose grassland vegetation makes admirable pasture. Its western surface is cut by four valleys which, added to a narrow coast strip, make an area of fertile agricultural land. The Plateau of the Shotts is an area of inland drainage, with numerous depressions which become shallow lakes, or *shotts*, after rain. Its surface is covered with alfa grass which yields good pasture and is used for making paper and straw hats. The town of Tunis, near the site of ancient Carthage, is on the coast at the east end of the Plateau. The rivers are of the wadi type, flowing only after rain, except on the Mesa where the streams are sufficiently stable in volume to be used for the production of electric power.

The soil on the coastal lowlands is very fertile and in ancient days formed the 'granary of Rome'. Irrigation, terracing, and careful cultivation are necessary. Cereals, especially wheat and barley, are among the chief crops, but tobacco, wine, and Mediterranean fruits are the chief exports. Besides these, however, there are all the various kinds of produce common to Mediterranean countries: olives and olive oil, cedar and pine wood, and cork, as well as such Saharan products as dates and gum. The higher mountains are of little use and are sparsely inhabited. Their rough surface has caused them to be the refuge of warlike bands who have until quite recently been a permanent source of disorder in the country. The lower slopes and the tablelands are used for grazing, horses, sheep, goats, and cattle being raised in large numbers. But this leads only to the export of a little wool. There is some mineral production of which phosphates are the chief item.

In Tripoli there is (1) a narrow, discontinuous belt of coast land of moderate rainfall and (2) a belt of semi-desert backed in places by (3) the uplands of Jebel Nefusa and Barka. The first is very fertile and produces dates, olives, and oranges. The second, which is gradually being settled by Italians and scientifically cultivated, produces chiefly barley and wheat, but also to a less degree olives and almonds. In the upland areas thrive the olive, the fig, and the vine, and sheep and goats are raised. Efforts are being made to cultivate the banana around Benghazi in the Barka Peninsula.

The coastal waters of the region are extremely rich in fish. In the Mediterranean the catch consists chiefly of tunny, sardines, and anchovies, while sponges and coral are collected for export. Along the Atlantic coast there is an abundance of tunny, cod, and mackerel. These waters are visited by the fishing fleets of France and Spain, but there is a local industry with its headquarters at Casablanca, where preserved fish is exported.

The people of the region are largely a mixture of Berber and Arab and are of the Mediterranean race, but there is also a fair admixture of negro stock. They have the Mediterranean tradition of town life in 'honey comb' villages and dress in flowing robes of cotton. Under Carthaginian rule two centuries before Christ and again under Arab domination in the eighth to the tenth centuries A.D. they played an important part in the history of the Mediterranean, but the region has now sunk into a political extension of France, Italy, and Spain.

The chief town on the Tell is Algiers (pop. 257,000 in 1931), whose importance is derived from a central position on the coast strip and the terminus of a natural route down to the sea from the interior. Orán and Bona are secondary ports. On the Mesa Marrakesh is the exchange market for coastal produce with dates, etc., from the Sahara, while on the lowlands is Fez with its outlet at Casablanca. At the other end of the region is Tunis. On the Saharan slopes are a number of settlements on oases which serve as starting places for caravans across the desert. Railways run from Algiers eastwards to Bona and Tunis and westwards through Orán, Fez, and Casablanca to Marrakesh.

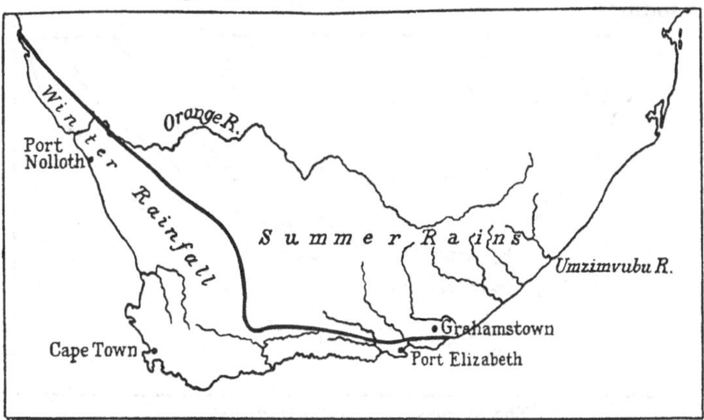

The Cape 'Mediterranean' region, as outlined by the rainfall régime

Another crosses the Atlas to Colomb Béchar, whence the trans-Saharan bi-weekly motor service starts for Gao on the Niger. Daily air services run from Marseille to Algiers and from Toulouse to Casablanca via Tangier, and there is a weekly service between Casablanca and Dakar.

The southwest district of the Cape Province in the extreme south of the continent has a similar climate. The area is bounded roughly by a line running from east of Port Elizabeth to Luderitz Bay (see map), but may conveniently be taken to include the whole of the Great Karroo. North of the Orange River, though the rainfall system is Mediterranean, the amount of rain is too small to redeem the land from desert, and this portion of the area must therefore be omitted from the region. Along the south coast the tendency of the African tableland to descend to the sea in terraces is developed to the full, and below the High Veld, as the tableland is called here, there are three distinct steps, each of which is bordered by high ground. The highest is the Great Karroo, a barren area somewhat similar to the Plateau of the Shotts. The vegetation is that of poor steppe, viz. scrub and the little dusty-grey karroo-bush. The latter is very nourishing for sheep, which are accordingly raised here. Formerly, ostrich farming was a flourishing occupation around Oudtshoorn, but the disappearance from fashion of the use of feathers has killed the industry.

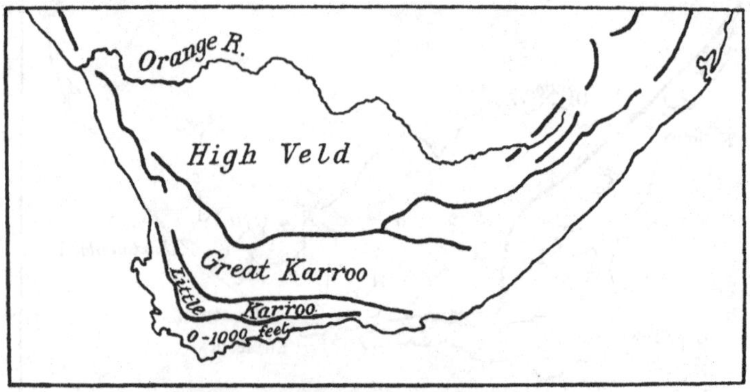

The terraced rise to the High Veld

The next terrace is the Little Karroo, on which there is sufficient rain to feed permanent streams. The drainage passes northwards into the Olifants River, a tributary of the Doorn, and southeastwards into the River Breede. Seawards the Little Karroo is bordered by the Lange Bergen and its continuations westwards and northwestwards, and between these ranges and the coast is the third and lowest terrace. Its surface is broken by huge tabular hill masses whose shape is due to the horizontal position of the layers of ancient rocks which compose the rock of the terrace. Table Mountain is a well-known example of the type. Between these tabular hills are fertile valleys, and along the coast there are strips of lowland which help to make up the really useful part of the region. The native vegetation is of the same type as that of the Mediterranean, though of course the species are different; and the olive, grape, orange, and other Mediterranean fruits flourish exceedingly when introduced. Sixty per cent. of the wheat produced in the Union of South Africa is grown here. Maize, however, is neglected and only 5 per cent. of the total production of the Union is derived from this region. Wine- and jam-making and fruit preservation have become definite industries. Since the fruit ripens at the opposite time of the year to that at which it matures in Europe, the fruit growers now export fresh fruit in cold storage to English markets to supply the needs of the British Isles when fruit has become scarce there. The imports received in exchange are manufactured goods, of which textiles, machinery, and other metal goods are the chief.

In this region lie Cape Town (European pop. 150,000 in 1931), one of the two capitals of the Union of South Africa, and Port Elizabeth (European pop. 44,000 in 1931), a growing port farther east. These are outlets not only for this region, but also for the adjacent ones on the tableland. It must not be thought that the edges of the Karroos are continuously scarped: breaks occur here and there to afford passage for railways and other routes. The main line of what was once hoped would be the Cape to Cairo Railway starts from Cape Town and zigzags up the Karroos to de Aar Junction on the High Veld. It connects Cape Town with the Orange Free State, the Transvaal, and the Rhodesias. Other shorter lines run along the grain of the land to Natal and to the outlying towns within the region. Cape Town is also the terminus for the passenger air service which was started in April, 1932, between Croydon and the Cape.

8. The Warm Eastern Region

The east coast strip below the tableland and between Port Elizabeth and Delagoa Bay continues the terraced formation described in the south of the Cape Province, though here the terraces are less well marked. There is a margin of coastal lowland, backed by a rather wider belt of broken, hilly country lying below the slopes which lead directly up to the edge of the continental tableland. At Durban on the coast the mean annual temperature is 70·8° F., with a range of 12° F. between the hottest month (February, 76·6° F.) and the coolest (July, 64·6° F.). Pieter Maritzburg, at a height of 2250 feet above sea-level, has a mean annual temperature of 66·6° F. and a range of 15·8° F. This is 7° F. cooler, with nearly double the annual range, than at Dar es Salaam. Owing on the one hand to the effects of the warm Mozambique Current which flows along the coast of Natal and on the other to the cold Benguella Current on the west coast, the East Coast Strip is far warmer than the west. Compare, for instance, the mean annual temperatures at Durban (70·8° F.) and Port Nolloth (57·4° F.).

The prevailing wind is the Southeast Trade which blows strongly in summer, but weakens in winter. Hence, 70 per cent. of the rainfall occurs in summer. The coastal lowland gets an average of 40 inches, while the slopes of the continental tableland receive as much as 45 inches. The intermediate belt is drier, with an average of 30 inches. Generally speaking, there is a ten-mile-wide fringe of tropical open forest along the coast with extensions up the river valleys. Behind this is temperate grassland, the dry winters being unfavourable to the growth of trees. The middle courses of the river valleys, being sheltered from the rain-bringing winds by the hills, are often too dry for grass and are covered with thorn bush and scrub. The valleys of the Pongola, Tugela, Umzimvubu, and Great Kei Rivers are striking instances of this. On the high slopes of the edge of the continental tableland, where one would expect forest, the prevailing vegetation is grass, owing to the effect of exposure to the Southeast Trades; but there is bush and forest on the banks of watercourses in sheltered valleys.

These upland forests produce pit-props and building timber for use on the treeless tableland beyond, while the grasslands are used for sheep and cattle. The rest of the region may be divided into three economic parts: (1) the grasslands of Swaziland and Zululand in the north which are almost given up to native use and are economically of little importance; (2) the backland of Durban which is the richest area and contains Pieter Maritzburg, the capital of Natal, and the smaller towns of Ladysmith and Greytown; (3) a large area of grassland and thorn bush between the Umzimvubu and Great Kei Rivers which is known as Transkeia or Kaffraria and is given up to native use; and (4) the backland of East London (European pop. 27,000 in 1931) which is a district of moderate fertility. Production is agricultural and mineral. Maize and other cereals are grown, and fruit is cultivated for export to England. Along the coast and in Zululand there are large plantations of sugarcane and tea, while in the drier parts acacia is grown for its bark which is used for tanning purposes. The region is rich in minerals, the number of metals exploited amounting to more than a dozen, though coal and gold by far exceed the others in quantity. A scientific whaling industry is carried on from Durban.

The native population is of Bantu race. Swaziland is a native protectorate, while Transkeia is largely inhabited by its original people. Western European civilisation has hardly raised the blacks as yet, although those who live among the European settlements have adopted the dress and some externals of the customs and religion of the colonists. In their own homes the natives are herdsmen by nature and tradition, keeping cattle and sheep on the grasslands; but like most of the Bantu tribes they till the soil to some extent and by primitive methods grow maize and kaffir corn from which they make 'mealies'. Owing to their inefficiency as labourers, large numbers of Indians have been imported to work on the sugarcane plantations.

Durban (European pop. 86,000 in 1931), the third largest town in the Union of South Africa, and East London are the two chief ports. The former is connected by rail with its backland, and the main lines cross the edge of the continental tableland to Johannesburg and Bloemfontein. Motor roads are being rapidly constructed and bus services started along them. Except for the railway from Ladysmith through Piet Retief to Johannesburg, all the public traffic in Swaziland consists of motor buses. East London is linked up with the railway network of the south of the Cape Province.

9. The Warm Highlands

The High Veld, or continental tableland south of the Tropic of Capricorn, forms a separate region by reason of its height. The surface of the tableland is tilted downwards to the west (see diagram below) from a height of about 6000 feet to one of 4000. The eastern edge rises up in the Drakensberg and Quathlamba Mountains, some parts of which are 10,000 feet above sea-level.

In the south, where the most prominent parts of the plateau edge are the Stormberg and the Nieuwveld, the general height is less, and in the north there is a gradual falling away in the Low Veld to the valley of the Limpopo. A knot in the Drakensberg forms the hilly country of Basutoland. Elsewhere, the tableland presents a monotonous surface broken by rocky hills, known locally as *kopjes*, and by *dongas*, or gorge-like river valleys. The southern portion of the region is drained by the Orange River and its feeder, the Vaal, and the northern portion by the Limpopo.

The region lies just outside the hot belt, and its height makes its mean temperature cooler than that of the east coast strips. Yet in summer the days are hot, since the clear air allows the sun's rays to act fully on the ground. For the same reason the summer nights are cool, radiation being assisted by the absence

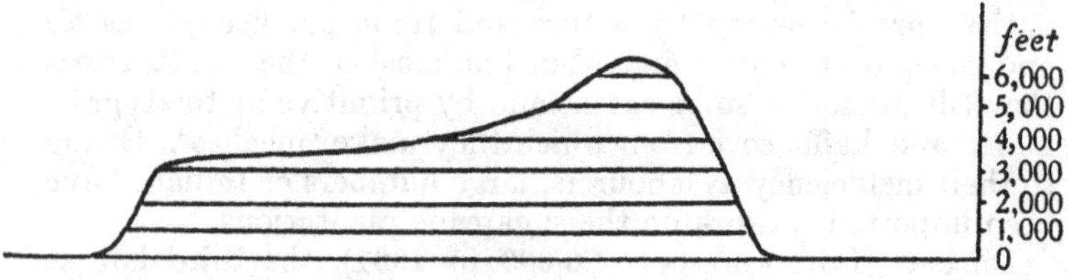

Section across the High Veld

of clouds and other atmospheric moisture. Winter is a period of cool days and nights, the temperature being between 20° F. and 25° F. lower than in summer. On the highest ground towards the south falls of snow are not uncommon, while in most of the region morning frosts occur every year. It is the winter temperatures that bring the annual mean below that of the east coast. But the influence of the Benguella Current keeps the temperatures of the west coast even lower. Thus,

	January	July	Year	Range
	° F.	° F.	° F.	° F.
Pretoria	71·7	51·7	63·5	20·0
Kimberley	75·8	50·6	64·8	25·6
Durban	76·6	64·6	70·8	12·0
Port Nolloth	59·9	55·2	57·6	6·5

It is the occurrence of a really cool season that makes the region fit for permanent settlement by Europeans.

The prevailing wind is the Southeast Trade, which blows strongly in summer. But in winter anticyclonic conditions are usual, and at this season the sky is clear and there is little or no rain. Summer is the period of greatest rainfall, precipitation being due partly to convection and partly to relief. The Southeast Trades deposit as much as 45 inches a year on the Drakensberg; but the rainfall decreases towards the west fairly quickly. The map below shows the isohyets for 30 inches and 10 inches

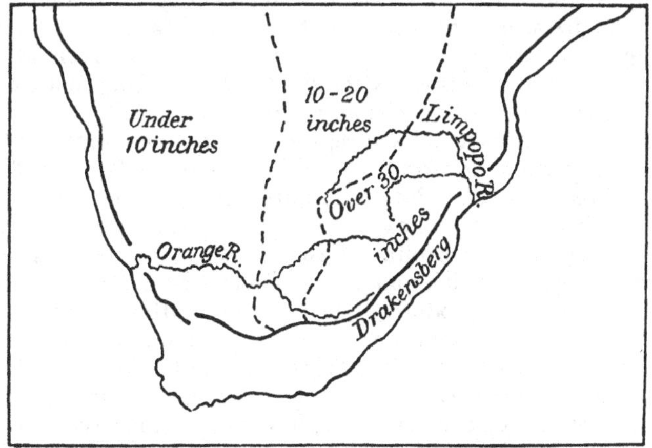

Rainfall regions of the High Veld

on the tableland, dividing the region into three parts. Only in the most easterly is agriculture possible, the vegetation being temperate grass. In the middle area, where there is a rainfall of 10 to 30 inches, the grass dries up in summer and is mingled with patches of scrub. West of the isohyets for 10 inches there is poor steppe where scrub prevails.

The rivers have a tropical régime, and greatly resemble those of Australia. Owing to the thinness of the soil, which causes the rain to run off as soon as it falls, and to the fact that most of the precipitation occurs in summer, the rivers rush along in spate in the latter season, but degenerate into brooks or even lines of waterholes in winter. Many streams actually dry up at times.

The Orange, which is by far the largest of the rivers, illustrates this régime perfectly. As a result of the low rainfall, the rivers tend to cut their beds below the surface of the tableland and to form the shallow canyons known as *dongas*. They are of course useless for either navigation or irrigation, and they are difficult to bridge when they are of any size.

The grassland vegetation gave rise to a plentiful supply of 'big game' of the same types as have been described as still existing on the savanas of the equatorial highlands. But the guns of the white farmers have cleared these animals from the region, except in the west, which being too dry for settlement still harbours a few antelopes, wart hogs, and jackals. There is also a good deal of game still left in the Low Veld, where in order to preserve the beasts from extinction the Kruger National Park has been set aside as a special game 'reservation'.

The European settlers have replaced the native animals with sheep and cattle, which form the chief wealth of the farmers. The Orange Free State and Basutoland are almost wholly given up to the raising of sheep and cattle, the centre of the industry being at Bloemfontein, the capital of the former state. Wool is exported in large quantities mainly to Great Britain, and a good many hides are also produced. The beef and mutton are used for food locally, but is insufficient in quantity, and a large amount is imported every year from the Argentine. In the east, where the rainfall is greater, agriculture tends to displace pastoral pursuits. The chief crop is maize, though wheat and tobacco are also cultivated. The districts around Pretoria and Potchefstroom in the Transvaal and Caledon and Aliwal North in the Orange Free State are the chief agricultural areas.

The wealth of the region, however, lies at present in its vast mineral resources. Gold, diamonds, coal, copper, and tin are all found in important quantities. The first two are mined in many places, the famous 'reef' at Witwatersrand in the Transvaal being the world's richest gold-producing area. The town of Johannesburg (European pop. 203,000 in 1931), the second largest in Africa, derives its importance wholly from its mines. In 1932 the gold output of the Transvaal was 11,553,564 oz., valued at over £50 million. The chief diamond centre is at Kimberley in Bechuanaland, from which comes a large proportion of the world's supply.

In general, though local factories are being slowly established, the foreign trade of the region consists of an exchange of primary products, whether pastoral, agricultural, or mineral, for textiles, machinery, and other manufactured goods. The ports through which these imports and exports pass are naturally outside the region and are reached by rail. Cape Town still holds first place, but the Portuguese town of Lourenço Marques on Delagoa Bay is being more and more used by the northern Transvaal, since the whole of the basin of the Limpopo is the natural backland of this port. Durban is not so clearly the outlet of the southern Transvaal and the Orange Free State, but a good deal of the traffic of these areas passes through this port and through Port Elizabeth.

Communication and transport, which formerly depended on ox-wagons, is now carried on chiefly by rail, though the use of motor vehicles is growing. The roads are poor on the whole, but in such a dry country roads of moderate quality are easily made and still more easily kept up. The railway system focuses to a large extent on Cape Town, from which two main lines separating at de Aar Junction run north, one through Kimberley and Mafeking to Rhodesia, the other through Bloemfontein and Johannesburg to Pretoria (European pop. 62,000 in 1931), the second capital of the Union of South Africa. But other lines reach the coast, after a shorter route, at Lourenço Marques, Durban, East London, and Port Elizabeth. Connexion with Southwest Africa is maintained by a line from de Aar Junction to Luderitz Bay and Swakopmund.

The natives of the region are Bantus, who thrive under European civilisation and greatly outnumber the whites. In their natural state they are cattlemen—as, for instance, in Basutoland—but they also cultivate in a primitive way sufficient maize and millet to feed themselves. They were formerly warlike and do not make good labourers. In the west of the region the tribes are of mixed race, and in the dry expanses of Bechuanaland they are pastoral nomads who have so far been hardly touched by European civilisation. The white settlers are either the descendants of the early Dutch farmers who first entered the region in about the year 1810, or later British immigrants whose main object was the exploitation of minerals. This difference in origin still largely remains, the Dutch being for the most part farmers (Boers), while the British are miners or business men in the larger towns. Economically, the two sections are complementary, but for historical reasons they are politically opposed to each other. The region forms the greater part of the Union of South Africa, since it includes a large part of the Province of the Cape of Good Hope, the whole of the Transvaal, the Orange Free State, and Basutoland, and some of the Bechuanaland Protectorate.

Exploration and Settlement. Although Egypt was the home of one of the earliest civilisations and the coast lands of North Africa are part of the Mediterranean region and have shared in its progress, Africa as a whole has been strangely behindhand in becoming known to the civilised nations of the world. The reasons are not far to seek: the great breadth of the Sahara and its eastern extensions were an effective check on travellers, the one possible crossing up the Nile was blocked by sudd, and the country to the south offered no obvious rewards for difficulties overcome. The desert could be avoided by moving along the coast in ships; but fresh obstacles presented themselves to such attempts. The African coast is singularly repellent, offering few havens, a barren or mangrove-lined shore, a fever-stricken approach, and hostile natives of little or no social and political organisation. Consequently, the Carthaginian and other Phœnician voyages proved a failure, and the later Arab attempts to effect a settlement along the east coast did little more than secure a foothold at Zanzibar and a few other places which were used as depots for the trade in slaves and spices. Hence, at the beginning of the fifteenth century little was known of the continent south of the Sahara. Nor had anything but the merest influence of civilisation penetrated even to the natives of the Sudan, Abyssinia, and Somaliland, in spite of considerable race drifts of superior peoples into those countries.

During the second decade of the fifteenth century the Portuguese under the stimulus of Prince Henry, called the Navigator, began to make systematic attempts to explore the west coast of Africa. Their progress seems to us nowadays to have been astonishingly slow, and it was not until 1497 that Vasco da Gama rounded the Cape of Good Hope and made his way along the east coast to the Arab settlements in what is now Kenya. His objective was India, and his voyage established a seaway between

Plate XV

(*South African Railways*)

CAPE TOWN AND TABLE MOUNTAIN

The mantle of clouds, which occurs frequently, is known as the 'table-cloth'

Plate XVI

(*South African Railways*)

VIEW OVER PRETORIA

The view is taken from Meintjes Kop. The Union Buildings are seen in the foreground

Western Europe and that country. The Portuguese were followed a century later by the Dutch, whose trade sought the

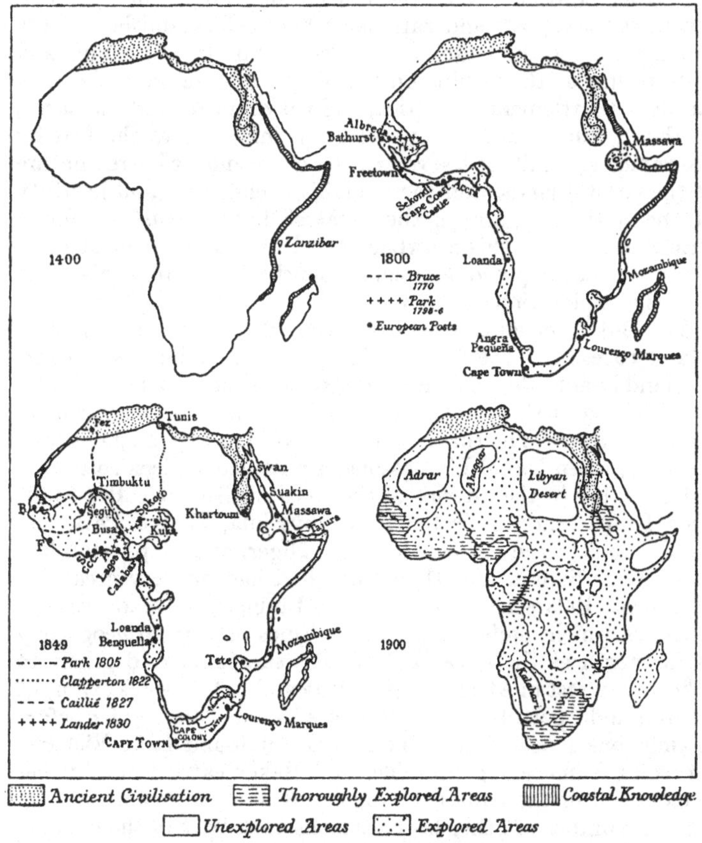

The opening up of Africa

In 1400 only the Mediterranean coast strip and Egypt were known to Europeans, and the east coast to the Arabs. By 1800 the coastal exploration was complete, but there was little penetration inland. The progress made between 1849 and 1900 is remarkable

spices of the East Indies, and by the English and French, who opened regular commerce with India.

Meanwhile, a traffic in negro slaves had led to the establishment of trading posts along the Guinea coast, where later gold and ivory were added to the objects of trade. For over two centuries these posts and various ports of call established by the Portuguese and the Dutch for the re-victualling, re-fuelling, and re-watering of their ships on the way to India were the only European settlements in Africa, and they represented the barest foothold. All attempts at penetration inland were checked by the rapids and falls of the rivers, the malarious or barren nature of the coastal lands, and the uncivilised character and hostility of the natives. Perhaps the greatest barrier was the fever-stricken coast. To this drawback the Dutch settlement at Cape Town was not exposed, and consequently it was from this point that the exploration of the interior began.

The Dutch settlers, impatient of official rule, gradually moved inland to escape interference. This movement increased when England bought the settlement at the end of the Napoleonic Wars, and little by little Natal and the High Veld were occupied by pioneer farmers. The turning point in African exploration came, however, when David Livingstone, a missionary, decided to find out what lay in the interior of the 'dark continent'. Before his time Bruce had penetrated into Abyssinia, Mungo Park and Lander had traced the course of the Niger, and Clapperton had crossed the Sahara, but their work had had no very great importance. Between 1849 and 1865 Livingstone made several journeys which made known the outlines of African geography as far north as the Great Lakes. His success stirred others to take an interest in African exploration and to follow his example. Burton and Speke together discovered Lake Tanganyika before Livingstone reached it, Speke and Grant found Lake Victoria, returning home along the Nile, and Baker carried out further exploration of the equatorial highlands and the eastern Sudan, while a number of other lesser persons took part in the opening up of Africa. When Livingstone was thought to be lost on his last journey, Stanley undertook to find him and, in doing this, crossed the continent from east to west, following the Congo River almost from its source right to its mouth. This practically completed the main survey of Africa.

This increased knowledge, the ever-growing needs felt by modern industry for raw materials which were to be had in the hot belt alone, and the desire for the expansion of national power now began to make the peoples of Europe realise the vast potential wealth of Africa, and those nations which had settlements in

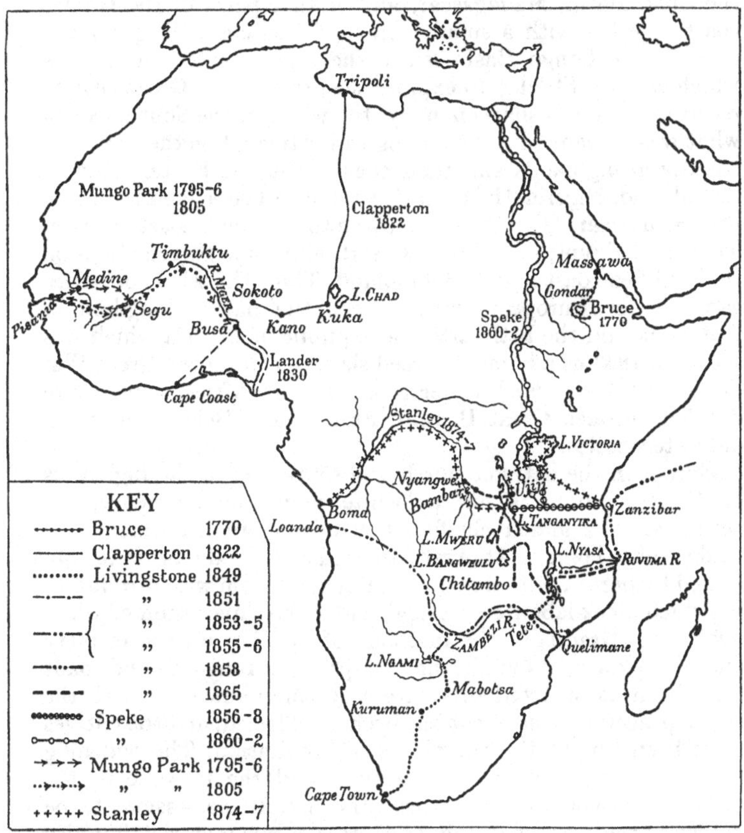

The exploration of Africa

the continent laid claim to the backlands as widely as possible. The British and French divided almost all West Africa between themselves. In 1883 Germany laid claim to the Cameroons, and this decided the Great Powers of Europe to hold a conference to settle claims quietly. This conference, which has been called 'the scramble for Africa', was held at Berlin in 1885. At it Germany was allowed the Cameroons, Togoland, and Tanganyika, while France was given Algeria, Tunis, the Sahara, Western Sudan, the Chad Basin, Madagascar, and certain parts of the Guinea coast together with a small area in Somaliland. Belgium was allotted the Congo Basin, since she had supplied the money which enabled Stanley to explore the great river. Great Britain received the lion's share, namely, the whole of the South, except what was recognised as belonging to Portugal, together with the equatorial highlands and their coast strip, the Eastern Sudan, Somaliland, Nigeria, the Gold Coast, Sierra Leone, and the Gambia. Some years later Italy laid claim to the north coast between Tunis and Egypt, the Libyan Desert, and some barren strips on the Red Sea coast and in Somaliland. Thus, the whole of Africa came under European rule, except the native kingdom of Abyssinia and the little artificial republic of Liberia which was set up in 1847 as a home for freed slaves. During the Great War of 1914–18 Germany lost her possessions in Africa, which were divided between Great Britain, France, and Belgium as mandated territories.

Except in the extreme north and south and in the highlands of Kenya, Africa is not a 'white man's land'; hence, the earlier settlers aimed chiefly at acquiring wealth which they could take home. At first, the native products were obtained by barter, but later the plantation system was introduced, by which black labourers worked under the supervision of white overseers. Gradually the methods of opening up the country and of exploiting its wealth have improved. Railways and roads have been constructed to secure easy communication with the coast, peaceful government has been established, and the natives have been taught the rudiments of civilisation. The widening needs thus acquired by the natives and the increase in the number of colonists have caused the African possessions to be valued nowadays not merely as sources of raw material, but also as markets for European manufactured goods. By degrees, however, the settlers and in some cases the natives have become more and more unwilling to be ruled from Europe. In 1910, the Union of South Africa was established as a Dominion of the British Empire and is united to Great Britain by its allegiance to the King. Since then Egypt, which had been a British Protectorate, has been set up as an independent kingdom, though with certain restrictions. Lastly, Rhodesia and Kenya are showing a wish to be independent also. France does not allow her possessions to govern themselves, but admits them to a share

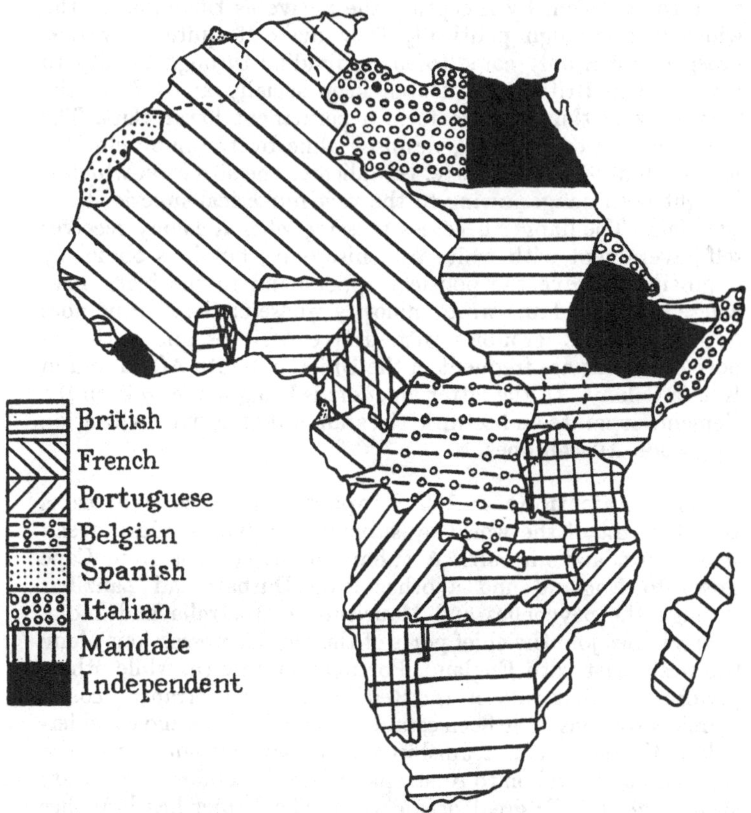

Political map of Africa

in her own government. Thus, Algeria is politically a part of France. Other countries have not developed their possessions sufficiently for the problem to arise.

One of the most difficult questions which confront European governments in Africa is the future of the native. The blacks thrive and multiply under European rule, but their lower intelligence and their traditionally different outlook upon life prevent them from assimilating an exotic civilisation, whose more primitive elements they readily absorb. The French try to solve the problem by accepting the native as the equal of the white man, though politically they make it quite clear that prosperity can only come to the individual through loyalty to France. The British and Dutch keep socially apart from the natives, who thus tend to form a permanent lower class. The presence of 'poor whites', despised alike by the more efficient European and by the native, complicates the situation which is fraught with danger owing to the vast numerical superiority of the black. The danger becomes pressing when a colony receives self-government with white franchise only; but it is obviously impossible to give the political upper hand to the black man where there is a large white minority whose civilisation he does not understand. Troubles with native Africans and with imported Indians are frequent in the Union of South Africa and in Kenya Colony. In the latter an effort is being made to keep the elements separate and to raise the standard of native civilisation on pseudo-African lines.

Communications and Transport. Africa communicates with the rest of the world by submarine cable, wireless, ocean steamships, and aircraft. A submarine cable runs from Cape Town to England and another from Durban and Zanzibar through the Seychelles and Mauritius to Australia and India. Coastal lines join the chief ports of the various possessions along the west coast with England, Portugal, or France, while others perform a similar duty on the Red Sea and Mediterranean coast. Wireless stations have been established at all towns and considerable settlements. The steamship routes from Southampton to the Cape (12 days) and on to Australia and from London to Suez are two of the world's great ocean ways. The former has branches to the chief ports on the west coast and touches at intervals at the oceanic islands of St Helena and Ascension Island. From Cape Town a branch runs along the east coast as far as Beira, the ports north of that town being served by lines which pass through the Suez Canal. Algiers and Tunis are connected by frequent boats to Marseille. A regular weekly airmail and passenger service has been established from Cape Town to Alexandria and Croydon, but there are signs that this may be replaced by the shorter way along the west coast to Lagos and across the Sahara to Algiers and Croydon. French airways connect Toulon with Algiers, and with Agadir and Dakar, the latter having a proposed extension to Port Natal in Brazil.

Internal communication is mainly by rail, road, river and lake steamer, and aircraft. There are also important overland telegraph lines from Alexandria to Khartoum and Port Sudan and

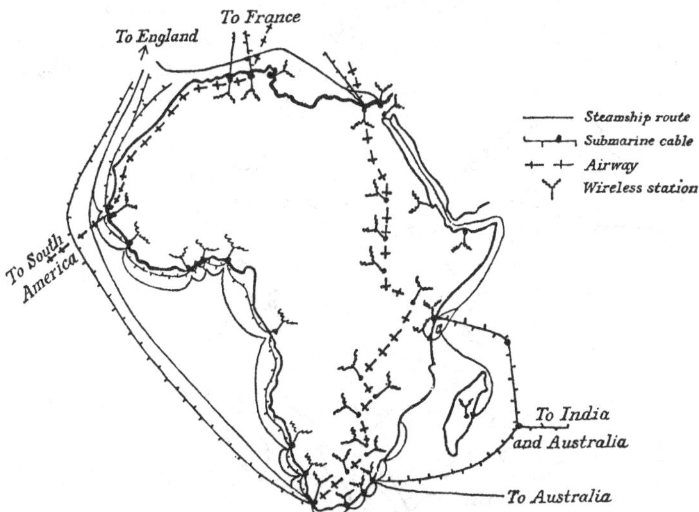

External communications of Africa

from Cape Town to Katanga with branches to Beira, Lourenço Marques, Durban, East London, and Port Elizabeth. At the end of the nineteenth century Cecil Rhodes conceived the idea of a great railway from Cape Town to Cairo; but geographical factors are against the construction of such a line. Most of the railways have been built from the seaports straight inland to their backlands; the most important being that from Cape Town to Salisbury, with its extension to Katanga, where it meets the new Katanga line from Benguella. The French have plans for building a railway from Colomb Béchar across the Sahara to

Lake Chad, but owing to the perfection of the motor vehicle there is much doubt whether railway expansion will continue. Roads are being constructed everywhere and are too numerous to mention.

The navigable systems of the Nile, Congo, Niger, Shirë, and Zambezi are used by river steamers and other small craft. Unfortunately, they are interrupted by rapids or by bars and

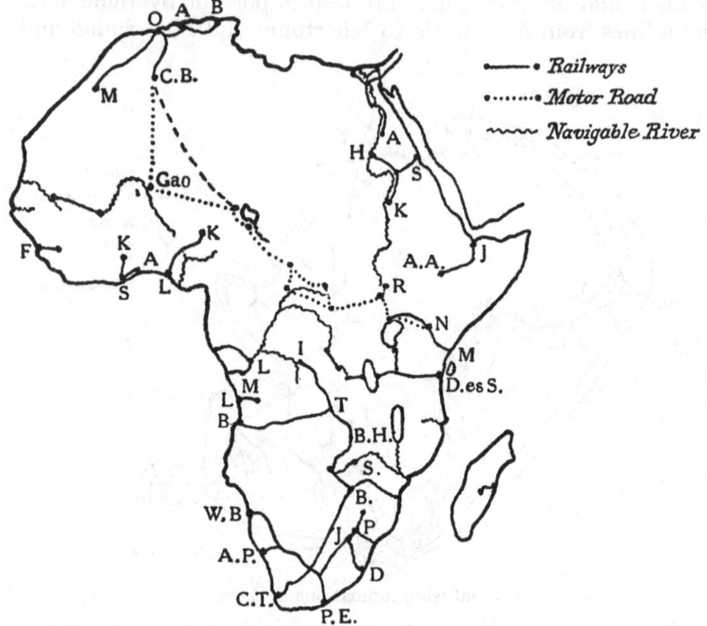

Internal communications of Africa

sandbanks. But, except for the breaks caused by their rapids, the Nile can be ascended to Rejaf, the Congo to Stanley Falls, and the Zambezi to the Victoria Falls. The swift current of the latter is a drawback to navigation. The lakes, too, are useful for traffic, especially Nyasa, Tanganyika, and Victoria. Owing to the difficulty of making landing places, the local use of aircraft is largely confined to the Union of South Africa, Egypt and Anglo-Egyptian Sudan, and the Atlas lands.

PART IV

NORTH AMERICA

IN the geography of North America the interest centres round two points: the compactness of the physical regions and the transplantation of Western European civilisation to a new environment. The continent includes all the physical regions known to the Old World, but packs them together so closely that, for instance, in the couple of hundred miles between New York and Savannah every kind of temperate and subtropical product can be had, while a few miles farther south brings in the tropical region of Florida. No other land mass offers such advantages owing to the closeness of climatic belts. Then, again, the peopling of North America by Europeans has been a great human experiment of which the results are not yet fully evident. Leaving their Old World homes while Europe was in the flush of vigour, the settlers occupied an almost empty land, transferred to it the culture and civilisation of their homeland, and built up those new nations overseas. But inevitably geographical and historical factors have influenced the growth of these 'young' peoples and given to them features which distinguish them from their parent stocks.

Position and Size. North America is joined by the Isthmus of Panama to the southern continent of South America, the two land masses combining to form the New World. This is conventionally regarded as being to the west of the Old World, since the prime meridional is by international agreement the one which passes through Greenwich. North America lies wholly in the northern hemisphere, extending from Lat. 8° N. in Panama to within a few degrees of the pole in Greenland and the Arctic archipelago of Canada. Roughly triangular in shape, it is washed by the North Atlantic Ocean on the east, the Pacific on the west, and the Arctic Sea on the north. The widest part of the triangle covers about 120° of longitude, but the average is 50°. New York is in Long. 74° W. and Vancouver in Long. 130° W., the distance between them being about 3000 miles. The great longitudinal breadth has necessitated the establishment of five belts of standard time, as shown in the accompanying map.

In the extreme north the continent approaches to within sixty miles of the coast of Asia at Bering Strait, while the distance

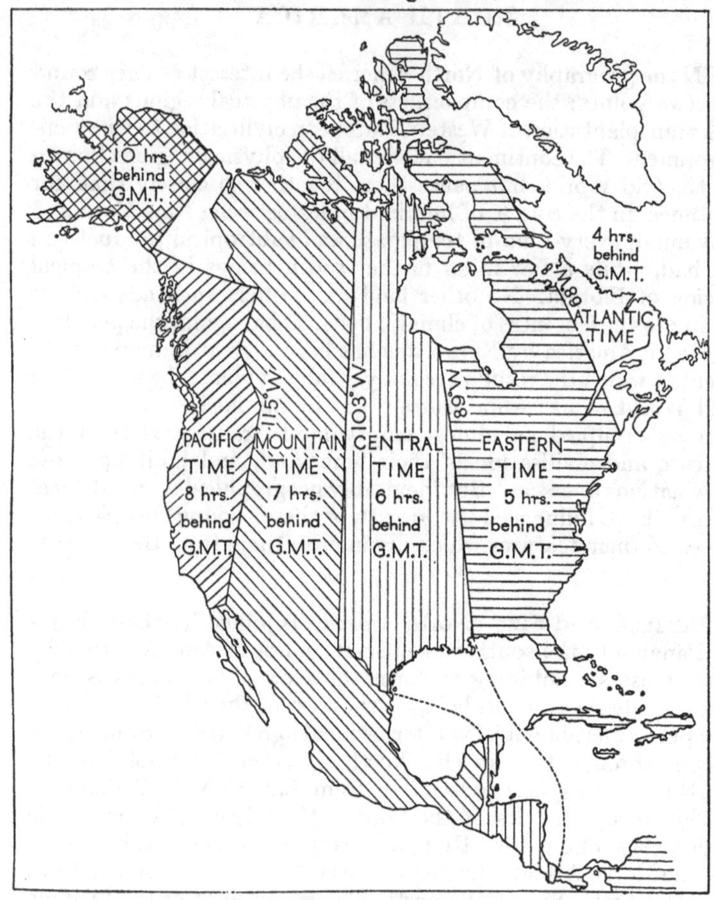

Time belts of North America

between Greenland and the European outpost of Iceland is about 200 miles across Denmark Strait. Farther south, North America is separated from its neighbours by the whole breadth of the Atlantic and Pacific Oceans. The closeness of the northern extremities to Asia and Europe has had a fundamental influence on human life in North America, since proximity to Asia led to early immigration of Mongolian peoples into the continent, and the relatively short distance from northern Europe has not been without influence on recent colonisation.

Build and Relief. North America has three main features of relief: a great series of fold mountains which follow the west coast nearly throughout its length, a far shorter and lower range in the east, and a vast area of lowland in between. The eastern range, which we shall call the Atlantic Mountains, consists of two ancient folds which have been so disrupted by earth move-

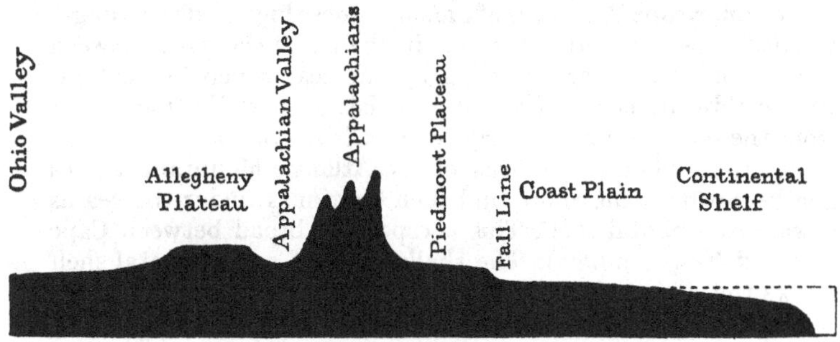

Section across the Atlantic Mountains. The broken line ----- marks the sea-level

ment and carved out by the forces of erosion that their original formation is almost obliterated and they have become just a mass of broken highland. The range is divided by gaps into several portions, the largest and most important of which is known as the Appalachian Mountains. It consists of a wooded line of hills bordered by low upland shelves on either side. That on the west is known as the Allegheny Plateau and is separated from the central hills by a longitudinal valley. The eastern shelf, or Piedmont Plateau, descends gradually to the coast plain. Since the coast plain is formed of softer rock, the rivers erode it more quickly than they do the older and harder material of the plateau, and hence they descend from the one to the other by cascades or falls. The junction of the hard and soft rocks, which is known as the Fall Line, may easily be traced on a map, since important towns have sprung up on the rivers at the break in the gradient in order to use the water power provided by the falls.

The Appalachians are separated from the more northern blocks by the Hudson-Mohawk Gap, a natural passage through the mountains which has played an important part in the settlement of North America and the opening up of the Great Plains, and which to-day offers the chief route from the east coast of the United States into the interior (see page 310). A branch of the gap runs north along the valley of the Richelieu River, making a highway between New York and eastern Canada and separating the Adirondack Mountains from those of the New England States and the Maritime Provinces of Canada. These highlands are rough, wooded, and of no great height. In the north the gaps between the various blocks are invaded by the sea, which has cut off Prince Edward Island, Cape Breton Island, and Newfoundland from the mainland and nearly severed Nova Scotia.

The coast plain to the east of the Atlantic Mountains slopes gently down to the shore and then continues under the sea as a wide continental shelf. This is especially broad between Cape Cod and Newfoundland. The shallowness of a continental shelf allows the sun's light to reach the bottom of the sea and to encourage the growth of seaweed and other fish food. Hence, this stretch of coast is the home of countless fish, and fishing is one of the chief occupations of the people. The Grand Banks of Newfoundland and the fishing grounds of Nova Scotia have long been famous for the large quantities of cod caught on them every year.

The Central Lowlands consist of vast plains stretching from the Gulf of Mexico to the Arctic without a break except in the Ozark Mountains, a small upland area supposed to be an outlying portion of the Atlantic Mountains. The Lowlands are divided into three parts: the Mississippi basin, the Northern Plains, and the Great Plains. The Mississippi valley comprises a wide area in which the rock has been laid down in nearly horizontal layers and has lain undisturbed for ages. On either bank of the great river and its tributaries there is an alluvial strip which grows wider down stream, ending in a large delta. To the west the valley gradually climbs to the Great Plains, which stand about 1000 feet above sea-level and are also formed of horizontal strata.

These plains, which are a kind of broad shelf or step leading to the Rocky Mountains, extend from Mexico to the Arctic Sea.

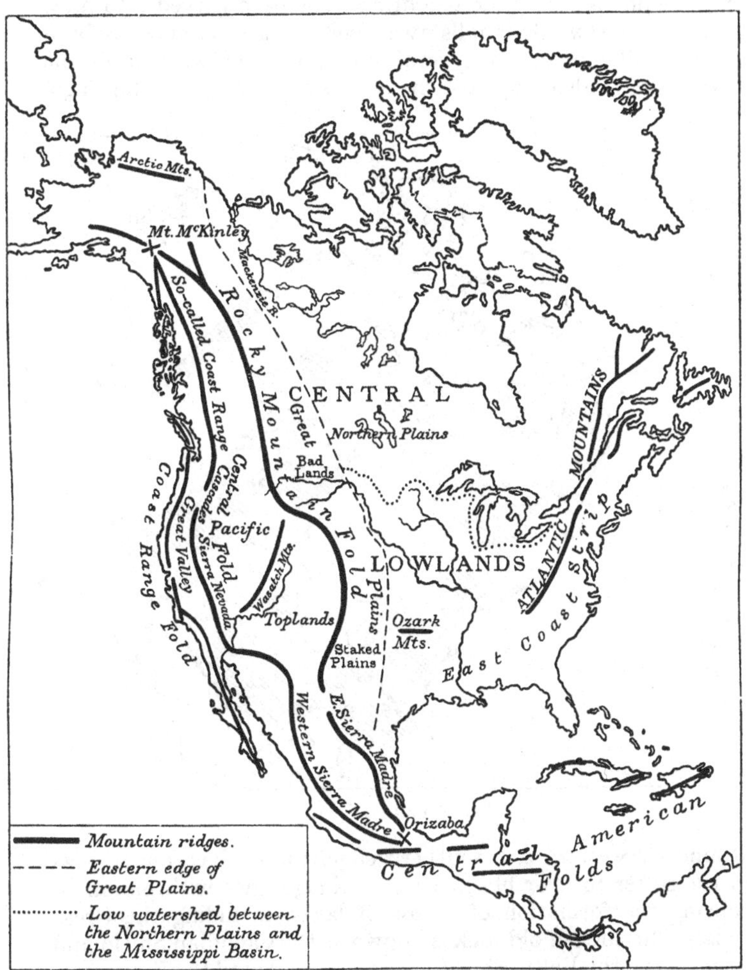

Main physical features of North America

Lastly, the whole of central and eastern Canada is a peneplain of archæan rock whose surface has been worn down, polished, and hummocked by the action of the ice-cap which lay over it during the Tertiary age. In the east, in Labrador and Quebec, the ground rises to some height and has been carved into hills, but northwestwards it falls away below sea-level and has been overlaid with horizontal layers of the younger rock which forms the margins of Hudson Bay and the islands of the Arctic archipelago.

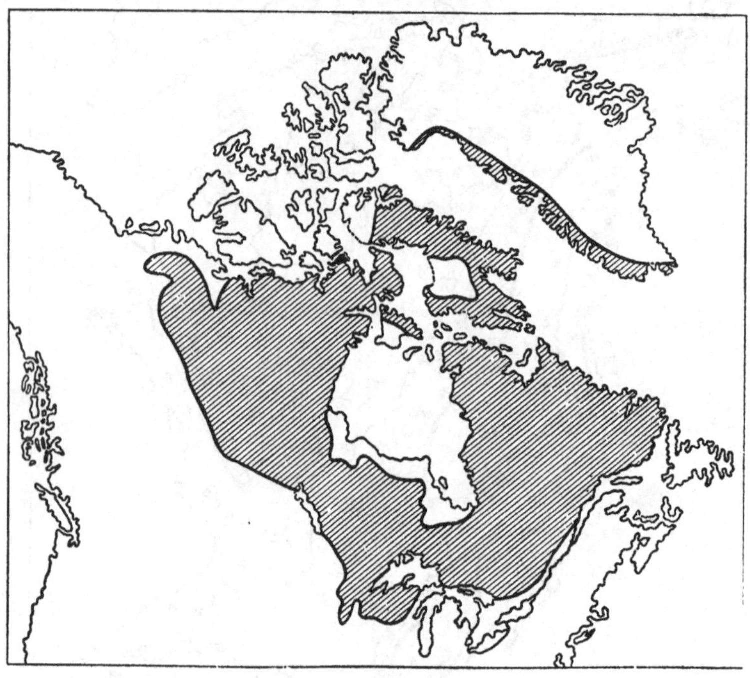

The Canadian Shield

Round its western and southern edge, the ice has dug hollows in the softer rock or blocked the drainage lines with moraines, causing the formation of a row of lakes, including the Great Lakes. This area of old rock is known as the Canadian Shield and is similar to the Baltic Shield which centres round the Baltic Sea.

The Pacific Mountains begin as a single range in Alaska, but east of Mount McKinley, the highest peak in North America, they separate into three folds. The eastern, or main range, is the Rockies, the western is the Coast Range, and the central is called the Cascades in the north and the Sierra Nevada and Sierra Madre in the south. In Canada the ranges are separated by great longitudinal valleys, but in the United States the Rockies and Sierra Nevada swing apart to hold between them the large topland of the Great Basin of Utah and, farther south beyond the crosswise Wasatch Mountains, the topland of Colorado and Arizona. Along the Mexican frontier the mountains are somewhat lower for a short distance, but they soon rise again as the Eastern and Western Sierra Madre which enclose between them the tableland of Mexico. They end in a knot of volcanoes, of which the best known are Orizaba, Popocatepetl, and Jorullo (pron. *Horoolyo*).

Subsidence has drowned all but the higher ridges of the Coast Range in British Columbia, but in Oregon and California the range is almost continuous. The most important break lets in the sea to form the Bay of San Francisco. Between the Coast Range and the Sierra Nevada is the Great Valley of California, which is continued northwards to Seattle by the valleys of the Willamette and Cowlitz.

The mountains of southern Mexico and central America belong to the ancient Antillean folds which have been broken and partly submerged, but which reappear as the backbone of the Greater Antilles.

Climate. The fact that North America extends from near the Equator to within a short distance of the North Pole gives the continent a wide variety of climatic conditions. Furthermore, since the continent tapers towards the Equator, the area within the hot belt is far less than that in the temperate belt. In this respect North America is the exact reverse of South America.

The maps overleaf illustrate the distribution of temperature. It will be seen that, broadly speaking, there is a decrease from south to north. The changes in insolation according to season, however, cause the isotherms to move great distances north and south. Thus, the isotherm for 60° F., which is thickened on the maps, runs through the north of Canada in summer, but through Mexico in winter. The isotherms do not pass straight across the land, for differential heating and cooling of the land and its adjacent oceans causes greater extremes of temperature in regions inland than along the sea coast where the modifying influence of the sea is felt most. Besides, where mild onshore

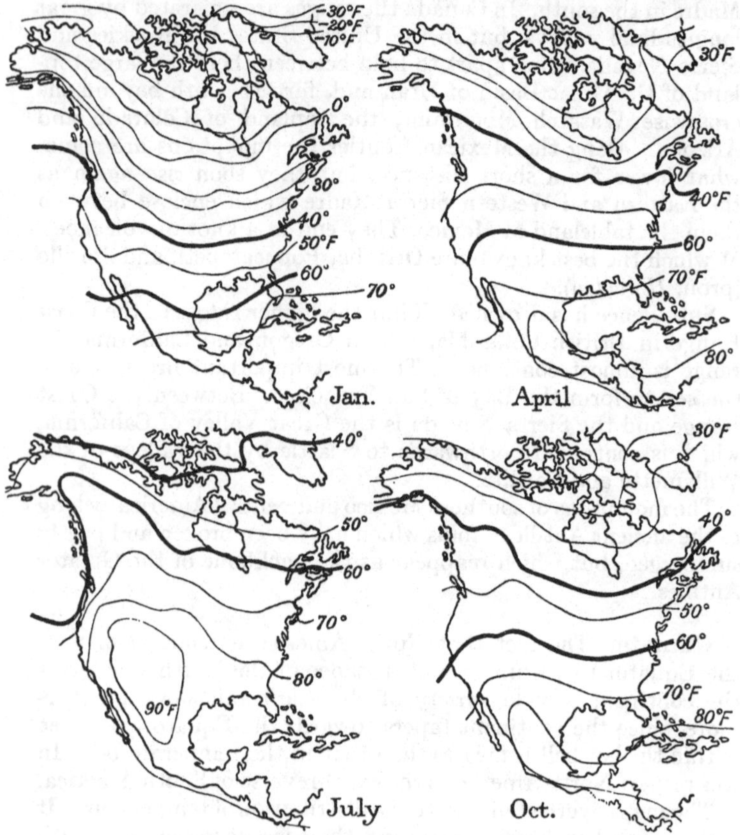

Distribution of mean temperature in North America

breezes are the prevailing winds, as in British Columbia, there the sea influence is greatest. The diagram on page 273 illustrates the influence in temperate latitudes of distance from the sea (Winnipeg), the regularity of the monthly rainfall at New York, the typical 'Mediterranean'. pattern of the rainfall at San Francisco, and the tropical régime at New Orleans. The great ranges of temperature at Winnipeg should be noted.

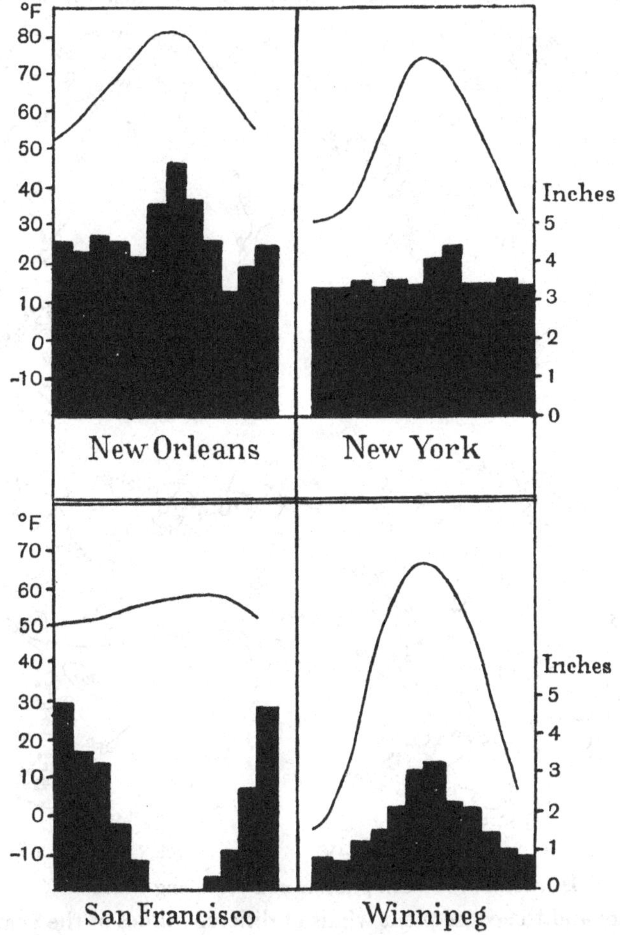

The isotherms on the maps on page 272 show temperature reduced to sea-level values. It must be remembered that height lowers the temperature about 1° F. for every 300 feet of ascent and that therefore there is in fact a great extension southward along the Pacific Mountains of the cold regions of the north.

The next set of maps (see below) show the distribution of

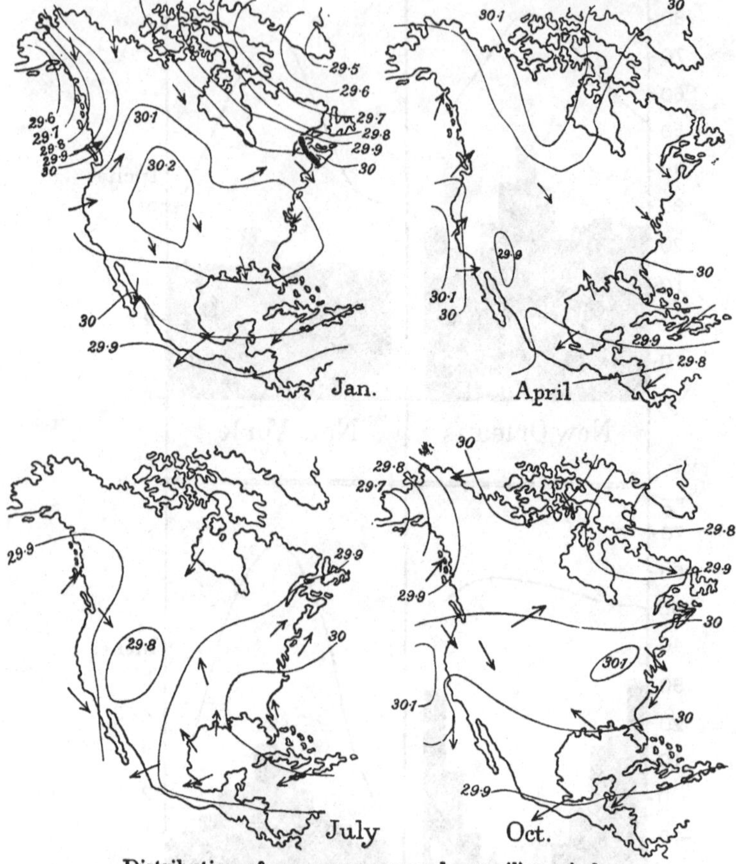

Distribution of mean pressure and prevailing winds

pressure and the consequent winds at different times of the year. The outstanding feature is the reversal of pressure conditions over the Great Basin in January and July, owing to changes in temperature. This upsets the normal wind system and causes a monsoon along the shores of the Gulf of Mexico. Generally speaking, there are variable winds in the north and east, a belt of regular trades in the south, and prevailing westerlies in British

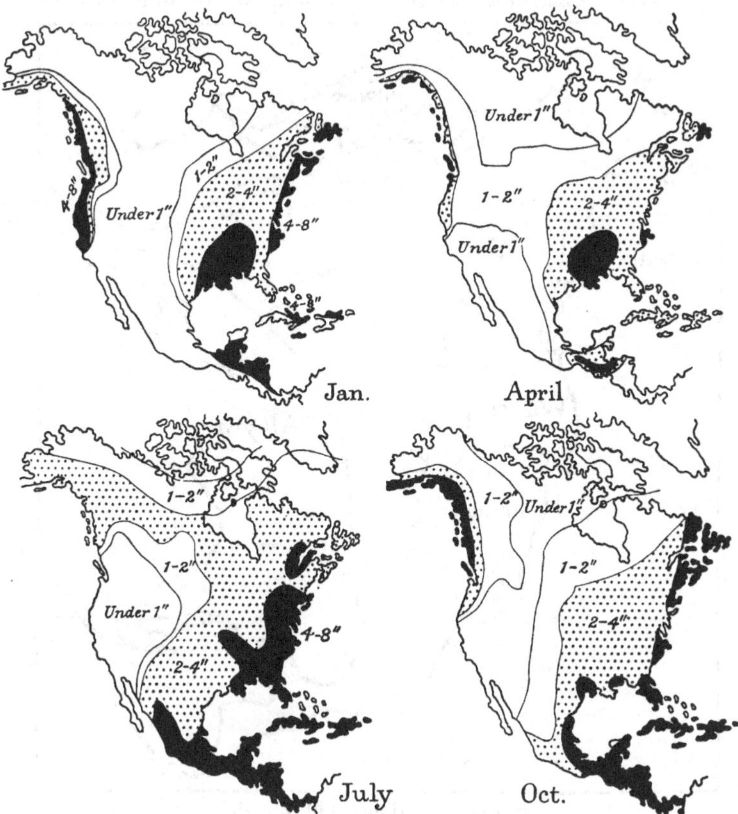

Distribution of mean rainfall

Columbia. Onshore winds reach California in winter and spring, but not in summer. Owing to the absence of high barriers of land lying across the continent, freezing blasts of wind blow from the Arctic regions from time to time in winter, bringing crop-ruining frosts as far south as Florida. These 'cold waves' usually pass from the west of Hudson Bay southeastwards along the Central Lowlands to the Gulf of Mexico.

The maps on page 275 show the distribution of rainfall in the four seasons, and the summary map below indicates the

Seasonal distribution of rainfall

extension of the six rainfall systems. Two are areas of drought: the Arctic coast fringe and archipelago, where the absolute humidity is too small for much precipitation; and the States of Nevada and Arizona, where there are desert conditions owing partly to the rain shadow of the surrounding mountains and partly to the high-pressure conditions which are normal in these altitudes. The prevailing westerlies give a well-distributed rainfall to the coast of Alaska and British Columbia, but the Pacific Mountains prevent the influence of the wind from penetrating far inland. In the same way the winter rains of California do not spread far inland, owing to the height of the Sierra Nevada, and the Great Basin of Utah is as a whole a region of scanty rain, though its maximum is in winter. The Atlantic coast from the mouth of the Rio Grande del Norte as far as Labrador has rain throughout the year, the southern portion having a summer maximum and the northern a winter maximum. This last is due to the low-pressure disturbances which frequent the Maritime Provinces of Canada, New England and Newfoundland, and give an abundant precipitation often in the form of snow. The central regions of the continent have a summer maximum, though not for the same reasons. Southern Mexico and Central America have a tropical régime with rain in the warm months when the Trade Winds weaken, while the lands to the northwest of the Gulf of Mexico owe their maximum to the summer monsoon. Farther north, in the temperate belt, most of the rainfall is due to convection.

Drainage. The drainage is simple. The rivers of the Pacific slopes are short and swift. Only the Skeena, the Fraser, the Columbia, and the Colorado have succeeded in eating their way through the Cascades-Sierra Nevada line. The last, fed by the snows of the Colorado tableland, flows through a region of scanty rain and has cut for itself a huge canyon more than 1000 feet deep at one point. In the Great Basin of Utah there is a large area of inland drainage, and another smaller one exists in Mexico. The Atlantic slopes are also drained by short streams which nevertheless have a navigable volume and gradient after crossing the Fall Line. The St Lawrence alone has been able to cut through the mountain barrier, which it does along the junction of the old rocks of the Canadian Shield and the newer rocks to the south. It carries off the water of the Great Lakes, but so slight is the watershed towards the south that the sewage of Chicago has been turned into the Illinois River so as to avoid fouling the lake on whose shore the city stands.

Half of the drainage of the Central Lowlands, except around the Great Lakes, flows off southwards in the great system of the

Mississippi. This river rises near the southwest of Lake Superior. At St Louis it is joined by the Missouri and 150 miles farther on by the Ohio, its two most important feeders. Farther south it receives the Arkansas and the Red River, both of which have a

Drainage system of North America. The dotted areas drain inland

tropical régime and carry far more water in summer than in winter. The lower course of the main river lies through a flood plain above whose level it has raised itself, containing its water by means of natural embankments known as *levées*. The summer floods often cause a breach in these banks, and serious damage is done by inundation in the surrounding country. Finally, the river reaches the Gulf of Mexico after a course of 4000 miles, building its famous 'goose-foot' delta out into the gulf.

The other half of the Central Lowlands drains northwards into the Arctic Sea through countless rivers. The watersheds are low, and it is possible to journey by canoe through most of this country, making short portages from one stream to another. The most important rivers are the Red River, which is known as the Nelson between Lake Winnipeg and Hudson Bay, and the Mackenzie, which drains the Great Slave and Great Bear Lakes and is a very large stream, but suffers from the drawback of having its mouth frozen for nine months in the year.

Vegetation. The map on page 280 shows the distribution of the various types of vegetation and should be studied in connexion with the temperature and rainfall charts so that the relation between climate and vegetation may be grasped. In the hot belt are the tropical forests of Central America and Mexico, which are nearly as dense as equatorial forests. They contain a number of sappy trees which are for the most part useless, but there are also hardwoods which, like the *lignum vitæ* and mahogany, yield excellent cabinet woods, or like the logwood provide vegetable dyes. The zapote tree of Yucatan gives the *chicle* which forms the basis of chewing-gum. Besides these, there are many fruit-trees like the guava, the cocoa, the banana, and the coffee. Large numbers of insects, especially mosquitoes and ants, fill the air and cover the ground, while on the trees live hosts of parrots, humming birds, snakes, monkeys, the cat-like raccoon, the opossum (the only pouched animal found outside Australia), and many huge bats and lizards.

The tableland of Mexico and the subtropical regions of scanty rainfall farther north are semi-desert or desert, the chief forms of plant-life being the sage brush, chaparral, and succulents, like the prickly pear, Spanish needle, and agave. Animal life is scarce and consists mostly of lizards and spiders. Above the 6000-foot contour, however, there are park-like woods of piñon and silver pine, while in the canyon bottoms and other well-watered spots are cotton woods, tropical trees, and undergrowth.

On the Central Lowlands open parkland formerly covered the ground between the Atlantic Mountains and the Mississippi before the natural vegetation was cleared away to make room

Distribution of vegetation

for cultivation. From the great river to the foot of the Rockies, and from the Rio Grande up to the Peace River in Canada are treeless grass plains, once the haunt of the American buffalo, but now the feeding-ground of multitudes of cattle. Only in Canada and in North Dakota has the land been brought under the plough. Marmots and jumping-mice still burrow in the plains, while the prairie hen and sage cock nest among the grass.

In the Mediterranean region of California there are evergreen trees, prominent among which are the Douglas firs and sequoias, giants which tower up to 300 feet on the slopes of the Sierra Nevada. When introduced, Mediterranean fruits like the orange, the grape, the peach, and the apricot, thrive in this region.

The east coast strip and the Atlantic Mountains continue the subtropical forests of the Gulf coast, but the species gradually change. The magnolia and judas tree of the southern states are gradually replaced by oaks, beeches, hemlocks, maples, and other leaf-shedders. Still farther north these are ousted by the conifers which occupy Quebec and run right across the continent. Beyond the July isotherm for 50° F., however, the trees die away, and their place is taken by dwarf species, mosses, lichens, and in summer by bright-coloured annuals. Here the musk-ox, the caribou, the Arctic fox, and a number of fur-bearing animals succeed in existing.

Natives. Before the arrival of European settlers, North America was inhabited by two races of men: the Eskimos, who dwelt in the tundra region, and the Red Indians, who sparsely occupied the rest of the land. The former are short and squat, with straight, black hair and the 'Mongolian' eye. They are wholly dependent on hunting not only for their food, but also for their clothing, boats, tents, tools, weapons, and even their fuel of seal blubber. Hence, they lead a nomadic life, following their prey from place to place. They often have permanent winter homes of half-underground houses roofed over with driftwood logs. In summer they visit their hunting grounds in sledges drawn by teams of dogs. Nowadays, they are adopting European dress and customs and have been found to make good mechanics.

The Red Indians were fairly tall and of a copper-brown colour, with straight, black hair. They were mostly hunters, but also cultivated maize and tobacco. The arrival of the European settlers found these people in a state of barbarism in Canada and the eastern United States, though the Iroquois tribes dwelling along the southern shores of Lakes Erie and Ontario had made some progress in social organisation and architecture. The birch-bark canoe was a principal feature in the life of these people of the northeast. In the barren tablelands of Colorado and New Mexico there were 'cliff dwellers' who had made much advance in social life, had learnt the potter's art, and built honeycomb villages along the ledges of cliffs on mountain and canyon sides. The most civilised of all, however, were the Aztecs of the Mexican tableland, who made houses and temples of stone, wove cotton cloth, and were divided into social classes under a king.

Only in Mexico, where there are some four million pure Indians, do these aboriginals survive in considerable numbers, many of them having interbred with the Spanish conquerors. In the United States and Canada they are now to be found only in reserves. Many died fighting against the European settlers, but for the most part the tribes simply melted away before the advance of civilisation and its new diseases.

Natural Regions. The various combinations of natural conditions enable the continent to be divided for study into eleven regions, namely—(1) the tropical south, (2) the sub-tropical west coast, (3) the Mediterranean region of California, (4) the great topland region, (5) the subtropical lands of the centre, (6) the subtropical east coast, (7) the temperate east coast, (8) the temperate lands of the interior, (9) the northern Rockies, (10) the temperate west coast, and (11) the Arctic coast. The map on page 283 indicates their positions.

THE REGIONS

The Tropical South. This includes the southern coastlands of Mexico, the extreme end of the peninsula of Florida, and Central America from the Isthmus of Tehuantepec to the eastern frontier of the Republic of Panama. The coastal lowlands bordering the Pacific are narrow, but those facing the Atlantic are somewhat broader, especially in the peninsula of Yucatan. Florida

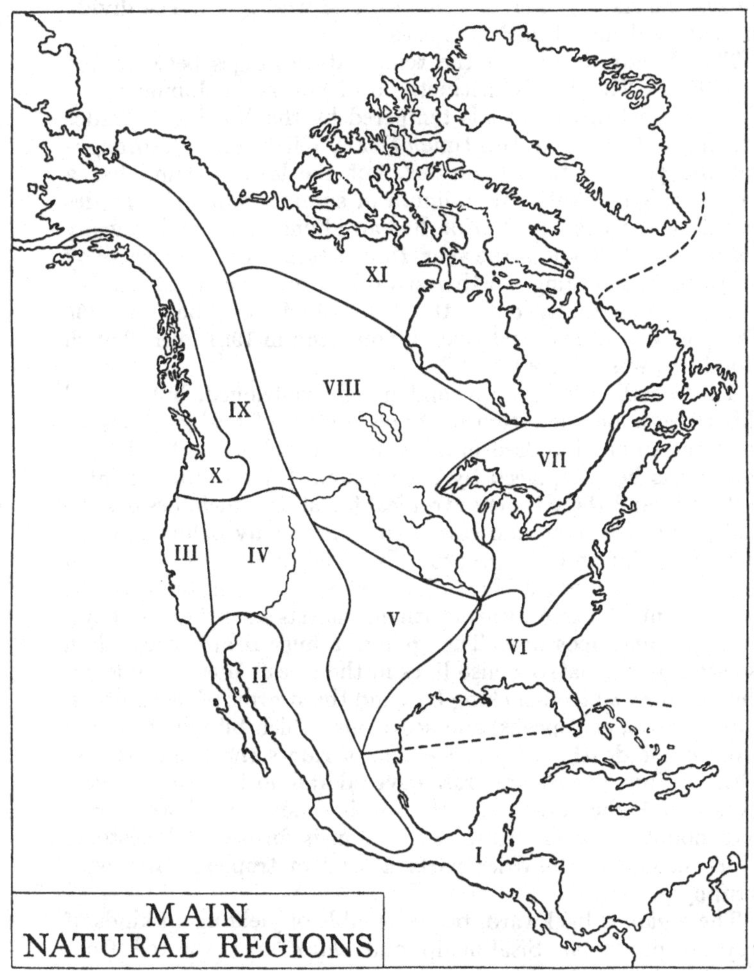
XI
VIII
IX
X
VII
III
IV
V
VI
II
I
MAIN
NATURAL REGIONS

is low-lying and swampy. Through the length of Central America runs a backbone of mountains belonging to the ancient Antillean ranges and having therefore an east-and-west grain. The divide formed by it hugs the Pacific coast.

The climate is hot, for the temperature ranges between 70° and 80° F. There is a distinctly cool season in January and February, when the heat is tempered by the Northeast Trades which are strongest at this time. Heavy relief rains fall throughout the year on the mountains, but the lower ground has a tropical régime, with a maximum in summer when the Trades are at their weakest. Florida has an abundant rainfall, evenly distributed throughout the year. Its coasts are often laid waste by hurricanes travelling northwestwards from the West Indies. These storms rarely enter the Gulf of Mexico, but in 1933 Tampico was almost destroyed by one, and in 1931 even British Honduras was ravaged.

The coastline is swampy and mangrove-fringed, but behind this there is on the Atlantic shores a strip of lowland varying from ten to one hundred miles wide on which the natural vegetation has been replaced by cultivation. On the hilly country farther inland there is dense tropical forest, in which, besides the mahogany, logwood, and carob, there are many palms, like the cabbage palm and the grugru, while the more open woodland bursts at times into a blaze of scarlet with the blossoms of the flamboyant. The trees swarm with marmosets and other monkeys, raccoons, and opossums. The iguana, a huge lizard whose flesh is eaten by the natives, also lives in the trees. More terrible are the vampire, a blood-sucking bat, and the swarms of mosquitoes, ants, spiders, centipedes, and scorpions, which bite or sting and often cause death. On the lee side of ridges there are savanas which at lower levels are grass-covered and dotted with tropical trees, but higher up, where the soil is sandy, are shorter grass and mountain pines. Most of Yucatan is formed of limestone, which causes in the drier parts a kind of tropical karst with scrubby growth.

The region is backward, but is capable of yielding all kinds of tropical products. Sisal-hemp and *chicle* are exported from Yucatan to the United States, while some coffee is grown in Mexico. Maize, yams, and sweet potatoes are grown as local food supplies, but sugar, cocoa, rubber, tobacco, and various tropical fruits, like the banana, coconut, pineapple, and plantain, are cultivated for export. The chief forest products are mahogany and logwood. The savanas provide good pasture, but there is no export of meat. The most valuable product is mineral oil which is found on the coast strip behind the port of Tampico and also to a less extent in Yucatan. The Tampico oilfield is the fourth richest in the world.

Five hundred years ago the region was inhabited by barbarous tribes, those near the Mexican tableland being under the rule of the Aztecs. In 1521 a Spanish expedition under Hernán Cortés landed at Vera Cruz, to-day the largest port in the region, and after deeds of incredible bravery conquered the Aztecs. Gradually, the whole region came under the yoke of the Spaniards, who as usual got out of it as much as they could, but did nothing to improve it. When the Spanish colonies in America broke away from the mother country a hundred years ago, Central America formed itself into the five little republics of Nicaragua, Guatemala, Salvador, Costa Rica, and Honduras, to which number was added later the republic of Panama, formerly a part of Colombia. There is also a small area (about 8600 square miles) which forms the crown colony of British Honduras. Its total population is some 50,000 persons, of whom 16,700 live in the chief town, Belize. The people of the region as a whole are either of Spanish or Indian descent or a mixture of both.

In recent years the region has increased in importance owing to the working of its oilfields and to the efforts of American business enterprise to draw from it the tropical products which cannot be grown easily in the United States. The oil is worked by British and American companies. Apart from the port of Vera Cruz (pop. 70,000 in 1931), the only towns of any size are the capitals of the six republics, namely,

Managua	cap. of	Nicaragua	pop. in 1931	32,000
Tegucigalpa	,,	Honduras	,,	40,000
Guatemala	,,	Guatemala	,,	166,000
San José	,,	Costa Rica	,,	56,000
San Salvador	,,	Salvador	,,	95,000
Panama	,,	Panama	,,	114,000

Communications are bad, being carried on chiefly by ill-constructed roads and cart tracks. A railway spans the Isthmus of Tehuantepec and is joined up with the Mexican system. Another line runs from Vera Cruz to Mexico City. Within the last few years airways have been established and are rapidly increasing. An air mail route which enters Mexico from the United States stops at Guatemala City, Tegucigalpa, San José, and San Salvador on its way to the terminus at Colon. Another goes from Florida through Havana to Belize and Guatemala City. Few regular ocean routes touch at any ports except Vera Cruz and Colon. The latter is a modern port on the Atlantic end of the Panama Canal which was constructed during the years 1904–14 to afford a quick passage between the Atlantic and Pacific Oceans.

The Subtropical West Coast. Northwards from Cape Corrientes as far as the Colorado Desert the west coast from the ridge of the Sierra Madre to the sea and the peninsula of Lower California is a barren region which becomes actual desert north of the latitude of Guaymas. Its summer temperatures are the highest in North America, the July mean being 75° F.; but there is a distinct cool season, as shown by the January mean of 55° F. in the north and 60° F. in the south. Since the winds blow off-shore or parallel to the coast throughout the year, the region is rainless and barren as a whole. Yuma, near the junction of the Gila (pron. *Hee-la*) and Colorado Rivers, has a mean annual rain-fall of 3 inches. Towards the south there is an improvement which is sufficient to redeem the land from desert conditions; neverthe-less, cultivation depends on the water-supply derived from the streams carrying off the drainage of the well-watered upper slopes of the Western Sierra Madre. Population, elsewhere scanty, is con-centrated near the mouths of these streams, where are situated the few small towns of the region. Mazatlán has some import-ance as a seaport and is on the main railway line along the west coast, but the largest town is Phœnix (pop. 48,000 in 1930), the state capital of Arizona. Although most of the region is semi-desert, there are large patches of true desert at the head of the Gulf of California, among them being the Mohave (pron. *Mo-háh-vay*) and Colorado Deserts. The only products are minerals, silver, gold, and copper, these being mined in consider-able quantities. A railway runs from Mexico City along this coast for the collection of this produce and connects up with the rail-way system of the United States.

The Mediterranean Region of California. The Great Valley of California, which is over 400 miles long and 50 miles broad, is a downfold between the Sierra Nevada and the Coast Range. It is the only productive part of the region, which also includes the western slopes of the Sierra Nevada and both slopes of the Coast Range. The valley is drained by the Sacramento and San Joaquín (pron. *Ho-ah-keén*), which flow from the north and south respectively and empty into San Francisco Bay. This inlet is a drowned portion of the valley and communicates with the sea of the mile-wide Golden Gate.

The winters in this region are cool, not only because of the latitude, but also on account of the cold California current which passes along the coast. San Francisco, for instance, has a January mean of 50° F. But the summer temperatures do not rise as high as they do in the corresponding region in Europe owing to the absence of a permanent store of warmth in an inland sea and to the influence of the California Current. Thus, the highest mean monthly temperature is 59° F. (in September), and the range is only 9° F. There is, however, the characteristic feature of onshore westerly winds in winter, and light, uncertain breezes in other seasons. The rainfall is typically Mediterranean, with a maximum in winter and practically nothing in June, July, and August. San Francisco, with an annual total of 23 inches, gets no less than 10 inches in December and January, and zero in the three summer months (see page 273).

The valley has a shrub vegetation of plants adapted to stand the long summer drought; but nowadays there are extensive irrigation systems which use the water of the many streams from the surrounding mountains. These are especially needed in the north, which is drier than the south. Some wheat is grown, but efforts are being more and more directed towards the cultivation of orchards of apricots, grapes, figs, oranges, lemons, and other Mediterranean fruits, and fruit-canning is an important industry. On the mountains on both sides there are great forests of the famous 'big trees'. On the Coast Range grows the giant redwood, while on the Sierra Nevada are the still bigger sequoia, which reaches up to 250 or 300 feet above the ground, and the Douglas fir. These trees are a valuable source of timber, but they have been wastefully felled in the past and are in danger of becoming extinct in this region.

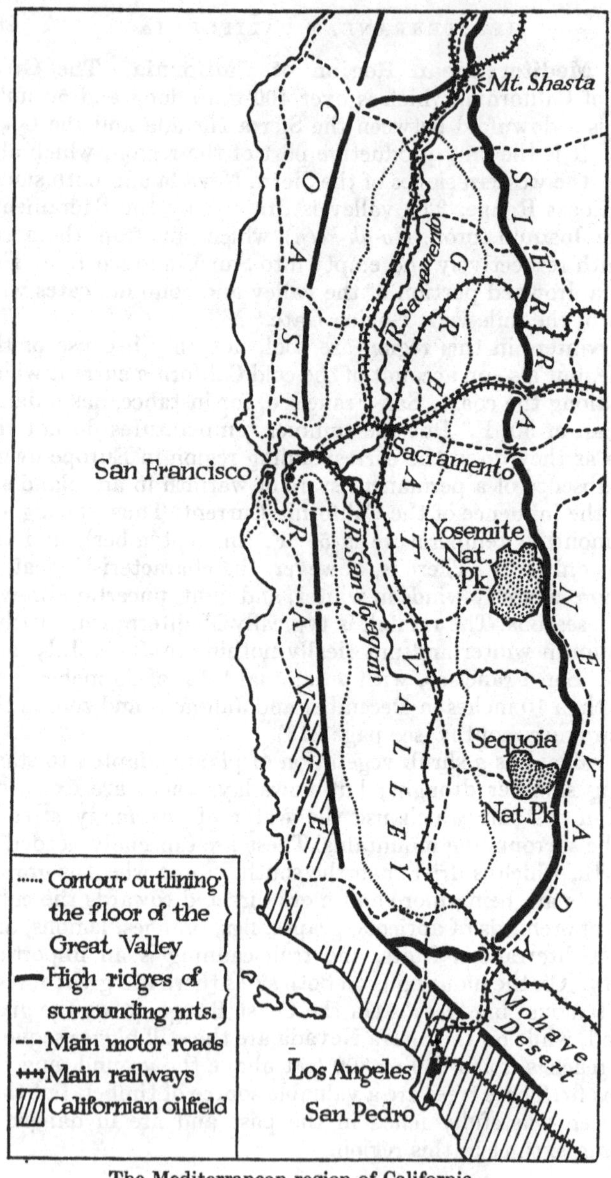

The Mediterranean region of California

Plate XVII

(*Canadian Pacific Railway*)

YOSEMITE FALLS, CALIFORNIA

One of the most beautiful cascades in the world

Plate XVIII

(*Mondiale*)

NEW YORK

The view shows Manhattan Island, with its skyscrapers in the foreground. A corner of Brooklyn appears on the right, while the main stream of the Hudson is on the left

The chief wealth of the region lies in the mineral oil which is raised in the neighbourhood of Los Angeles. This is the second richest oilfield in the world and has made Los Angeles (pop. 1,238,000 in 1930) the fifth largest town in the United States. Better known for its film studios, which are situated in the suburb of Hollywood, it is twice as big as San Francisco, the collecting centre and port of the region. This town is the Pacific terminus of the main railway routes across the United States, and steamers run from it to Japan via Hawaii, and to Australia and New Zealand via Samoa. A large sea-borne trade is also carried on through the Panama Canal with the eastern United States in canned fruit and oil. Pipe lines lead from the oilfields to San Francisco as well as to the lesser ports of Monterey and San Pedro.

The region must not be left without a word of praise on its scenery. A delightful climate and a background of mountains clad in forest and snow-capped on the east form an ideal setting for the carpet of beautiful flowers which bloom in winter. To the indigenous species have been added many exotic blooms from Europe. The land is by no means crowded as yet, and comfortable homesteads and country houses dot the valley. Good motor roads run everywhere, thus rendering accessible all the beauties of the region.

The Great Topland Region. Between the Sierra Nevada and the Rockies lies a vast highland area consisting of the inner ranges of the Pacific Mountains together with the toplands between them. The transverse Wasatch Mountains separate the Colorado topland in the south from the Great Basin in the centre, while other ridges divide the Great Basin from the Snake or Columbia Basin. To the east of the source of the Snake River the Rockies become lower and sprawl out into a highland mass rather than a mountain range. This forms the plateau of Wyoming. Broadly speaking, the surface of the region is very rough, being seamed with high ridges and pitted with hollows. Death Valley at the foot of the lofty Sierra Nevada is more than 50 feet below sea-level. The horizontal strata of the rocks have weathered into blocks locally known as *mesas*, and this is the characteristic land-form. The rainless nature of the region causes the rivers to carve deep, trench-like canyons through the toplands.

The high altitude gives the region a climate which is cool in relation to the latitude. The air is bracing, healthy, and clear. As usual in elevated areas, there is a wide daily range of temperature, owing to the clear skies and active radiation, and a great annual range. Thus, Salt Lake City has a January mean of 29° F. and a July mean of 76° F., with a range of 47° F. Towards the south there is a rise of temperature, especially in summer, where a mean of 90° F. is common. The wind system is light and variable, though in the Snake Basin and even farther south the westerlies succeed in crossing the Cascades in winter.

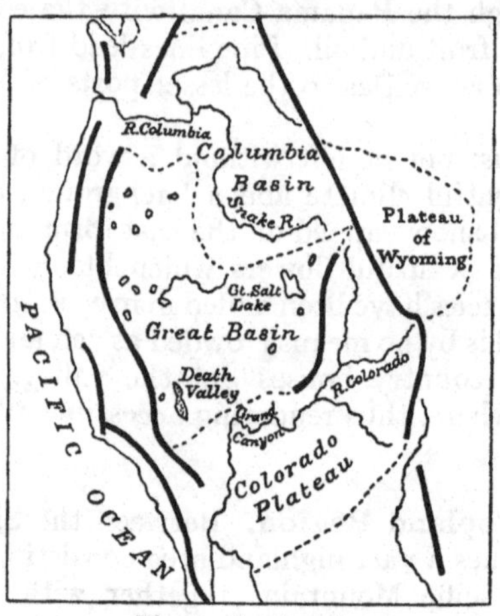

The topland region

Local winds are the rule, e.g. those which tend to blow along the canyons and the chilly breezes that blow down into the valleys from the mountain tops, causing an inversion of temperature. The rainfall is scanty, and large areas have less than 10 inches a year. What rain falls comes mostly in winter to the west of the Wasatch Mountains, but summer maxima are the rule in the areas east of the Rocky Mountains divide.

The whole of the Great Basin drains inwardly, not into one centre, but into scores of salt lakes and pans. Of these the largest is the Great Salt Lake, which is so shallow that a railway runs across it. The salts and other chemicals collected on the shores of this and other salt lakes are a valuable product of the region. In the north the drainage of the Snake Basin is carried off by the Snake River into the Columbia. The surface of this area is largely composed of an ancient flow of basaltic lava which has been much worn by the forces of erosion. In the south the drainage runs southwestward through the Colorado River and its tributaries which flow through the canyon country, and southeastward into the Rio Grande del Norte.

The vegetation of highlands above 6000 feet is forest, the piñon and southern pine predominating in the south and the silver pine in the north. High valleys carpeted with grass and dotted with conifers are known locally as 'parks'. Below the 6000-foot contour the sage brush predominates, except where salty soil or unusually bad conditions of drought lead to rock or sand desert. When the rivers do not fill the bottoms of the canyons, a dense subtropical vegetation of cottonwood and bush covers the dry ground. In the south sage brush often gives way to or mingles with the agave, prickly pear, and Spanish needle, and with thorn bushes like the dreaded chaparral.

In such a region mining is of course the chief occupation. Gold and silver are found in the Sierra Nevada, and lead and copper and other metals are found everywhere. Denver (pop. 287,000 in 1930) on the Colorado topland is almost entirely a mining town. Efforts are made to raise cattle and sheep, and large numbers of these animals feed on the sage brush; but drought is a severe handicap. Agriculture is carried on in the more favourable places. In the Snake Basin, where the volcanic soil is rich and holds moisture well, wheat is grown without irrigation, but elsewhere the crops depend on artificial watering. The Mormons of Utah have been most successful in wheat-farming. Farther south maize takes the place of wheat, and cotton is cultivated in Arizona, New Mexico, and Mexico.

Towns are naturally small and far apart. Besides Denver, there are Salt Lake City (pop. 140,000 in 1930), the focus of Mormon farming in Utah; Spokane (pop. 115,000 in 1930) in the Columbia Basin at a point where several routes through the Rockies meet; and Mexico City (pop. 960,000 in 1930) in the far south of the tableland of Mexico. A line of little towns has sprung up at the eastern foot of the Sierra Nevada, owing to the acknowledged healthiness of the climate, the abundant supply of water brought down by snow-fed streams, and the wealth of minerals in the rocks. The Mexican tableland forms part of the Republic of Mexico, but the rest of the region lies within the United States and includes Idaho, Wyoming, Utah, and Nevada with parts of Oregon and Washington in the north and Colorado, New Mexico, and part of Arizona in the south.

The region is the 'wild and woolly west' about which so many strange tales have been told. It was settled only some forty years ago, and the pioneers were adventurous and often bad men. A more settled and orderly life is gradually dawning, but it should be remembered that even now the wild scenes described by Zane Grey and Clarence Mulford still occur occasionally. The scenery of the country is magnificent. The bare rock is often bright red or yellow, while the clear air allows the sight to cover vast distances. Among the chief difficulties of the inhabitants are the long journeys which must be made to reach the towns. The mountains, canyons, and deserts are hard to cross, and the traveller must often make a wide *détour* to avoid them. Amazing feats of engineering were needed to take the great railway lines of the United States across the region.

The Warm Lands of the Centre. These include a roughly triangular area contained between the Black Hills of South Dakota, the junction of the Mississippi with the Ohio, and the town of Monterey at the foot of the Eastern Sierra Madre in Mexico. The region falls into three clearly marked divisions: (1) the shelf of high plains averaging 1500 feet above sea-level which stands against the Rockies and which is known as the Great Plains, (2) the lowlands east of this reaching to the bottom of the Mississippi valley, and (3) the Ozark Mountains. The Great Plains are deeply trenched by transverse rivers which carry off melt-water from the Rockies and find their way into the Mississippi. The chief are the Canadian, the Arkansas, and the Platte. Otherwise, they are undulating and fall gradually away to the east. The Ozark Mountains are a boat-shaped mass of old worn mountains which are supposed to be related to the Appalachians. They rise from the newer rocks of the lowlands around them. The lowlands of the region are a continuation of the Great Plains, but average less than 1000 feet above sea-level. Their undulating surface falls gradually away to the southeast, the drainage running off into the Mississippi and the Gulf of Mexico.

The climate of the region is largely controlled by the alternating centres of high and low pressure which form over the great toplands in winter and summer respectively and which give a kind of monsoon to the region. In winter the prevailing winds are from the north and northwest and are very cold, while in summer warm breezes from the Gulf of Mexico cause high temperatures. Normally, there is a steady rise in January from 30° F. in the north to 60° F. in the south. But from time to time cyclonic conditions sweep masses of cold Arctic air over the region right to the Gulf of Mexico. Such periods of cold are known as 'cold waves' or, in Texas, as 'Northers'. When they occur, severe frosts are carried all through the region. They are often followed by quick reverses of temperature, so that a range of over 50° F. has been known in January. In July the temperature is more even and averages 70° F. throughout the region. In the hottest days of summer the contact of cold and hot surface currents of air often gives rise to *tornadoes*, a small, but violent form of rotatory storm which sweeps unobstructed across the plains at a speed of thirty to fifty miles an hour, destroying everything in its path. The devastated area usually measures some twenty miles long by one or two hundred yards wide.

Owing to the monsoon effect of the wind system, the rainfall shows a summer maximum; but no month is dry. San Antonio in Texas, for instance, has a minimum of 1·5 inches in January and a maximum of 4 inches in September, with a mean annual total of 28 inches. A similar régime is observed at Omaha in the north of the region. The Great Plains are dry, and, though their régime is like that of San Antonio, their total rainfall is as low as 12 inches in some districts. The northern portion in Nebraska is almost semi-desert and is known, on account of the scarcity of water, as the Bad Lands. The central portion is better, but deterioration occurs again in the Staked Plains of Texas and New Mexico. These are so-called because early settlers erected lines of stakes to serve as guides across this featureless country. Owing to the rainfall régime, the rivers have a strongly marked summer rise which begins in spring with the melting of the snows in the Rockies.

Except in the Ozark Mountains, where the higher ground is

forested, this region is an extensive grassland whose vegetation becomes gradually scantier towards the west. On the Great Plains not a tree is seen for miles and miles, though the more barren parts are sometimes dotted with low bushes. The lowlands sometimes have fringing woods along the watercourses, but even here trees are rare. The region was formerly the home of the bison, or American buffalo, which roamed the plains in herds numbering thousands. During the middle years of the last

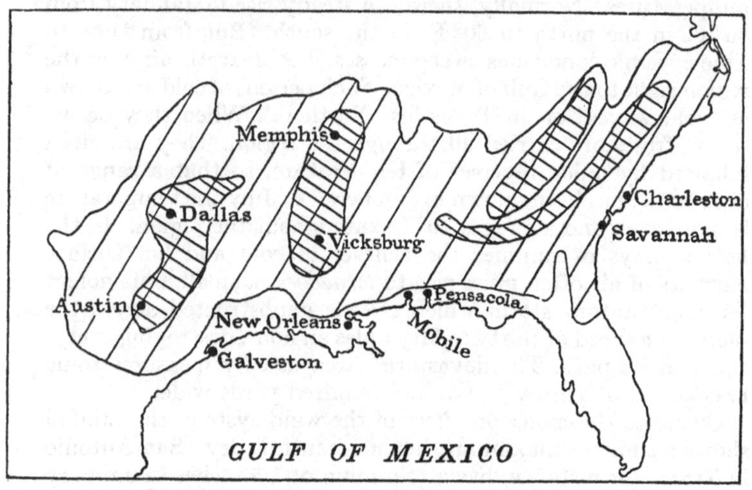

Distribution of cotton crops in the United States
The whole of the shaded area produces cotton, the chief districts being more heavily shaded

century the animals were ruthlessly slaughtered for their warm coats and were all but exterminated. A few small herds survive in reserves in the United States and Canada. Apart from the bison, America was without the many kinds of large hoofed animals which are found in Africa and Asia. But, when introduced from Europe by the Spaniards, the horse quickly made itself at home on the plains and on the toplands, where the original breeds degenerated into a small, though hardy beast known locally as the mustang.

Formerly a pastoral country in which large herds of cattle

were raised, the region has rapidly become agricultural. The chief crops are maize, wheat, and cotton. When the ravages of the boll-weevil destroyed large areas of cotton on the east coast strip from 1892 onwards, the cultivation of this plant gradually spread westward, the planters using an upland variety instead of the famous sea-island plant, and one of the areas of most intense production lies between Dallas and Austin in Texas (see map on page 294). In 1932 Texas was by a wide margin the chief cotton-growing state, the yield amounting to 4½ million bales (20 million lb.). Wheat is confined to the north of the region and is cultivated on a large scale in Missouri, Nebraska, Kansas, and Oklahoma. It is now being planted even on the Great Plains wherever irrigation is possible. The mildness of the region enables a variety to be grown which is sown in autumn, while farther north the seed must be sown in spring. In 1932, Kansas was the second wheat-producing state, with a total yield of 106 million bushels.

In the south and centre maize either replaces wheat or is grown side by side with it, and the crop is found in Missouri, Nebraska, Kansas, Oklahoma, and Texas. In the last state it is cultivated side by side with cotton. The United States produce more maize than any other country, and the use of the grain as human food is commoner there than elsewhere; yet it is as fodder for cattle and pigs that the crop is usually grown. Animals raised on one of the huge ranches of the Great Plains are generally taken to the maize-growing lowlands to be fattened before being turned into tinned meat at Kansas City or one of the smaller meat-canning towns.

From the first settlement of the region until recently pastoral and agricultural production has been carried out on the largest scale and on the one-crop principle. The rapid growth of industrial towns in the eastern states and in Europe has enabled the produce to find a ready sale in an expanding market, and large fortunes have been made. The financial depression of 1931 and the consequent contraction of the markets have demonstrated the weakness of the system of dependence on a single crop. Many of the farmers have been ruined, and those who have weathered the storm have wisely taken to growing vegetables and to raising supplies of food for local consumption.

Perhaps the chief wealth of the region lies in its minerals. The

West Central coalfield (see map on page 309) centres round Kansas City and has a long extension into Oklahoma and Texas. The fuel derived from it is inferior, but is used for driving the meat-canning machinery of the factories. Farther south is the

Distribution of oilfields in North America

oilfield of Kansas, Texas, and Oklahoma, the most productive in the world. Like cotton, the oil finds its outlet to the sea through the Gulf port of Galveston, which is now second only to New York as a commercial port in the United States. The quantity of oil amounted in Texas alone in 1931 to 171 million barrels of petroleum and 426 million barrels of natural petrol. Besides this there was a yield of a vast quantity of natural gas. Production is so easy that it has had to be regulated by government orders so as to prevent a glut on the market.

Great industrial development has not yet come over the region; hence, it has surprisingly few large towns. Besides Kansas City (pop. 122,000 in 1930), which has already been mentioned for its coal and meat-canning, the only other place of any size in the north is Omaha (pop. 214,000 in 1930), which stands on the peninsula formed by the Platte River as it flows into the Missouri. In the south are Dallas (pop. 260,000 in 1930) and San Antonio (pop. 231,000 in 1930), besides the far smaller, but yet important, cotton centre of Austin. The region contains only four states, namely, Nebraska, Kansas, Missouri, and Oklahoma.

Less than a hundred years ago the region was the home of nomadic Red Indians, the chief tribes of whom were the Comanches, Apaches, and Pawnees. These people naturally resented the invasion of their hunting-grounds by the pioneers who moved westward from the Mississippi in their 'prairie schooners', or huge tilted wagons, and began to establish cattle ranches on the plains. Hence, many fights ensued. Gradually, however, the Indians were exterminated, and the few survivors were placed in reserves in what is to-day the State of Oklahoma. No roads led the early pioneers over the plains, but by degrees 'trails' were worn by the waggon wheels and became easy to follow. The best known were the California and the Santa Fe trails, both of which started from the settlement of Independence near Kansas City. The former led northwestwards towards South Pass, where it crossed the Rockies, while the latter went southwestwards to the town of Santa Fe in the mountains of Colorado. A realistic description of pioneer life and of these trails is given by Zane Grey in his *Roaring U.P. Trail.*

The Warm Lands of the East Coast. These include the coast plain from Chesapeake Bay to the Rio Grande del Norte. They consist of the Atlantic coast strip, Florida, the Gulf coast, and a portion of the Lower Mississippi valley in the States of Tennessee, Arkansas, and Mississippi. In the last area it extends north as far as the town of Cairo at the junction of the Ohio with the Mississippi.

The Atlantic coast plain comprises four strips running parallel with the coast. The shore line is fringed by sandspits which have blocked the inlets and straightened out the coast. Consequently, important harbours occur only where considerable rivers enter the sea, as at Charleston and Savannah. Behind the sandspits is a chain of lagoons and salt marshes, on the inner side of which is a narrow belt of sandy soil which was once covered with forests of southern pine. Behind this comes a strip of fertile soil known as the 'cotton belt' owing to its chief crop. A broad, sandy belt follows, which is forested with conifers, among which the southern

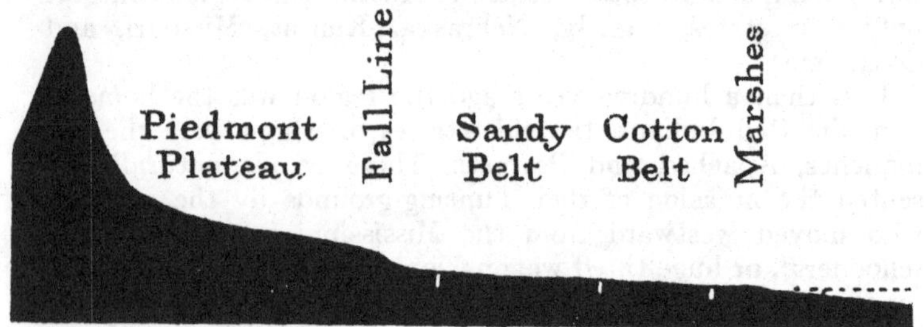

Section across the Atlantic coast of North Carolina
The broken line ----- denotes the sea-level

pine and the red cedar are the chief trees. The 'fall line' separates this belt from the Piedmont Plateau.

Florida is a low-lying and badly drained peninsula, whose interior is marshy, especially in the Everglades district. It is forested, with progressive density towards the south, where real tropical forest exists. The characteristic native tree is the swamp cypress. The Gulf coast is also low-lying, much of it consisting of the delta of the Mississippi. The interior lowland includes the flood plain of the river together with the lower spurs of the Allegheny Plateau in Tennessee.

The climate is one of cool winters and hot summers. In winter the temperature decreases from 65° F. at Miami in Florida to 40° F. at Raleigh in North Carolina, while in summer it is 82° F. and 79° F. at the same two places respectively. On the Gulf coast and the Mississippi valley the readings are slightly higher latitude for latitude. The coolness of the winters alone marks off the region from the tropics. The 'Northers' which occur from time to time in that season emphasise the difference by bringing a period of sharp frost which often does great damage to the crops. These winds are felt on the east coast and not infrequently descend on Florida. Normally, the winds blow from the north, except along a narrow belt in the west of the Appalachians, where, owing to a local high-pressure centre on those mountains, a stream of air passes from the Gulf through Alabama and Tennessee. In summer the direction of the winds is from the Gulf. This is therefore the season of maximum rainfall, though most of the region receives abundant rain all the year round.

The plant life has special features in that it is, with the exception of China, where the natural vegetation has been almost entirely removed, the only large region of its kind. The striking characteristic is the meeting and mingling of several different kinds of forests. The more northern portions of the region are clad in a woodland of magnolias, hickories, walnut trees, tulip trees, judas trees, etc., many of which are leaf-shedding. These subtropical species are replaced in poor and sandy soils by cedars and pines, while in the south they merge gradually into tropical forest. Except where the woodland has been preserved for one reason or another, the trees have been cleared away and replaced by cultivation. The Gulf shores are mangrove-fringed, and the delta of the Mississippi and, as has been said above, much of Florida has a swamp vegetation.

The surviving woodland provides a good deal of timber, especially on the sandy east coast strip, where the southern pine yields wood, turpentine, and resin. Red cedar is used for casing lead pencils, while hickory is made into axe-hafts and other articles which need a combination of strength and flexibility. The almost tropical Gulf coast and Florida have had much of their forests replaced by fruit trees, especially the orange, grape fruit, lemon, and banana.

Apart from fruit, the chief crops are maize, cotton, sugar, rice, and tobacco. Rice and sugar are cultivated on the warm, swampy lands on or near the Mississippi delta. So much of the former is produced that no rice has to be imported into the United States. The centre of the cane-sugar industry is New

Orleans (pop. 460,000 in 1930), where there are refineries. The production does not meet the needs of the United States, and large quantities are imported into that country, chiefly from Cuba and Puerto Rico. In 1931 the output of sugar from Louisiana amounted to over 135,000 tons. Tobacco is grown throughout the region, but North Carolina and Tennessee are the chief producers. Virginia, whose name is connected with American tobaccos, produced in 1931 only one-fifth as much as North Carolina.

Cotton and maize are grown farther inland. Maize is used largely as fodder here, as elsewhere in the States, and its distribution is wide. Cotton, however, is the most important crop of all. In the early days of settlement it was grown on the low, sandy islands which fringe the Atlantic coast and consisted chiefly of the 'sea island' variety which, owing to its long staple, is still regarded as the best kind. In the last years of the nineteenth century the plants in this area were attacked by a weevil which ate the 'boll', or seed-pod, and so ruined the crop. Hence, the area in which cotton is grown has shifted westwards, and an upland variety has taken the place of the 'sea island'. The whole region except the coastal marshes now produces cotton, the cultivation of which has even spread into the drier parts of Texas. In fact, the chief producing area lies outside this region and centres round Dallas and Austin. But within the region there are three districts where cotton is cultivated intensively, namely, the flood plain of the Mississippi between Memphis and Vicksburg, the Piedmont Plateau, and the inner coast belt. The outlets are at New Orleans for the Mississippi area, and at Charleston and Savannah for the east coast districts. Besides these larger ports there are secondary outlets at Mobile and Pensacola on the Gulf coast. See map on page 294.

The industrial revolution in England in the eighteenth century led to the growth of cotton plantations along the Atlantic seaboard of the southern states. The raw cotton was shipped to the Lancashire mills for manufacture into cloth. This economic interdependence partly explained English sympathy with the south during the American Civil War. Raw cotton still forms the chief export of the United States, the value of the quantity shipped amounting in normal years to over £60 million. Of this the greater part goes to the United Kingdom. But towards the end of the last century local factories sprang up rapidly, and to-day the United States supplies the bulk of its needs. The industry

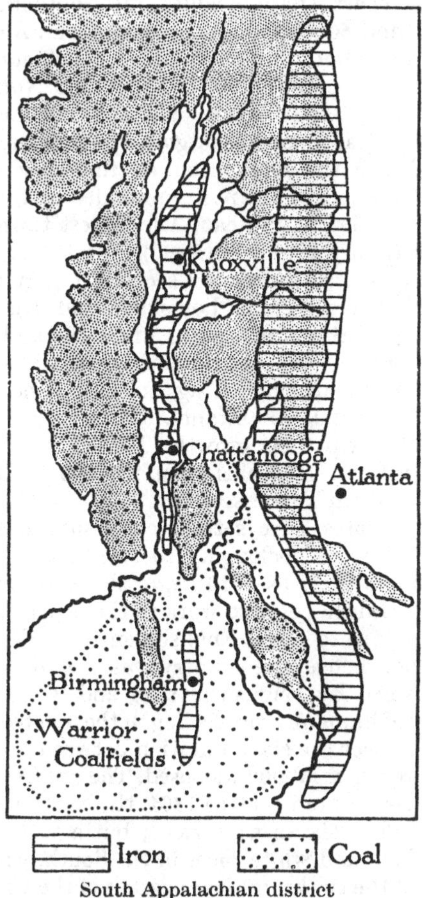

South Appalachian district

This is the second most important industrial area in North America

centres round the towns of Birmingham (pop. 259,000 in 1930) in Alabama and Chattanooga (pop. 119,000 in 1930) in Tennessee, where the presence of coal and iron has helped to establish manufactures. By-products of the crop are cotton-seed oil and cotton-seed cake, the latter being used for feeding stock.

The Atlantic coast strip includes the southern states of Virginia, North and South Carolina, Georgia, and Alabama. The two largest towns, both in Virginia, are Richmond, the old capital of the South (pop. 182,000 in 1930), and Norfolk, a port at the mouth of Chesapeake Bay. The old 'fall line' towns of Raleigh, Augusta, Macon, and Montgomery are at the head of the tidal waters—hence the navigation of their several rivers—and have the advantage of the power provided by the falls of the rivers from the Piedmont Plateau. The largest town in Florida is Jacksonville (pop. 129,000 in 1930), but better known are Miami and Palm Beach. Tampa is a secondary port on the west. The Mississippi states include Louisiana and Mississippi, together with large portions of Texas, Arkansas, and Tennessee. New Orleans is by far the largest town, but Galveston is the chief seaport. As a rule, in this agricultural region the towns seldom exceed 100,000 inhabitants, the larger ones having some industry to help their growth. Thus, Houston in Texas (pop. 292,000 in 1930), Memphis (pop. 258,000 in 1930), Nashville, and Knoxville in Tennessee.

The southern states were settled early in the seventeenth century as 'plantations' belonging to companies or noblemen in England, and as early as 1616 negro slaves began to be imported from the Guinea coast to work in the fields. The summer is too hot and damp to allow Englishmen to work on the land with any comfort; hence, the country was soon divided up into large estates owned by English settlers and worked by black labour. The rise of humanitarian feeling in the nineteenth century caused a dispute over the slave question between the South and the North, where white labour was used. The quarrel culminated in a war which lasted from 1860 to 1864, the North winning after a hard struggle. The slaves were freed, but a 'colour bar' still separates the black and white races in the South, where live the great majority of the twelve million negroes of the United States. The presence of this inferior people is one of the problems of the Government of the States, a problem which seems no nearer solution than ever. The forced transplantation of these blacks and their reaction to their new environment is no less interesting than the white colonisation Some idea of the spirit of the people may be gathered from reading the tales of *Uncle Remus* and from listening to jazz music and the well-known plantation songs. Harriet Beecher Stowe's novel, *Uncle Tom's Cabin*, draws an ugly picture of the old days of slavery, but should be read.

The Cool Lands of the East Coast. These lie between latitude 38° and 50° N., but have extensions towards the north, south, and west. The height of the Appalachian uplands carries the temperate climate southwards into the heart of Tennessee and South Carolina, while the moderating influence of the sea redeems much of Labrador from the Arctic. On the west the Great Lakes act as an inlet and carry the equability of the coast far into the Central Lowlands. The main features of the relief are the Atlantic Mountains, the Canadian Shield, and the two great gaps of the St Lawrence and Hudson-Mohawk which lead from the Atlantic coast to the Great Lakes and the Central Lowlands.

The temperature is cold in winter and warm in summer, but there is a decrease from south to north. At Washington the January mean is 33° F., while at Nain in Labrador it is – 7° F. The coldness of the winters as compared with those of corresponding latitudes on the west coast is due partly to northwesterly winds which are set in motion by the low-pressure centre over Iceland, and partly to the land winds which reach the lakes from the southwest. The presence of the Labrador Current off the coast as far south as the mouth of the Delaware River is not without its effect, and the Gulf of St Lawrence is blocked by ice for four months in the year. The significance of this will be realised when it is remembered that the gulf is in the same latitude as the English Channel.

In summer the weakening of the Icelandic depression enables the Southwest Antitrades to blow steadily, and these winds, coming from heated interior plains, give rise to high temperatures. At times when the anticyclone near Bermuda is strengthened and a low-pressure centre forms in the Mississippi valley, warm moist winds blow from the south and southeast, causing a 'heat wave' which is very unpleasant. At such times hundreds of people are overcome by the heat and faint or even die in the streets. It is to this feature of the climate of this region that Americans owe their habit of taking iced drinks and of regarding a refrigerator as an indispensable fixture to be provided by the landlord of a house to let. The following table gives the main facts about the temperature of typical points distributed over the region:

	January mean	July mean	Range
	° F.	° F.	° F.
New York	30·4	74·5	44·1
Halifax	24·1	64·6	40·5
Montreal	13·2	69·0	55·8
Toronto	23·0	69·0	46·0

The rainfall is spread through the year, the even distribution being striking in the north. Nova Scotia, the east of Newfoundland, and the coast of New England have a winter maximum owing to the prevalence of low-pressure disturbances at that season. Elsewhere, there is a summer maximum which is slight in the north, but marked in the south. On the whole spring is driest, and the late summer months of August and September are the wettest. And naturally the Atlantic coast is the rainiest part of the region, as the following table shows:

Place	Régime	Jan.	April	July	Oct.	Total
		in.	in.	in.	in.	in.
Halifax	Nova Scotia	6·0	4·6	3·7	5·5	57·3
Toronto	north, inland	2·8	2·4	3·0	2·6	31·4
New York	north, coast	3·3	3·3	4·1	3·4	42·5
Pittsburgh	south	2·5	3·0	3·9	2·7	36·2

Owing to its good rainfall, the region was heavily forested when European settlers first arrived, but now the lowlands have been cleared for agriculture, and even in the uplands the trees have been widely cleared away by timbermen. In the south are found leaf-shedding species of the same types as occur in England, namely, oak, elm, ash, beech, sugar maple, and hickory. Farther north these trees gradually give way to birches, pines, firs, and larches. Owing to the elevation, this last type of forest extends farther south in the Appalachians than on the lowlands.

The region is too varied and too important to be treated as a whole, for it contains the heart of both Canada and the United States. It falls naturally into five subregions: (*a*) the middle Atlantic coast, (*b*) New England, the Maritime Provinces of

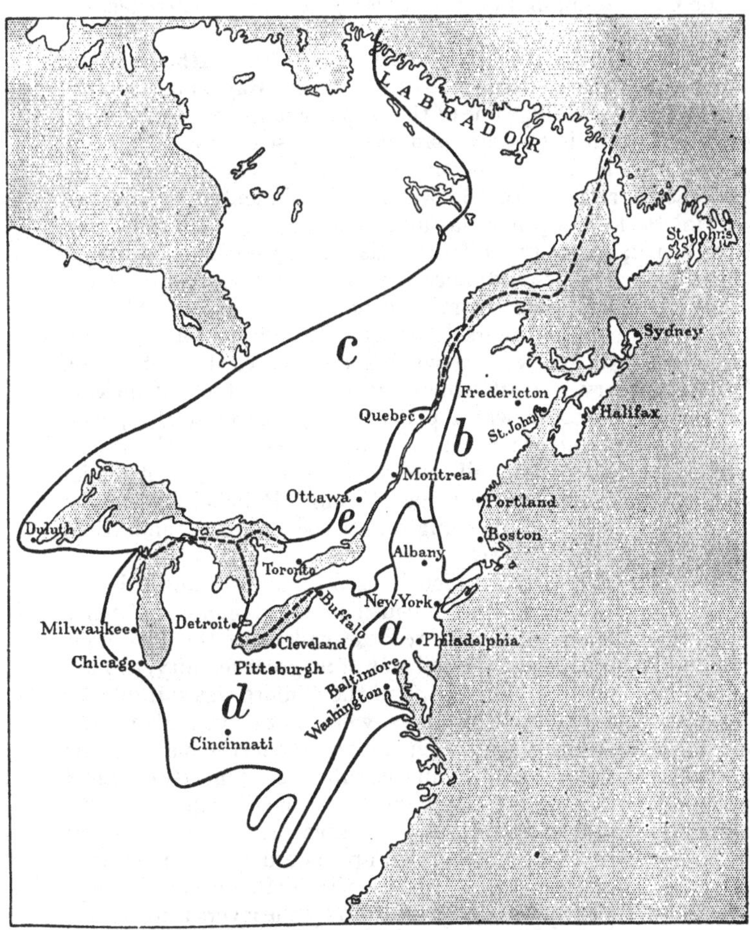

Subregions of the temperate east coast

These include the wealthiest and most densely peopled parts of North America

Canada, and Newfoundland, (*c*) part of the Canadian Shield, (*d*) the Ohio basin, and (*e*) the lowlands of the St Lawrence.

(*a*) *The Middle Atlantic Coast.* This runs from Rhode Island southwards to the entrance to Chesapeake Bay and stretches inland to the main ridge line of the Appalachians, where its length north-and-south is greater than along the sea coast. The same series of belts in the relief occurs here as in the region next south. The coast is rather more drowned and is indented by the two great inlets of Chesapeake Bay and Delaware Bay; and the action of the sea in smoothing off the coastline by means of sandspit formation is in a less advanced stage. But the spits offer excellent sites for seaside resorts and holiday towns, like Atlantic City. The lagoons and marshes which lie inside the sandspits are surrounded by a sandy strip which is devoted to the cultivation of the cranberry, a fruit which Americans are fond of making into tarts. The coast was repellent to the early settlers, who pushed on up the drowned estuaries to the head of the tidewater. Here they reached the 'fall line' with its advantages of settlement, a line which can be traced through Philadelphia on the Schuylkill, Baltimore, Washington on the Potomac, and Richmond on the James River.

The belt of fertile soil which is used for cotton and tobacco farther south is of no great breadth in this region, and hence the early settlers soon penetrated the valleys of the Piedmont Plateau. But even there the soil was of no great fertility, though it could be made to yield crops, and the climate was suitable for European field labour. Nowadays, when wheat and other main food crops can be grown more cheaply on the fertile plains of the Central Lowlands, agriculture in this subregion is mostly confined to market gardening (or truck, as it is called locally) and the growing of potatoes and other perishable produce required by the large towns which have grown up. The denuded Piedmont Plateau is succeeded westwards by the main line of the Appalachians. These mountains are cut up by the rivers into several blocks, the largest of which is the Blue Ridge, though the range has its highest peak in Mount Mitchell (6700 feet) in the Smoky Mountains of North Carolina. In the north the low hill mass which overlooks the Mohawk Gap is known as the Catskill Mountains, while on the northern side of the gap are the wild

Adirondacks. The eastern side of the Hudson-Richelieu valley is formed by the Green Mountains.

The streams which penetrate the Appalachians are of great importance, since their valleys offered routes which were followed by the early pioneers and which are used to-day by railways and other routes. In the north the mountains are completely breached by the Mohawk, whose gap presents the easiest gradient for routes from the Atlantic coast to the Central Lowlands. A canal now connects the stream with the Niagara at Buffalo, while an eightfold railway threads its way from Utica to New York. The next two are the Delaware and Susquehanna, the last of which has all but breached the mountains. The passages offered are moderately easy, yet not so convenient as that of the Mohawk. Next south comes the Potomac, whose upper valley leads to that of the Monongahela, at whose junction with the Ohio stands Pittsburgh. The James and Roanoke lead up to the Cumberland Gap, the gateway into the valley of the Tennessee, but also connecting with the Kanawha valley which leads to the Ohio. The point at which the streams break through the main ridge and reach the Appalachian valley beyond can easily be detected on a map by noticing where the longitudinal head-waters turn at right angles to flow eastwards or southeastwards to the Atlantic. See map on next page.

The Atlantic Mountains yield minerals almost throughout their length, but the richest area is the East Pennsylvania coalfield (see map on page 309), whose annual output places Pennsylvania easily first of all the states in mineral production. In 1930 some 70 million tons of anthracite and 125 million tons of bituminous coal were extracted from this field. It is largely owing to the presence of this supply of coal that the subregion has become part of the great industrial area of the United States. The factory towns are mostly on the Piedmont Plateau, but they have the great advantage of being near the big tidewater ports. Iron and steel goods form the chief class of manufactured goods, but textiles (especially cotton goods), chemicals, and paper are also largely made. The markets for these products are chiefly in Europe and South America, to the principal ports of which run important ocean routes. The great circle route to Europe begins more or less parallel to the Atlantic coast, so that the more northerly ports have

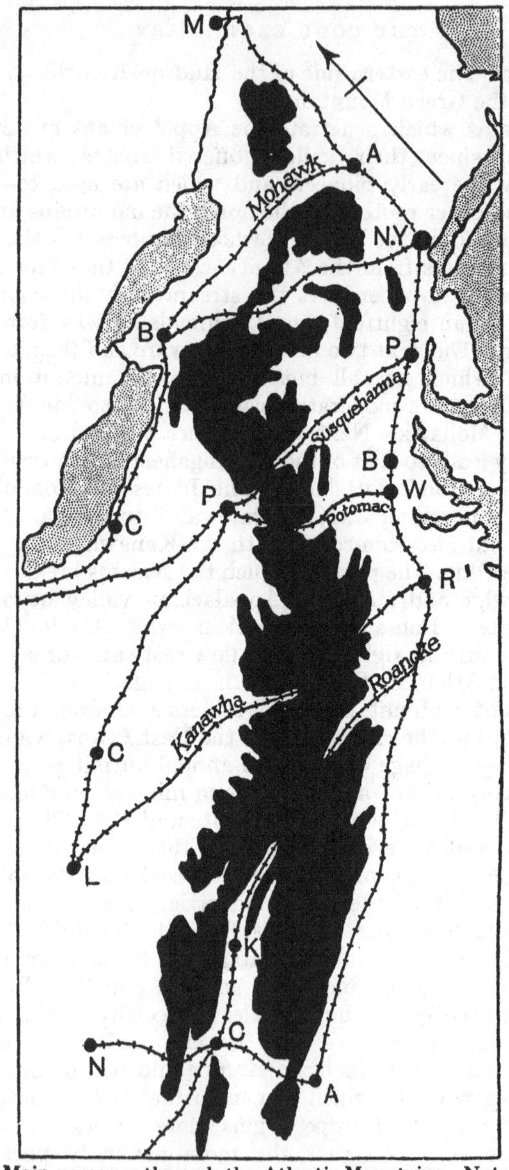

Main passages through the Atlantic Mountains. Note the names of the rivers marked

the advantage of shorter distance to England, France, and Germany.

New York, the largest and most important city in the sub-region, is built on Manhattan Island at the mouth of the Hudson, but overflows on to the mainland north and south and on to Long Island. With Jersey City it contained 7,300,000 inhabitants in 1930, which makes it the second most populous city in the world. In area, however, it is relatively small, owing to the restrictions caused by its island nucleus. In order to house a large population

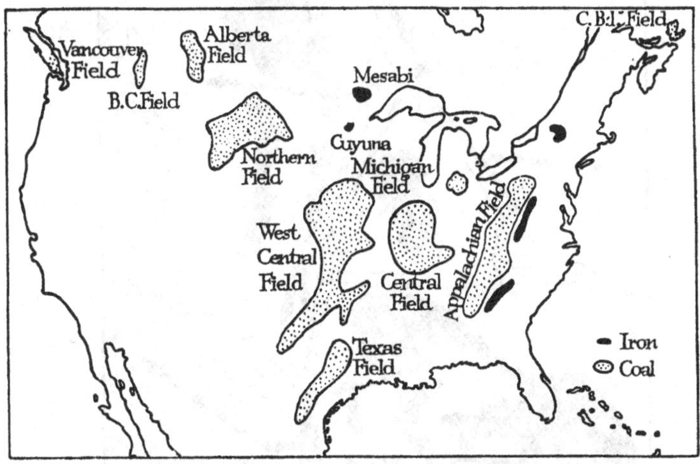

Main coal and iron fields of North America

Note the relation of the Mesabi and Cuyuna ironfields to the northern area of the Appalachian coalfield

on a comparatively small area, a new form of architecture has been invented which comprises lofty 'skyscrapers' towering up to 900 feet. The natural importance of the city is due to its position at the eastern end of the Hudson-Mohawk route, its general focal relation to other routes, and to its being the closest Atlantic ice-free port to Europe, which is easily accessible from the interior of the continent. The next largest town is Philadelphia (pop. 1,950,000 in 1930), a tidewater port on the fall line, with moderately easy communications through the valleys of the Delaware and Susquehanna. It shares the advantages possessed by New York, but has them in a less degree. Baltimore in Maryland (pop. 805,000 in 1930) is on an arm of Chesapeake Bay and has access to the west through the Potomac valley. It also commands the longitudinal coast route from north to south. Washington (pop. 486,000 in 1930), the capital of the United

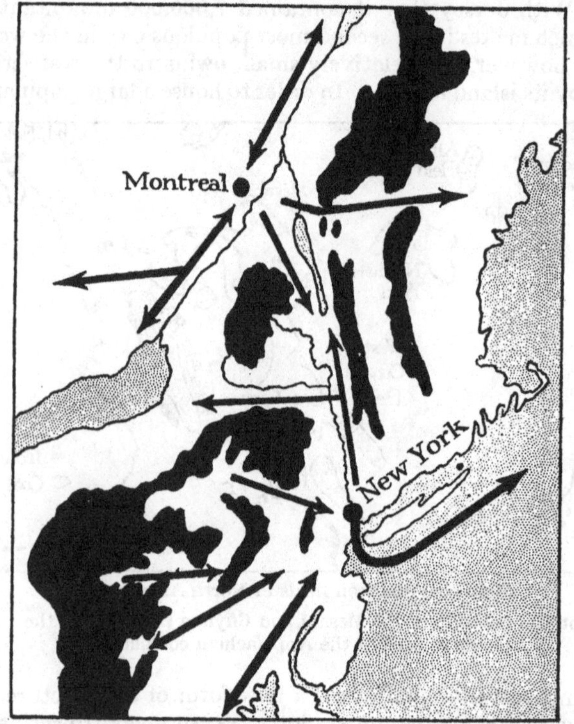

The position of New York and Montreal. Note the converging routes which make focal points of the two towns

States, is an artificial city in the sense that it owes its growth and importance to its selection as the headquarters of the Government of the U.S.A. At the head of the tidewater of the Potomac it has advantages similar, but inferior, to those of Baltimore. Richmond (pop. 183,000 in 1930), the old capital of the South, is on the James River and commands the routes across the Atlantic Mountains along the valley of that river and that of the Roanoke; but it is less of a manufacturing town than its more northern neighbours.

(*b*) *New England, the Maritime Provinces of Canada, and Newfoundland.* In this subregion the Atlantic Mountains have deteriorated into a low, wide plateau whose coastal area has been drowned and forms a broad continental shelf. The Green Mountains define it sharply on the west, while the Gulf of St Lawrence marks it off in the north. Fluvial denudation has cut deep troughs, some of which have been drowned. Hence, Cape Breton and Prince Edward Islands have been severed from the mainland, while Nova Scotia narrowly misses the same fate. It is supposed that the Gulf of St Lawrence lies along an old fault line which has been deepened by fluvial action and then drowned by subsidence. In the southwest the valley of the Connecticut River separates the Green Mountains from the hills of New Hampshire. Farther northeast the divide lies close to the St Lawrence, the ground sloping steeply to the north, but gently and unevenly to the south and east. The coastline is ragged and fringed with islands. In Maine there is a well-developed ria coast with long headlands and deep inlets. The Bay of Fundy, which nearly severs Nova Scotia from the mainland, is famous for its tides, whose amplitude is as much as 70 feet in places. The St John River which empties into it is obstructed by a bar of hard rock which gives rise to the almost unique phenomenon of a waterfall whose direction is reversed at high and ebb tides (see photograph facing page 320).

The broad continental shelf, 200 miles at its widest, makes fishing an important occupation. The Grand Banks of Newfoundland have long been famous for the large quantities of cod caught on them every year. Over a century before the arrival of permanent settlers, fishermen from England, France, Spain, and Portugal used to visit the Banks in the summer. The French still have fishing rights, but most of the industry is now carried on from Newfoundland, Nova Scotia, and Massachusetts. Cod is still the most important fish, but lobster, herring, halibut, and haddock are caught in large numbers. In Prince Edward Island there is a growing oyster production. Most of the fish is 'cured', and, besides being used for home supplies in various parts of eastern North America, it is shipped abroad, especially to the Roman Catholic countries of Europe and to the West Indies. In the last, the 'salt fish' (dried and salted cod) on which the slaves were formerly cheaply fed is one of the principal foods of the negroes.

The subregion is one of mixed forest in which the most important trees are white pine, spruce, hemlock, balsam, birch, cedar, oak, maple, beech, ash, and linden. The stands of good timber are enormous. In the days of wooden ships, the presence of material for shipbuilding combined with good harbours and the practice gained on the fishing banks caused Massachusetts to become the maritime centre of the United States. To-day, the spruce is largely used for making pulp from which newsprint and a multitude of other articles are produced. Birch yields plywood, and maple provides syrup and sugar. Newfoundland and the Maritime Provinces manufacture a great quantity of their own pulp, but they also export much raw material to New England. Trapping was once a profitable occupation in these forests, but the animals are now almost exterminated. However, in Prince Edward Island there is a flourishing industry in the rearing of silver foxes on farms.

Agriculture is carried on in the valley bottoms and on the coast, but the soil is not fertile on the whole and the opening up of the Central Lowlands has ruined all but the best farms. Cereals are grown throughout the subregion, oats predominating in the north and maize in the south. Farming is mixed, the production of potatoes, hay, and dairy products going side by side. Fruit, especially apples, is grown nearly everywhere, but tobacco is confined to the southern portions of Massachusetts and Connecticut.

Mining is important in New Brunswick, where many kinds of minerals are found. The chief metals in the province are iron, copper, and antimony. Large quantities of natural gas are obtained near Moncton and contribute to the importance of that town. Cape Breton Island, which is politically a part of Nova Scotia, has a large coalfield near Sydney from which coal, coke, and tar are produced. Maine and Vermont yield much good stone, the latter supplying half the marble used in the United States. Granite, various kinds of building stone, and lime for cement are also quarried.

While fuel is provided by the few local coalfields, by the woods of Vermont, and by the East Pennsylvanian anthracite field, the many streams and broken ground afford ample supplies of hydro-electricity. Hence, manufactures have assumed some importance in New England, where cotton goods are produced in various towns in Rhode Island, New Hampshire, Massachusetts, Maine, and Connecticut. The little state of Rhode Island is almost wholly given up to manufactures. Other important industries are the production of paper, artificial silk, linoleum, etc., from pulp, and the manufacture of boots and shoes and woollen goods. Among the chief industrial towns are Boston (pop. 781,000 in 1930) in Massachusetts, Providence (pop. 253,000) in Rhode Island, and Hartford and New Haven in Connecticut. Portland in Maine is a seaport much used as an outlet by Canada and joined to Montreal by part of the Canadian National Railway system. Halifax, the capital of Nova Scotia, has one of the finest harbours in the world. It is ice-free and is much used in the four winter months during which the Gulf of St Lawrence is obstructed by ice. St John in New Brunswick is used in a similar way.

(c) *The Eastern Canadian Shield.* This subregion includes Labrador, which is politically attached to Newfoundland and which, owing to the severity of its climate, has no human settlement beyond some fishing and missionary stations along the coast. Seal-fishing is the chief occupation and is carried on mostly by boats from St John's, Newfoundland, which visit the numerous fjords in summer. From Labrador the subregion runs southwestwards to the Great Lakes. Its relief is that of a low tableland whose frayed eastern edge in Labrador is tilted up to form a range of mountains, but whose surface slopes away unevenly to the northwest. The hard archean rock has been smoothed and the eminences rounded by the ice-cap which formerly covered it, and the greatest irregularities are caused by the piles of rock waste deposited by the retreating ice. Moraines which have been laid across the drainage lines block the rivers and give rise to innumerable lakes, cascades, and falls. The soil is poor, but is covered with forests of fir, pine, spruce, and larch, wide stretches of which are of poor size and quality.

In the more favourable areas where the trees are of good growth, lumbering is an important occupation. This is chiefly in the valley of the Ottawa River, whose stream is used for floating down thousands of logs annually to the saw and pulp mills of the city of Ottawa (pop. 125,000 in 1931), the capital of Canada. The old city of Quebec (pop. 130,000 in 1931) near the mouth of the St Lawrence is one of the largest timber ports in the world. It also has large pulp factories. The early French settlers used the subregion chiefly as a hunting ground whence they procured furs by hunting and trapping. This method of securing pelts is fast going out of fashion, for the animals are being raised on farms in various parts of the Province of Quebec. In recent years many valuable minerals have been found, chiefly in central Ontario. Sudbury, Rouyn, and Timmins, which are small places with a big future, are the chief centres, and gold, copper, nickel, and cobalt are the most important minerals worked. An extension of the subregion to the west of Lake Superior contains the most valuable iron mines in North America. The ores, which are mostly red hematite, are found in the Mesabi, Vermilion, and Cuyuna hills and are shipped at Duluth on the west end of Lake Superior to be landed at Chicago, Cleveland, Buffalo, and other ports on the south shores of the lakes, which are within easy access of coal. In 1931 the production of iron from these mines amounted to over 17 million tons.

(*d*) *The Ohio Basin.* This subregion centres on the Ohio, but includes rather more than the basin of that river, since it extends westwards into Wisconsin and northwards over the Michigan peninsula. Except for small areas which are mostly in the north-east, the land is below the 1000-foot contour. It is monotonously level, being broken only by the shallow troughs cut by the rivers. Originally, these 'bottoms' were wooded, while the plain itself above was covered with grass. Hence, the name 'prairies' was given to it by the early French settlers. Nowadays, most of the natural vegetation has been removed and the land brought under the plough. The rivers are sluggish, owing to the low gradient, and they are therefore navigable by small boats nearly to their sources. Except in the Michigan peninsula and in the immediate vicinity of the lakes, the drainage is southwards into the Mississippi. The Illinois and the Ohio rise so close to Lakes Michigan and Erie respectively that a short portage was all that La Salle and other explorers needed to make their way by canoe from Canada to the Gulf of Mexico.

The subregion lies in the corn belt of the United States, and the four States of Ohio, Indiana, Illinois, and Michigan are

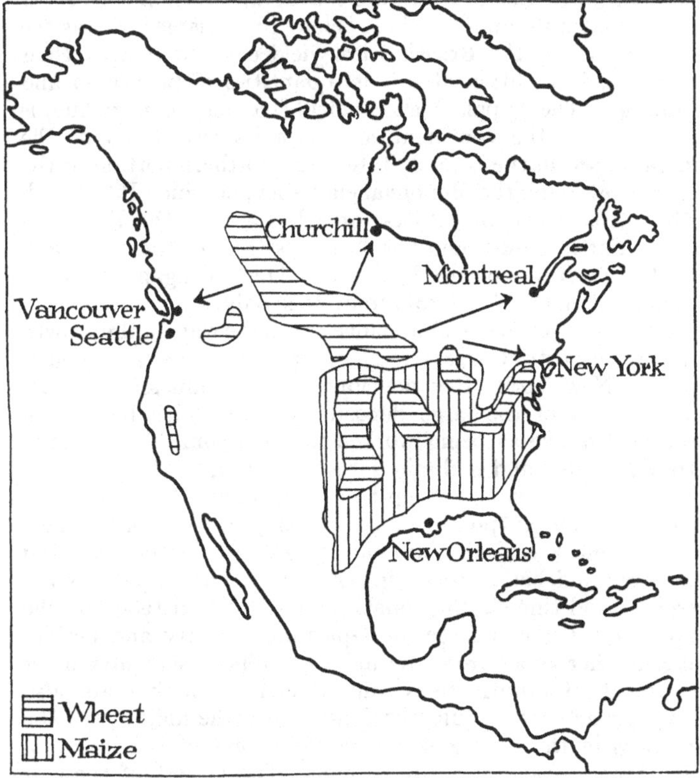

The wheat and maize areas of Canada and the United States

largely agricultural. Ninety per cent. of Indiana is under cultivation. Corn (i.e. maize) is the chief crop, the grain being used for the fattening of pigs, cattle, and sheep. Meat-packing (i.e. the preserving of meat in tins) is a principal industry. Oats and hay are also important, while wheat is grown as a winter crop, and potatoes and tobacco are widely cultivated. In fact, the subregion is one of the greatest suppliers of food for the world market. Illinois, Indiana, and Ohio together produced in 1931 a cereal crop of 1027½ million bushels.

Besides its immense agricultural wealth, the subregion is rich in minerals, the chief of which are coal and petroleum. The rocks of the Atlantic Mountains, which belong to the same series as the coal measures of the British Isles, yield coal throughout their length and in addition give large quantities of petroleum and natural gas. The Appalachian coalfield (see map on page 309) is the richest in the continent and supplies two-thirds of the mineral raised in the United States. The northern portion of the field centres round the Monongahela valley, of which Pittsburgh is the focus, while the central area lies astride the Kanawha valley. Farther west another field lies astride the Ohio and stretches from Illinois to Kentucky. The Michigan peninsula provides a small field of local importance only.

The presence of coal in abundance paved the way for the growth of a vast industrial area stretching from the Mississippi eastwards to New York. But development was impossible before the discovery of the large deposits of iron which have been mentioned above as being worked in Minnesota. The distance between the iron and coal mines is a drawback, but this is partly compensated for by the cheap transport provided by the Great Lakes. Pittsburgh (pop. 670,000 in 1930) on the Monongahela has been able to establish large manufactures of steel goods and machinery and has factories in which fruit and vegetables are preserved and tinned. Cincinnati (pop. 451,000 in 1930), on the Ohio coalfield, has a large meat-packing industry and besides makes leather goods from the hides of animals slaughtered for their meat. Clothing, tools, and electrical supplies are also among the articles manufactured here. But the majority of the big towns in the subregion lie on the shores of the lakes, to which the coal and iron are brought. The largest of these is Chicago (pop. 3,376,000 in 1930), the second city in the United States. Called 'Porkopolis', the town is the world's greatest producer of tinned beef, but it also turns out iron goods, printed matter, and clothing. Next in size comes Detroit (pop. 1,568,000), the city connected with the name of Henry Ford and the chief producer of motor vehicles. Like most other places in the subregion it has meat-packing factories. The towns of Cleveland (pop. 900,000), Buffalo (pop. 573,000), and Rochester (pop. 328,000) form a group on the southern shore of Lake Erie. Besides the usual meat-packing, they are engaged in making motor vehicles, iron goods, electrical supplies, flour-milling, and the preparation of 'feed' for stock. Other large towns are Milwaukee (pop. 578,000) and Indianapolis (pop. 364,000), but there are numerous others containing between 50,000 and 200,000 inhabitants and engaged in similar occupations.

The subregion is the great supply area for the trade of the east coast, and this gives great importance to the routes through the Atlantic Mountains. Much of the heavy stuff is shipped by barge down the Ohio and Mississippi to the Gulf of Mexico.

(*e*) *The Lowlands of the St Lawrence.* These are the fertile margins of the St Lawrence and the peninsula of Ontario. The heart of Canada, it was the earliest field of settlement and still contains the three largest towns and the densest population in the Dominion. It is mainly an agricultural area, in which dairying and mixed farming are carried on. Apples, pears, and plums are cultivated, especially on the peninsula. Toronto (pop. 631,000) is the focus for this produce and has growing manufactures. The Great Lakes provide a good deal of fishing, the chief fish being trout, whitebait, and herring. At the foot of the Adirondack Mountains to the south of Montreal there are important asbestos deposits whose yield forms a large proportion of the world's supply.

The importance of the subregion is largely due to its character as a natural corridor from the west to the Atlantic. In 1931 no fewer than 27,651 vessels with a total capacity of 17½ million tons passed along the St Lawrence carrying chiefly grain, timber, iron ore, and coal. Unfortunately, the waterway is interrupted at three points by falls and rapids and, though small ships can reach the Atlantic from Duluth or Port Arthur, ocean-going vessels cannot pass the Lachine Rapids near Montreal. This obstruction has been avoided by a canal navigable by 'whalebacks' (i.e. grain barges) and other small vessels. A further obstacle occurs between Lakes Ontario and Erie, for the Niagara River here leaps an escarpment to form the celebrated falls. To circumvent this the Welland Ship Canal has been constructed. The third break in the waterway comes at the Sault

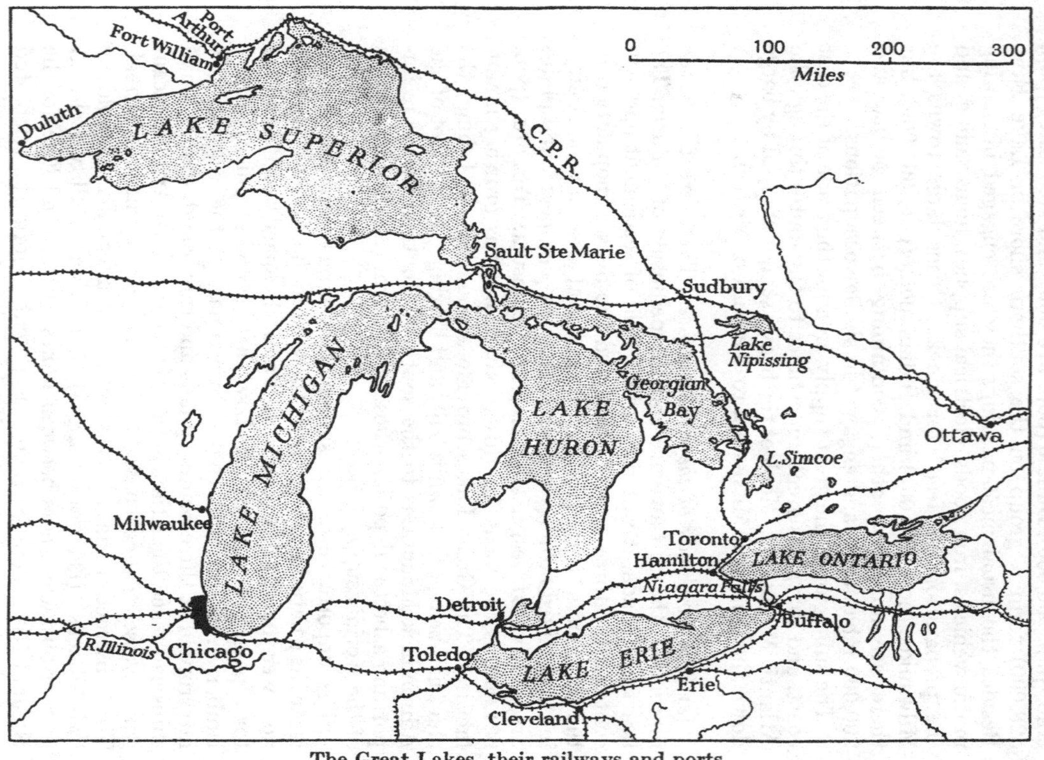

The Great Lakes, their railways and ports

Sainte Marie, where there is a fall in the stream connecting Lakes Superior and Huron. The St Mary's Falls Ship Canal (the 'Soo' Canal) has been built to avoid this fall.

Over £40 million have been spent on the improvement of the waterway, and schemes have been suggested for further betterment. The building of a ship canal along the French River to Lake Nipissing and so to the Ottawa has been suggested, but this purely Canadian route seems to have been shelved for ever by the provisional treaty arranged between Canada and the United States in 1932 for the enlargement of the longer passage through Lakes Erie and Ontario.

Railways run from the lake towns to Montreal and Quebec. The latter stands at the head of the Gulf of St Lawrence on a fine defensive site. It is the oldest town in Canada, but is comparatively small, the number of its inhabitants being 130,000 at the census in 1931. Montreal, with a population of 818,000, is the largest town in Canada. It owes its growth and importance to the breadth of the lowland strip and to the junction of a route down the Ottawa valley and of the Richelieu route from New York with the main St Lawrence waterway (see map on page 310). The rapids above the town supply electric power for lighting, heating, and driving the machinery of the factories that are growing up. Eastern Canada as a whole is rich in water power. Besides a large share in the Niagara Falls, it has the many rapids of the Ottawa and Saguenay Rivers as well as the Lachine Rapids. Much of this power is as yet undeveloped.

Since this region lies near the sea in the temperate belt, it not only attracted settlers from among the vigorous peoples of northern Europe, but it also maintained their energy. Whereas in the southern states negro labour was necessitated by the hot, damp climate, white labour is possible and profitable in the north. Eastern Canada, New England, and the middle Atlantic coast was the scene of some of the earliest settlements of the French, Dutch, and English, and it has many curious historical survivals. Thus, the Province of Quebec is largely peopled by the descendants of French settlers, who still speak French and use French methods of farming. By a curious paradox, this element of the population of Canada is the most averse to any form of political union with the United States. New England, which was settled by English Puritans and Quakers, still bears marks of those first pioneers. While the western areas of the continent are still young in civilisation, this region has grown to manhood, for it has filled with people and is fully developed in its resources. It is, in fine, the most important part of North America.

The Cool Lands of the Centre. This region lies between the Great Lakes and the Rockies and is an area of wide plains forming part of the Central Lowlands. Part lies in Canada, part in the United States, the political boundary being an artificial one along the parallel for 49° N., though this follows fairly closely the low watershed which divides the Arctic drainage basin from that of the Mississippi. The northeastern corner is occupied by the western part of the Canadian Shield which is here very low, but has the same rough, moraine-strewed surface as exists farther east. Low watersheds, many lakes, rapids and falls characterise it, and an infertile soil limits its usefulness. Westwards from the Red River valley the surface is more even, but rises by an escarpment to the Great Plains which form a shelf against the Rockies. In the United States the drainage is carried eastward into the Missouri and Mississippi and then southwards to the Gulf of Mexico. The Red River, which assumes the name of Nelson after leaving Lake Winnipeg, and the Mackenzie are the northern counterparts of the Mississippi, receiving the drainage of the Great Plains through the eastward-flowing Saskatchewan, Athabasca, and Peace Rivers.

The climate is one of extremes, the mean July temperature being about 65° F., while the mean January temperature falls as low as 0° F. Thus:

	January	July	Range
	° F.	° F.	° F.
Winnipeg	−13·2	66·2	79·4
Calgary	11·3	60·7	49·4
Fort Chipewyan	− 3·5	61·9	65·4
St Paul	11·9	72·0	60·1

It will be seen from the table that the temperature range is greater in the north than in the south. In the Canadian Province of Alberta the Southwest Antitrades often blow strongly across the Rockies, descending on to the Great Plains as a warm wind known as the Chinook. Their occurrence explains the compara-

(*Canadian Pacific Railway*)

REVERSIBLE FALLS, ST JOHN, NEW BRUNSWICK

The barrier of rock which lies across the mouth of the river causes a fall down-stream at ebb tide. At flood tide the rise of the sea and the rush of the tide up-stream causes a fall in the opposite direction

Plate XX

(*Canadian Pacific Railway*)

SASKATOON

A typical prairie town of Canada. Note the American influence on the architecture and the flatness of the landscape

tive mildness of the mean January temperature at Calgary. On the other hand, cold Arctic winds sweep down two or three times during the winter and lower the temperature of Montana and South Dakota to −50° F. These are the winds which are known farther south as 'Northers', but here they are called 'blizzards'. Accompanied by storms of wind and snow, they are much feared, since they cause men and animals caught in the open to die of cold.

The rainfall is everywhere low, 20 inches a year being a good average. Alberta is in the rain shadow of the Rockies and has a smaller total than the other parts of the region. The rainfall régime shows a maximum in summer, but there are no dry months. In winter practically all precipitation is in the form of snow. A large part of Montana has a scanty rainfall and is barren.

Over the whole of the region in the United States and a triangular area in Canada between Winnipeg, the Peace River settlements, and the eastern foot of the Rockies at the frontier, there is temperate grassland. The wide, rolling plains are treeless for mile after mile. In summer they are grass-covered, but this vegetation turns brown and dies down in autumn. In winter there is a monotonous expanse of level white snow, but as spring returns the green grass shoots up mingled with multitudes of brightly coloured wild flowers. Towards the north the grass gives way to the Arctic forest of conifers.

Up to some seventy years ago, there were but few settlers in this region. Most of those in Canada were fur-trappers engaged in collecting pelts for barter at the Hudson Bay Company's 'forts'. One of these, Fort Garry, has now grown into the city of Winnipeg (pop. 219,000 in 1931), a town as big as Leicester and the capital of the Province of Manitoba, formerly the Red River settlement. Winnipeg owes its importance to the near approach at this point of the infertile Canadian Shield to the United States border, which causes a bottle-neck through which all east and west routes must pass. When the C.P.R. (Canadian Pacific Railway) was built from Montreal to Vancouver, settlers thronged to these plains, which are known as the prairies of Canada. They are not to be confused with the prairies of the United States, which are different in character (see page 314). At first, the settlers on the low prairies planted wheat, while those on the Great Plains raised stock, chiefly cattle. Nowadays, wheat has invaded the stock country, which has come under the plough except in southwest Alberta and in Montana. Gradually, too, wheat cultivation has penetrated northwards till it has reached the Peace River valley. Meanwhile, the older wheat lands of Manitoba have been given over to mixed farming.

Across the American border, settlement spread gradually up the Mississippi and Missouri. The portions of Illinois, Wisconsin, and Minnesota which fall within the region, together with the whole of Iowa and North Dakota, are agricultural lands. North Dakota shares with Manitoba the fertile, alluvial soil which was originally laid down at the bottom of Lake Agassiz, a large body of water of which Lake Winnipeg is a mere survival. The vast quantities of wheat grown are milled chiefly at the twin towns of Minneapolis and St Paul (pop. 735,000 in 1930), which owe their origin to the water power supplied by the Falls of St Anthony on the Mississippi. Farther south St Louis (pop. 822,000 in 1930), at the junction of the Missouri and Mississippi, has large manufactures of boots and shoes.

The outlets for the American portion of the region are the ports of Duluth and Milwaukee on the Great Lakes, and Chicago both on the Lakes and the Mississippi waterway. In Canada the grain from Manitoba and Saskatchewan goes by train to the twin ports of Port Arthur and Fort William on Lake Superior, and is shipped thence eastwards along the lakes. Alberta finds a cheaper outlet westwards by train to the Pacific coast at Vancouver and Prince Rupert.

While in the United States the urban population is concentrated in a few large towns, in Canada it is distributed among a greater number of smaller places. Besides Winnipeg, there are Regina, the capital of Saskatchewan, and Saskatoon, both of which are centres of wheat-growing. Edmonton, the capital of Alberta, is in a district of mixed farming farther north. Calgary, which has some natural gas, oil, and coal in its neighbourhood, lies on the extreme western edge of the farming area. Medicine Hat in southern Alberta is in the ranching country. These towns are all modern and have not yet developed manufactures, but have remained collecting and distributing centres for the districts around them.

The Northern Rockies. From the Columbia valley northwards the Pacific Mountains comprise the three lines of folds

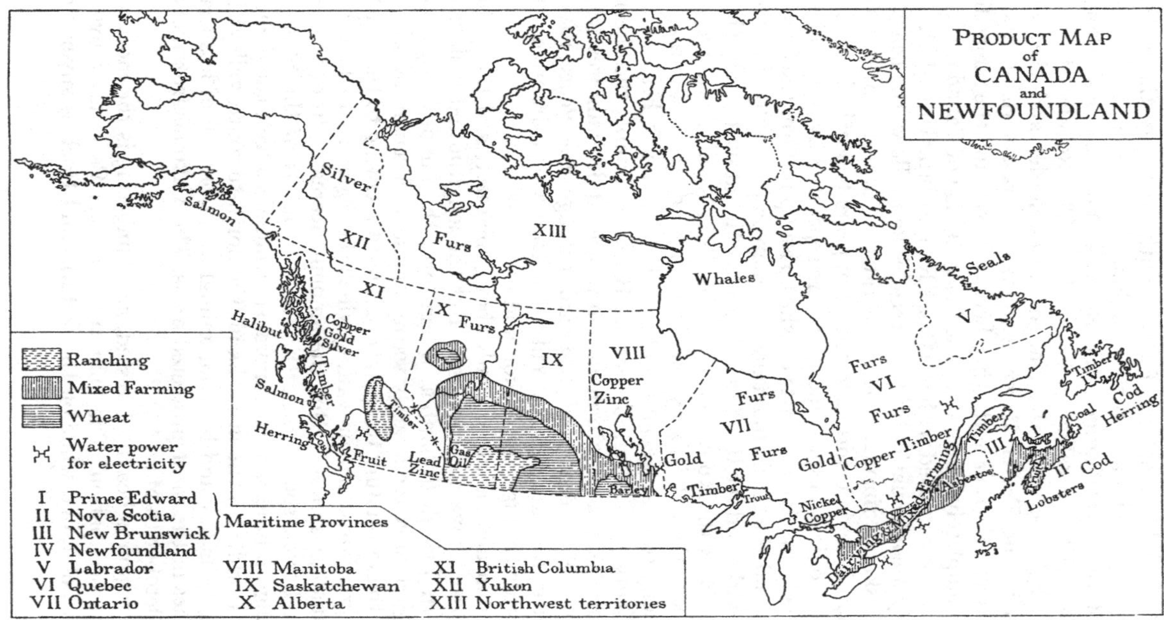
PRODUCT MAP of CANADA and NEWFOUNDLAND
Ranching
Mixed Farming
Wheat
Water power for electricity
I Prince Edward
II Nova Scotia
III New Brunswick
Maritime Provinces
IV Newfoundland
V Labrador
VI Quebec
VII Ontario
VIII Manitoba
IX Saskatchewan
X Alberta
XI British Columbia
XII Yukon
XIII Northwest territories
Silver
Salmon
XII
Furs
XIII
Whales
Seals
XI
X
Furs
Halibut
Copper
Gold
Silver
Timber
Salmon
Herring
Coal
Fruit
Lead
Zinc
Gas
Oil
IX
VIII
Copper
Zinc
Barley
Gold
Timber
Trout
VII
Furs
Furs
Gold
Nickel
Copper
VI
Furs
Furs
Copper Timber
V
Timber
IV
Cod
Herring
Coal
Timber
III
I
II
Cod
Lobsters
Asbestos
Dairying & Mixed Farming

which have been noticed farther south, but only the eastern and central ranges are on the mainland. The present region consists of these two ranges, the Rockies and the Coast Range, together with the valleys between them. The Rockies attain to some height between the United States border and the head waters of the Fraser River, a number of peaks rising above 10,000 feet. Farther north they become lower, and the range dies away completely in Alaska. The Coast Range of Canada, which must not be confused with the Coast Range of the United States, is really a continuation of the Cascades and follows the coast to the Alaska Peninsula, where the ridge is broken up into the Aleutian Islands before it finally disappears beneath the sea.

The two lines of fold mountains are separated by deep valleys and blocks of plateau, parts of which rise up to form the Selkirk Range and the Gold Mountains (see map on page 326). Between the Selkirks and the Rockies is the Rocky Mountain Trench, which is drained by the Kootenay River. The Gold Mountains are separated from the Selkirks by the deep valley of the Columbia. A third longitudinal trench is followed by the middle Fraser from Fort George to Lytton. In central British Columbia the Skeena Interior Basin is an intermountain tableland draining westward through the Skeena and eastward through the Peace River. Similar in formation is the Yukon plateau which drains northwestwards through the Yukon and its tributaries.

A mountain climate prevails above 5000 feet, and the higher peaks rise above the snowline. But the longitudinal valleys of southern British Columbia, which form the important part of the region, have moderately warm summers. Even in the Yukon valley the mean July temperature is 60° F. Owing to the southwesterly origin of the prevailing winds the region as a whole gets an abundant rainfall. But the amount decreases from west to east and from south to north. The mountains of the State of Washington have a mean annual snowfall of 70 feet, the highest in the world. On the other hand, the north-to-south valleys lie in a rain shadow and have an annual mean rainfall of less than 20 inches in the most protected areas. The Yukon is frozen over from October to May.

The region is heavily forested, and lumbering is the most important occupation. The 'big trees' of the Cascades grow in British Columbia, while on the Selkirks and Rockies silver pine and spruce are predominant. The rain shadow areas, however, are grasslands and are much used for ranching, especially on the uplands between the Fraser River and its feeder the Thompson. There is also a good deal of wheat and fruit cultivation in the great longitudinal valleys of southern British Columbia. Farther south, in Washington, wheat, oats, potatoes, and apples are produced in large quantities.

But naturally mining is an important industry. A great many of the early settlers entered the country in search of gold, which is found to some extent almost everywhere. The 'gold rush' to the Klondike valley in 1896 is famous and has been graphically described in Elizabeth Robins' novel, *The Magnetic North.* Dawson, in the Yukon territory of Canada, owes its growth to the development of mining. Gold is still worked, but the silver mines at Mayo are now of more importance. Alaska is rich in minerals, though the supplies have not been fully worked as yet, gold, copper, and silver being the chief metals. Farther south gold is also found at Stewart in the Coast Range of Canada, where there are also valuable copper mines. But the chief mining area is in southern British Columbia, where copper is found at Princeton and very rich mines of zinc and lead are worked in Nelson. Much coal exists in British Columbia and Alaska, but it is not yet worked to any extent. Nevertheless, in 1931 it headed the list of minerals raised in British Columbia. Electric power is plentifully supplied by the many rivers, the only source which has been much used so far being the Fraser River.

The region is transitional between the Pacific coast and the Central Lowlands, and hence the passes through the ranges are very important. Spokane in Washington is a junction of routes through the passes in the Rockies and Cascades. In the Canadian Rockies the Crow's Nest Pass affords a passage for the C.P.R. from Medicine Hat to the Kootenay valley, while farther north the Kicking Horse Pass enables the C.P.R. main line to reach the Fraser valley. Still farther north is the Yellowhead Pass, used by the C.N.R. lines from Edmonton to Prince Rupert and Vancouver.

The Temperate West Coast. This region lies between the lower slopes of the central folds and the Pacific and consists of the Coast Range proper together with the longitudinal valley separating it from the central fold. In this region the Coast Range proper (i.e. the range which outlines the coast, as distinct

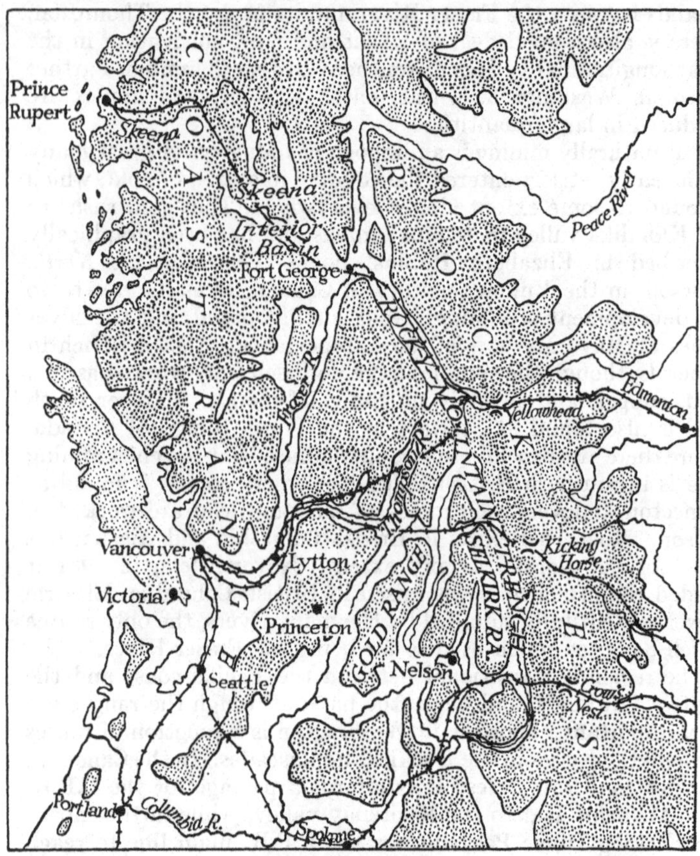

The longitudinal valleys of British Columbia and the routes through the Rockies

from the so-called Coast Range of Canada) is part of the mainland in the States of Oregon and Washington only, since farther north it has been drowned to such an extent that its lower ridges and inner valleys have been lowered beneath sea-level, while the higher ridges stand up as islands, the largest of which are Vancouver and Queen Charlotte Islands. In Alaska the region gradually narrows and finally disappears. The only lowland area of any size consists of the delta of the Fraser and of the undrowned portion of the longitudinal valley drained by the Cowlitz and Willamette.

The coastline is cut by fjords and presents the unique phenomenon of a combination of the fjord and Dalmatian types. The Alaskan boundary is purely arbitrary and violates geographical principles by giving to the United States the coast of British Columbia from Mount Cook to Dixon Entrance. The drowned portions of the Coast Range form a continental shelf which abounds in fish. The salmon and halibut catch is of great importance and normally amounts to a value of £5 million. The former are also caught in large numbers in the Skeena and Fraser Rivers, and large canneries have been built at Prince Rupert on the Skeena and at Vancouver and New Westminster on the Fraser.

The climate is of the temperate western or maritime type, though with many local variations due to the broken nature of the land. Victoria, on Vancouver Island, has records of temperature which closely resemble those of London, but towards the north there is a gradual decrease with latitude. Thus, Sitka has a mean January temperature of 30° F., and a July mean of 55° F., the range being 25° F. The prevailing winds are the Southwest Antitrades which not only come from the warmer south and from over the moderating ocean, but also from off the warm continuation of the Kuro Siwo, the Pacific counterpart of the Gulf Stream Drift. Precipitation is heavy on the mountain ridges and on the western slopes, but the lowlands are mostly in a rain shadow. Hence, while the exposed Sitka receives no less than 81 inches a year, Victoria has only 32 inches. There is a good distribution through the months, but a maximum in autumn and winter.

The mountain slopes are heavily forested with Douglas fir, red pine, and red cedar in the south and sugar pine and spruce in the north. Lumbering is an important industry in Oregon and British Columbia, and, besides timber mills, pulp factories have now been established at Vancouver, New Westminster, and Seattle. The soil of the lowlands is very fertile, and mixed farming is carried on everywhere in them as well as at both ends of Vancouver Island and in the lower Skeena valley. In the Puget Sound district and in the valleys of the Willamette and Cowlitz plums and other fruit are largely cultivated, and prunes are an important export. On the delta of the Fraser River apples and hops are grown.

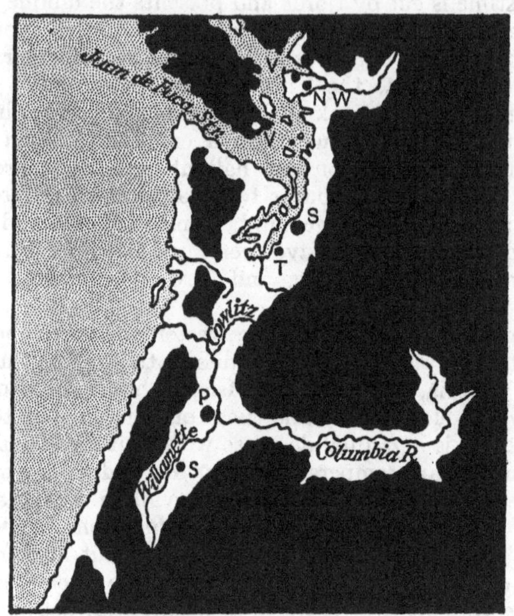

The lowlands of Oregon and British Columbia. Ground above 600 ft. is blackened

The chief towns in the region are Portland in Oregon, Seattle in Washington, and Vancouver and Victoria in British Columbia. Portland is the focus of the Willamette-Cowlitz valley and is at the breach in the Cascades formed by the Columbia River. It can be reached by steamer from the Pacific. In 1930 it had a population of 300,000. Seattle (pop. 365,000 in 1930) is an important port on Puget Sound. Vancouver (pop. 300,000 in 1931) is the western terminus of the C.P.R. and C.N.R. main lines. It has a fine, deep harbour, and its backland includes not only the productive valleys of British Columbia, but also the wheatlands of Alberta. Victoria is a relatively small place, with a population of 60,000 in 1931, but it is the capital of British Columbia. It owes its importance to the coal mines of the southeast of Vancouver Island.

The Cold Lands of the North. This region consists of the northern strip of the continent together with the Arctic archipelago of Canada and the large island of Greenland which belongs to Denmark. Except along the north coast of Alaska, where run the Arctic Mountains, the mainland slopes gently away northwards. The uneven, moraine-strewn surface reappears again in the islands to the north. Greenland is a tableland deeply covered in an ice-cap. Human settlement is only possible along the coast, where there are some villages of which Upernivik is the chief. The native Eskimos are largely nomadic and live by hunting the seal and by fishing.

The winter temperature of the region is always below freezing-point, while in summer it seldom reaches 60° F. Much of the ground is frozen constantly, only the surface thawing in summer. Hence, the vegetation is that of the tundra: dwarf birches and willows, annuals, and mosses. The caribou (a kind of reindeer) and the musk-ox contrive to find enough food, while the sea has countless seals, polar bears, walruses, whales, and fish. The Eskimos lead a precarious, nomadic existence, depending on the animals and fish which they hunt for food, clothes, fuel, and other necessaries of life. European settlement is confined to trading posts. Plans have been suggested for introducing the European reindeer, large herds of which could be reared on the tundra, and the flesh could be used as tinned meat. Some success has already been obtained in Alaska.

HUMAN GEOGRAPHY

Discovery and Settlement. It has been said that Columbus did not discover America, but only the way there and back. This is true to the extent that in A.D. 986 a Norwegian settler in Iceland named Bjarni Herjolfsson was blown by a storm on to a coast which has been identified as that of Nova Scotia. Four years later,

Leifr Eiriksson set out to explore the land seen by his friend Bjarni and during his voyage certainly went as far south as Cape Cod and perhaps to the coast of Virginia. The land was named Wineland, because of the grapes which were found growing there. But the Norsemen found richer and easier fields for adventure along the coasts of western Europe, where they so exhausted their resources that the Wineland settlement was left unsupported. Their discovery of the western continent was forgotten by all save the remote Icelanders, who recorded it in their sagas.

At the end of the fifteenth century, when the spirit of adventure was high and men were eager to discover what lands lay beyond the narrow confines which then comprised the world, Christopher Columbus succeeded in persuading the King and Queen of Spain to send out a voyage of discovery to the west. Columbus said that his purpose was to reach the Far East by a short sea journey, but it has been said that while on a voyage to Iceland he had heard of the existence of Wineland and knew beforehand of the existence of the land which he 'discovered'. However this may be, to him is due the credit for having found out and put into practice the sailing route westwards along the Trades from the Canary Islands to the West Indies and back eastwards along the Southwest Antitrades from Bermuda to Portugal. Columbus's discoveries were at once followed by the Spanish conquest and occupation of the West Indies, Central America, and Mexico. It is interesting to notice that the Spaniards, who were accustomed to a Mediterranean climate in Europe, did not penetrate northwards beyond the subtropical regions of America.

The discovery of Newfoundland by John Cabot in 1497 and the exploration of the Gulf of St Lawrence as far as the Lachine Rapids by Jacques Cartier in 1535 led to no immediate settlement, but at the beginning of the seventeenth century efforts were made by the French, English, and Dutch to establish colonies. Again, it is interesting to note that these more northern peoples sought familiar climatic regions in which to develop their settlements. At first the French were more successful than the English. The open door of the St Lawrence quickly led them to the Central Lowlands, where Marquette found the Mississippi and La Salle navigated the great river to its mouth. The establishment of a French colony at New Orleans is kept in mind by the French names in the State of Louisiana, itself called after

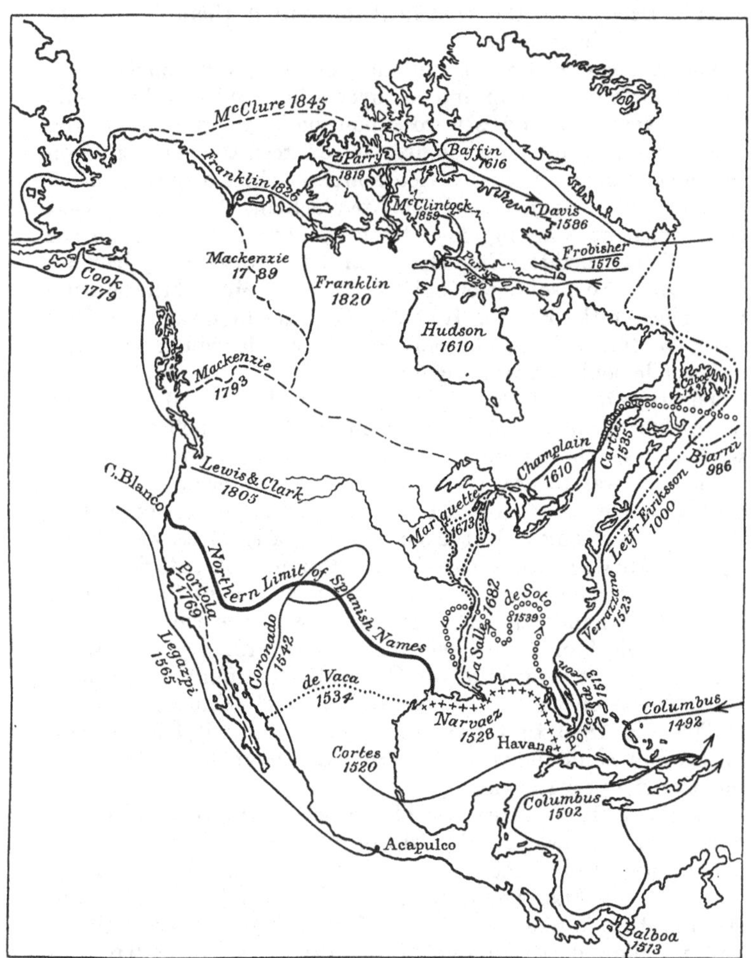

Exploration of North America

the then King of France. The very ease with which the French could spread over the continent prevented them from consolidating their positions.

The English settlements, on the contrary, were made along the Atlantic coast strip and for over a hundred and fifty years were prevented from westward expansion by the presence of the Appalachians. Thus, the colonists were forced to concentrate on the development of the restricted area at their disposal. When they eventually penetrated the Appalachians, they found that the French were laying claim to the Central Lowlands. In the war which ensued the English gained the victory largely because of the greater consolidation of their strength. Most of the fighting took place in the Hudson-Richelieu Gap, the highway from New York to Montreal and the only area in which regular troops of either side could advance against the enemy.

The surrender of Canada to the English in 1763 meant no more than a change of government for the French settlers in what is now the Province of Quebec, the only area in which settlement was to any extent close. The American War of Independence which followed in 1775 ended in the breakaway of the English colonies. But large numbers of loyalists were given land in Canada in what is to-day the Province of Ontario.

In the gradual western expansion of settlement Canada now found itself handicapped as compared with the United States. The infertile Canadian Shield discouraged ordinary drift westward, while the prairies of the United States offered a tempting field. It was only after the construction of the Canadian Pacific Railway in 1886 that the grass plains of Manitoba and Saskatchewan could be reached. From this moment the development of the prairies was rapid.

On the United States side of the border the removal of the French was followed by the drift of settlers into Kentucky, and the prairies were gradually peopled as far as the Mississippi by agriculturalists. The drought conditions of the Great Plains resisted the farmers until the development of the railway systems allowed cattle-ranching to be carried on. In 1849 reports of the discoveries of gold in the Sierra Nevada caused an exodus of adventurous spirits into California. This western expansion, which was temporarily checked but ultimately increased by the American War in 1861, brought about a dispute between Great

Britain and the United States over the boundary of Canada between Lake Superior and the Pacific. Eventually, the 49th parallel of latitude was agreed upon westward from the Lake of the Woods. Though artificial, this boundary more or less followed a main divide on the Central Lowlands and can be defended as a practical line in an area where the relief of the land offers no visible landmarks. From the Rockies to the Pacific, however, it cuts off the heads of the longitudinal valleys from their lower portions and from their natural outlet to the sea. Geographical considerations would suggest the inclusion in Canada of the State of Washington and much of Oregon.

The eastern boundary between Canada and the United States is equally unfortunate. After following the median line of the Great Lakes and the St Lawrence, it cuts across all geographical features along the 45th parallel of latitude as far as the watershed of the Atlantic Mountains. It runs along this divide up to the St John River, which it follows for a while, but afterwards leaves it for the St Croix River. Political considerations made inevitable the inclusion of the State of Maine in the United States, yet its loss to Canada has deprived the dominion of a short outlet to ice-free ports. Portland is, as far as commerce is concerned, largely a Canadian seaport.

In the south the spread of American settlement into Texas caused a war with Mexico, as a result of which a readjustment of the frontier was made whereby the Rio Grande del Norte became the boundary as far as El Paso. Thence westwards the border is marked by a series of artificial lines. On the whole this frontier is probably as satisfactory as any other, since it passes through mostly barren country. But the whole question of frontiers in North America illustrates the difficulty of drawing boundary lines across the natural features of a country.

Other interesting geographical problems arise out of the vast size of the United States and the inclusion in them of different natural regions where conflicting ideas are fostered by the diversity of conditions. The Civil War was the outcome of the opposite views taken by the northern states, i.e. those in the temperate belt where white labour is the most efficient, and the southern states in which the climate forbids white labour and favours the employment of blacks. The influx of foreigners into the United States has divided American opinion between pride in the power of absorption by the nation and the fear lest the 'Anglo-Saxon' character of the people should be weakened by too great an infusion of alien blood. To avoid this danger, a quota system has been adopted for the control of immigration, whereby a far larger proportion of northern Europeans is admitted than any other race. Meanwhile, the United States are saddled with a negro population of over twelve millions, which it has admitted to legal and political equality with the European elements. At the moment there is no sign of a possible absorption of the blacks by the whites, and 'colour prejudice' leads to much social difficulty. On the Pacific coast both of Canada and of the United States the immigration of Chinese and Japanese coolies whose low standard of life enables them to undersell white labour has led to trouble, both domestic and international, and steps have been taken to check the entry of Asiatics into the two countries.

The Pacific coast areas present another problem. Different interests and diverse points of view due to differences in geographical conditions have combined with great distance from the east and difficulty of transport to cause a feeling of opposition between British Columbia and eastern Canada and between the three Pacific states and the rest of the Union. The question was finally settled in Canada by the construction of the C.P.R., but it is still much alive in the United States. The building of the Panama Canal was more an affair of American domestic politics than of general ocean commerce, and its success removed for a time the ideas of secession entertained by Cailfornia and her fellows. Nevertheless, politicians in those states are often known to toy with the proposal to leave the Union.

The United States is divided into forty-eight states, which may be grouped as follows:

New England	*Middle Atlantic*	*South Atlantic*
Maine	**New York**	**Virginia**
New Hampshire	**New Jersey**	**West Virginia**
Vermont	**Pennsylvania**	**North Carolina**
Connecticut	**Delaware**	**South Carolina**
Massachusetts	**Maryland**	**Georgia**
Rhode Island		

Gulf
Florida
Louisiana
Mississippi
Alabama
Texas

Central
Ohio
Indiana
Illinois
Michigan
Wisconsin
Minnesota
Iowa
Kentucky
Tennessee

Middle West
Missouri
North Dakota
South Dakota
Nebraska
Kansas
Arkansas
Oklahoma

West or Mountain
Montana
Idaho
Wyoming
Colorado
New Mexico
Arizona
Utah
Nevada

Pacific
Washington
Oregon
California

The capital is Washington in the District of Columbia, a small area given by Maryland to form the seat of the Federal Government. The population in 1930 was nearly 123 million persons, of whom nearly 12 millions were negroes and 2 millions of other non-European races. The distribution is uneven, the middle Atlantic coast strip between Boston and Washington having the densest population and Nevada the sparsest. Four-fifths of the people live east of the Mississippi. Contrary to what might be expected, 52 per cent. of the population is urban.

Canada is divided into nine provinces, the territory of Yukon, and the three Northwest Territories of Mackenzie, Keewatin, and Franklin. The provinces are

Nova Scotia (including Cape Breton Island) } Maritime Provinces
New Brunswick } Maritime Provinces
Prince Edward Island } Maritime Provinces
Quebec
Ontario
Manitoba } Prairie Provinces
Saskatchewan } Prairie Provinces
Alberta } Prairie Provinces
British Columbia

The capital is Ottawa in Ontario. The country is settled and developed only along a relatively narrow strip on the southern border, and vast expanses of the north are practically uninhabited. The most densely peopled area in 1931 was the St Lawrence lowlands. The total population was short of 10½ millions, of which some 60 per cent. were of British and 30 per cent. of French descent.

Mexico is divided into thirty-one states and territories and has its capital in Mexico City. In 1930 the population totalled 16½ millions, the greater part of which was concentrated in the south of the tableland near Mexico City, while the barren areas of the north were sparsely peopled.

Communications. North America is specially favoured in natural waterways. In Canada east of the Rockies it is possible in summer to move about almost everywhere by birch-bark canoe, but there are two really important systems: that of the St Lawrence and that of the Mississippi. The former connects the Great Lakes with the Atlantic and so admits direct communication with the heart of the continent. Breaks in the natural waterway occur in three places: between Lakes Superior and Huron, at the Niagara Falls, and at the Lachine Rapids. The first are avoided by the 'Soo' Canal, the second by the Welland Ship Canal, and the third by the Lachine Canal. This important system is separated by a short, low divide from that of the Mississippi. This great river and its chief feeders, the Missouri and the Ohio, are navigable almost from their sources. A canal from Lake Michigan at Chicago to the Illinois River affords a passage for boats from the lakes to the Gulf of Mexico. Besides these great systems there are other rivers, like the Columbia, which are navigable for some distance. A canal connects Lake Ontario with the Hudson through the Mohawk Gap.

The eastern United States and Canada have a moderately close network of railways, and there are in all six main lines crossing the continent from east to west, two being in Canada and four in the United States. The C.P.R. runs from Quebec through Montreal, Ottawa, Winnipeg, Regina, and Calgary to Vancouver, passing the Rockies by a tunnel through the Kicking Horse Pass. The C.N.R. line is farther north and runs from Halifax through Quebec, Winnipeg, Edmonton, and the Yellowhead Pass to Prince Rupert and Vancouver. In the United States the first transcontinental line to be built was the Union and Central Pacific Railroad which runs from New York through Pittsburgh, Chicago, and Omaha to San Francisco. The other three routes use the same terminals, though the Northern Pacific

Main railways of North America

Railroad has a branch to Seattle. The map above shows these routes.

Internal communication was originally carried on by means of 'trails', or wheel tracks, across the grass plains. Most of these started from Independence near the modern Kansas City. The famous Santa Fe Trail led through the town of that name and thence on to California and Mexico. Another led northwestwards through South Pass, after crossing which it forked into

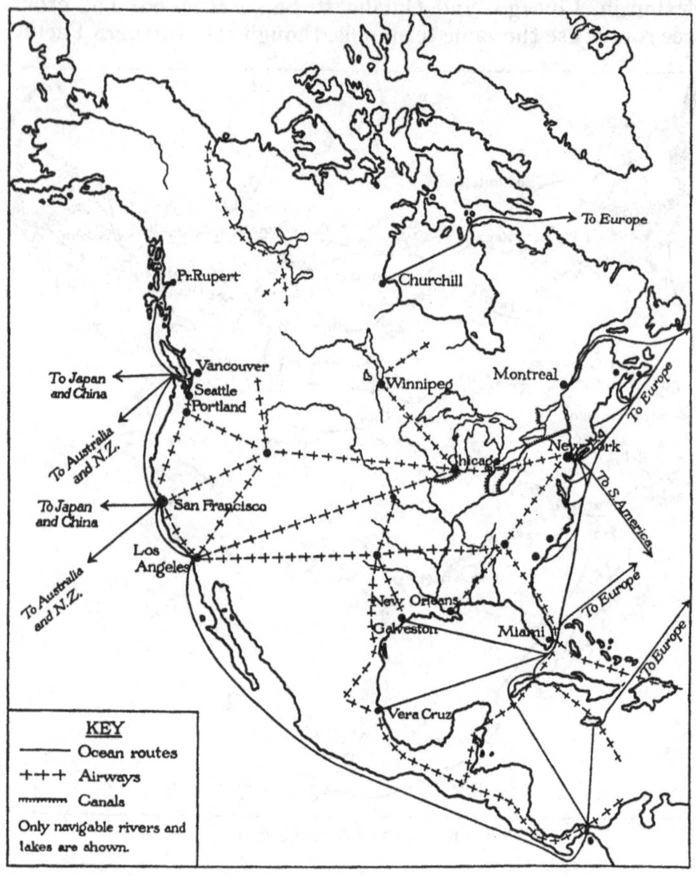

Waterways, airways and ocean routes of North America

the Oregon and California Trails. Nowadays, there has been a return to road traffic, and great motor-ways have been built. At intervals of about 400 miles these new roads have comfortable rest houses where long-distance travellers may spend the night.

Recent years have seen a great development in airways. Besides a number of main local services, Canada has established regular mails between Saskatoon and the mouth of the Mackenzie and between Winnipeg and Chicago. In the United States there are three regular transcontinental lines, besides others which run from north to south. The map on page 338 shows the chief lines on the continent.

External communication by aircraft has also been established, and regular lines now run from the United States to the West Indies and South America. The survey of a proposed route from New York through Newfoundland, Greenland, Iceland, the Faeroe Islands, to Copenhagen has just been completed, and the service may soon be in operation. An alternative route through Bermuda and the Azores has also been suggested, since here the weather is more reliable, though the 'hops' are far longer. There is, however, serious talk of a direct route from England dependent upon floating stations in mid-ocean.

Great ocean routes run from the Atlantic coast to western Europe, from New York to Buenos Aires, from the Gulf ports to Europe, and from Vancouver and San Francisco to Australia and New Zealand. There is also the Panama Canal route which gives cheap water communication between the east and west coasts of the United States and also lessens the expense of shipping wheat from Vancouver and Prince Rupert to England. In 1933 a new sea way was opened from Churchill in Hudson Bay to the British Isles. This route shortens the journey both on land and sea from the Prairie Provinces to their markets, but has the disadvantage of being open during no more than two months in the year.

PART V

SOUTH AMERICA

SOUTH America and North America together form the New World, a name formerly appropriate to these continents not so much because of their late discovery, but in order to emphasise the immense differences in climate, plant and animal life, and the conditions of human existence between the 'new' continents and the regions with which the European discoverers were familiar. The two great land masses are joined by the Isthmus of Panamá, the real geological boundary being marked by the valleys of the Atrato and San Juan Rivers which separate the northern Andes from the ranges of Central America. Eastwards, the bulge of South America in Brazil approaches to within 1800 miles of the west coast of Africa.

Position and Size. About one-fourth of South America lies in the northern hemisphere, the Equator passing through the continent at the mouth of the Amazon, and the 60th meridian of west longitude running nearly midway from north to south. The mainland stretches from 12° N. to 56° S. Lat. It is roughly pear-shaped with a length of 4800 miles and a breadth of 3200. Its area measures some 7 million square miles, or $2\frac{1}{2}$ times as large as Australia and 56 times as big as the British Isles. The portion outside the hot belt tapers to a point and is rather less than a third the size of the area within the Tropics.

South America has often been compared with North America and Africa. With the former it has many points of likeness. Both continents are strikingly similar in shape, have a backbone of fold mountains on the extreme west, with little or no Pacific coastal lowlands, and as a consequence an outlook towards the Atlantic. In both, also, the presence of old, worn highlands on the east concentrates the drainage of wide central plains into three great river systems. But important differences exist, all of which tend to give greater importance to North America. While South America tapers away from the Equator, North America tapers towards it. Hence, most of the northern continent lies within the temperate belt—a fact of supreme importance—and the broad northern end approaches Asia and

Europe. The last feature would have been of great historical import had the eleventh-century voyages of the Northmen been followed up.

The likeness to Africa depends on two facts: both continents contain large fragments of the ancient Gondwana Land, and both

are largely 'hot' lands. The latter is due to the position of the two continents astride of the Equator, from which it follows that they alone have true equatorial regions and show a duplication of the climatic belts. But Africa is more centrally divided by the Equator and does not go beyond the 'Mediterranean' climatic region, while South America contains a cold temperature area.

Above all, there is no likeness at all in the relief of the two continents.

Build and Relief. The map of South America shows three outstanding features, namely, the great chains of the Andes which outline the west coast, two large areas of rough upland

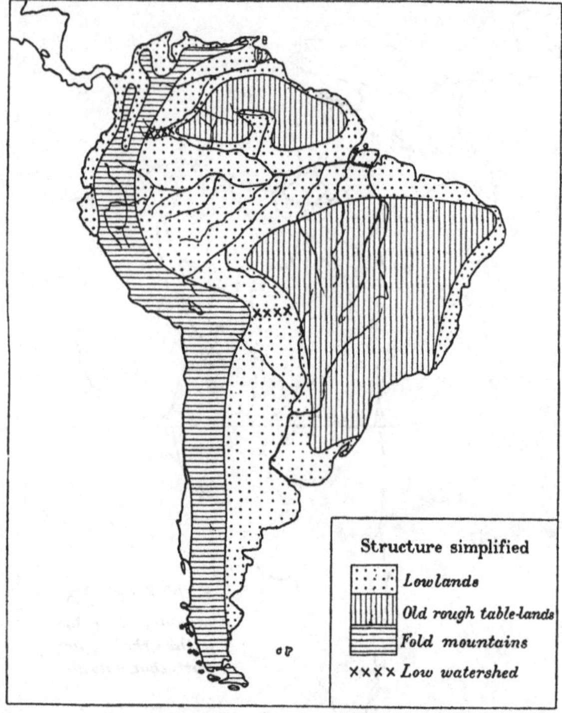

on the east, and a vast stretch of lowland filling up the space between the other two features.

From Tierra del Fuego northwards as far as the lofty peak of Aconcagua the Andes consist of a single cordillera (pronounced *korrdeelyaira*), snow-capped and uninhabitable, which forms the boundary between distinct climatic regions on either side. North of Aconcagua the range divides into two main branches, the Royal and the Western Cordilleras, which contain between them the great toplands of Bolivia and Peru. After reaching its maximum breadth and height, the whole chain swings westwards. At this point a third cordillera, the Central, rises between the others. At Cerro de Pasco the three ranges meet in a great 'knot', only to separate once more. Soon, however, the Royal

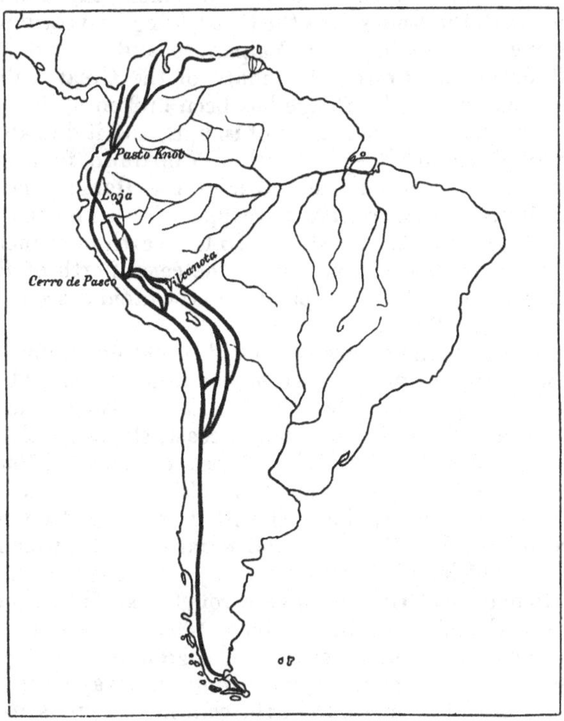

Main fold lines of the Andes

Cordillera grows lower and dies away, while the Central is pierced by intermontane streams which find their way eastwards to the Amazon. At Loja and again at Pasto there are 'knots' where the cordilleras come together, but from the latter four ranges emerge, the chief of which curves eastwards to outline the north coast of Venezuela and end in the island of Trinidad. The other three enclose between them the Maracaibo lowland and the valleys of the Magdalena and Cauca.

This great system of fold mountains has been formed by pressure from the Pacific against the resistant blocks of highland on the east. The movement is hardly finished, and volcanoes like Cotopaxi and Chimborazo rise at intervals along the cordilleras.

Parallel with the Andes runs the Coast Range, whose development is perfect only between Valparaiso and Puerto Montt, where it forms the western boundary of the Great Valley of Chile. Farther south the Range has been broken up by glacial erosion and fracture into a chain of islands, and it dies away on the west of Tierra del Fuego. North of Valparaiso, faulting and subsidence have reduced the width of the Range, and near Mollendo it disappears entirely, to reappear farther north only in slight projections of the coastline. To this earth movement are due the even, harbourless nature of the coast north of Puerto Montt, the absence of a coast plain, and the rapid descent of the land to the ocean floor.

The eastern highlands are composed of old crystalline rocks and show no sign of disturbance by earth movement. Those of Brazil form a tableland whose surface, after rising in a sharp escarpment along the South Atlantic coast, slopes gently away to the north. Most of the tableland is between 500 and 1000 feet above sea-level, but broken ranges of hills cross its surface from northeast to southwest. Parallel with them is a deep trough occupied by the San Fernando and Paraná Rivers, which thus divides the tableland into two parts.

The Highlands of Guiana have a rougher surface, a greater proportion of their area being above 1000 feet and the river valleys being more deeply scored. The ground rises from the Amazon basin in a steep escarpment, but falls away more gently to the north. In the far south of the continent there is another, but far smaller, tableland in Patagonia. It is separated from the Andes by a narrow trough and slopes very gently to the east.

The lowlands which fill the gaps between the high features comprise the Llanos, or grass plains of the Orinoco, the Selvas or forests of the Amazon, and the valleys of the Paraguay-Paraná together with the Pampas of the Argentine. The last are of æolian formation, while the others are due to fluvial deposition. The character of these lowlands is largely due to the constriction of their areas by high ground which gives them bottle-mouths and confines the drainage of each to one large river. A relief map clearly shows that the lowlands of the Orinoco are almost enclosed by the Andes and the Highlands of Guiana, that the Amazon basin has a narrowed mouth owing to the approach of the Highlands of Brazil and Guiana, while the River Plate is a comparatively small gap between the uplands of Tandil and the Uruguayan extension of the Highlands of Brazil. The connecting saddles between the three lowland areas are very low; indeed, during the rainy season boats can pass from the headwaters of the Paraguay to those of the Madeira, a large feeder of the Amazon.

Climate. Three factors dominate the climate of South America, namely, the great breadth of the continent at the Equator, the tapering towards the south, and the existence of the lofty barrier of the Andes. The first gives to South America the largest area of equatorial climate in the world; the second subjects the more temperate regions to the moderating influence of the sea and thus reduces their annual range of temperature; and the third prevents the circulation of surface winds across the mountains, except in the extreme north and south. The climate of the Pacific coast is also greatly affected by certain movements of the water of the Pacific. Not only does the Humboldt or Peruvian Current sweep its cold Antarctic waters along the coast from Valparaiso to the Gulf of Guayaquil, but there is also a belt of even colder water between this Current and the coast, owing to the upwelling of water from the depths of the ocean. In consequence, onshore winds do not yield moisture, except to the higher parts of the Western Cordillera, and latitude for latitude places on the west coast have a lower temperature than those on the east.

The lowlands of the Amazon basin, being traversed by the Equator, have a steady temperature of about 80° F. throughout the year, a feature which is shared by the west coast strip between the Gulf of Guayaquil and the Isthmus of Panama. The Andes at this point and the southern portion of the Highlands of Guiana share the equable temperature, though naturally the air is cooler owing to the altitude. On each side of this equatorial belt there are areas with a noticeably cool season. In the low-

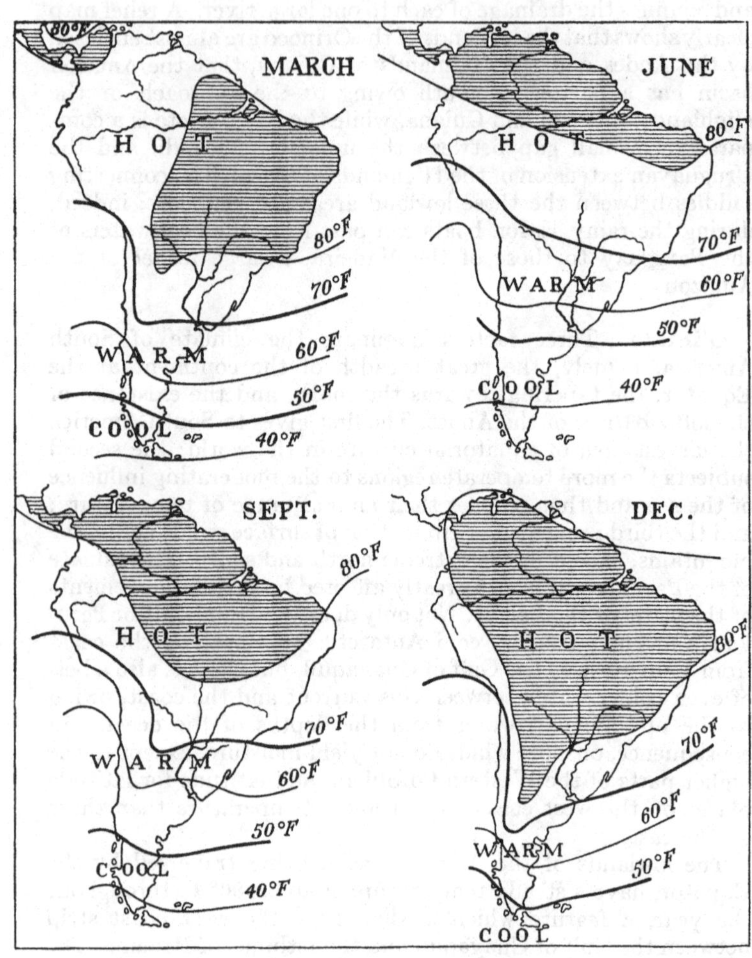

Distribution of temperature

lands of the Orinoco and on the north coast of Colombia, Venezuela, and the Guianas the cool season occurs in January, when the midday sun is overhead at the Tropic of Capricorn. On the other hand, in tropical Brazil it occurs in July, when the midday sun is overhead at the Tropic of Cancer. In this belt the Andes and topland of Bolivia are so high that they have a special mountain climate with low temperatures. South of the Tropic of Capricorn the mean annual temperature becomes less, though the clear skies of the northern provinces of the Argentine cause hotter summer days than are known elsewhere in the continent. Here too is found the greatest annual range, namely, 25° F. Farther south the summers are cool and the winters mild, until sea influence ends by giving a bleak, raw climate to the extreme south.

The great heat of the Amazon valley causes the low-pressure belt at the Doldrums calms to be very well marked, and the Trade Winds blow in from northeast and southeast. In July when the low-pressure centre is south of the Equator the Trades become very strong on the north coast and penetrate up the Maracaibo lowlands and the valleys of the Magdalena and Cauca, and along the west coast strip of Colombia. They weaken considerably as the Doldrums belt moves north. The high-pressure belt of the Horse Latitudes runs across the continent from northeast to southwest, its swing giving a Mediterranean climate to the west coast of Chile between Coquimbo and Puerto Montt. But within this belt, in the northwest of the Argentine, a weakened form of monsoon develops in summer, causing an indraught of wind from the east or northeast along the coast of Brazil and Uruguay. This development is strengthened by the powerful high-pressure system which persists throughout the year in the South Atlantic. The extreme south of the continent is covered by the Northwest Antitrade belt whose winds are more constant than they are in the northern hemisphere.

These Northwest Antitrades are moisture-bearing onshore winds, which on reaching the Andes drop a large amount of rain, especially in autumn (March). But the tableland of Patagonia on the lee side of the mountains lies in a rain shadow and has only 10 inches of rain a year. In the Mediterranean region farther north the Antitrades blow in winter only; hence, that is the season of rain in the Great Valley of Chile, which at other times

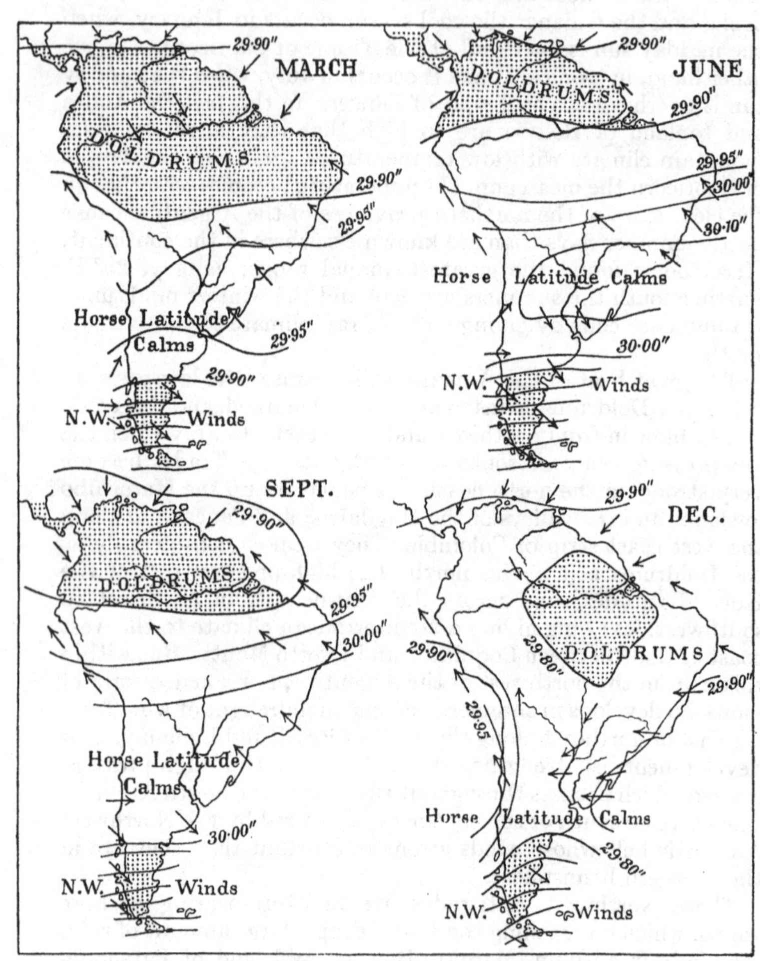

Distribution of pressure and winds

suffers drought. On the east side of the Andes here, there is a typical 'Chinese' climate, with rain from the summer monsoon chiefly, but with showers at other seasons as well. On the equatorward side of the Mediterranean and Chinese belts there is a subtropical desert on the west coast, which persists northwards as far as the Gulf of Guayaquil owing to the fact that the prevailing wind blows parallel to the coast and also to the chilling of the air by cold waters off the shore. But across the mountains there is no desert—though the Argentine province of Santiago del Estero comes near to being one—because of the monsoon winds from the South Atlantic in summer and also because the soil is watered by snow-fed streams from the Andes. In Brazil the Southeast Trades reach as far south as Lat. 30° in summer and give abundant rain to the east coast and to the valleys of the Paraná. But the valley of the San Fernando lies in a rain shadow, and so does the northeast coast of Brazil. The Amazon valley receives a large convectional rainfall. There is a slight north and south swing in the period of maximum rainfall conforming with the apparent movement of the sun; but heavy rain occurs throughout the year. In the tropical belt north of the Amazon valley the rains fall in summer, when the Trades weaken and the Doldrum belt is near; but the highlands, both in Guiana and the northern Andes, force the winds to give up their moisture even in the cool season. As a consequence, the snow line descends to 13,000 feet in Ecuador, though, owing to drought, it is never below 15,000 feet in Bolivia, and indeed the peaks bear surprisingly little snow.

Vegetation. The distribution of plant associations in South America is closely connected with the relief and climate. The constant high temperature and humidity of the Amazon valley and the Pacific coast of Colombia and Ecuador give rise to a dense tangle of equatorial forest with the usual storeyed arrangement and the accompanying binding of lianas, epiphytes, and parasites. A belt some miles wide along the Amazon is swampy and contains a thick growth of trees of moderate height with no shrub undergrowth. As the ground rises above the swamp level, the trees become taller and the normal equatorial forest begins. In such moist conditions the plant life is hygrophilous in the extreme, the majority of trees being of sappy species. On the

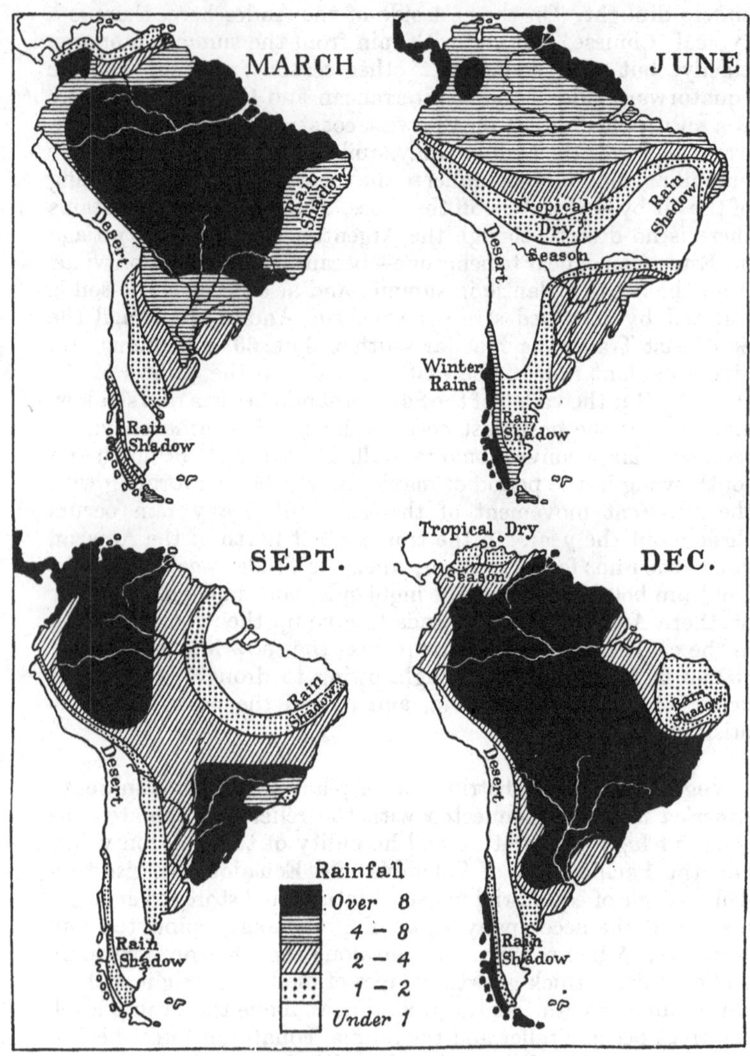

Distribution of rainfall

lower slopes of the Andes the forest changes slightly, retaining the same species, but losing much of its undergrowth. This belt is known as *montaña* in Colombia and Peru and as *yungas* in Bolivia. On the Pacific coast the equatorial forest forms a narrow belt along the shore, thinning out as it climbs the Andes.

In the well-watered tropical areas along the southeast coast of Brazil and along the north coast, especially in the Guianas and the Magdalena valley, there is luxuriant tropical open forest in which hardwoods are interspersed with the less woody species. This is the home of the mahogany, the greenheart, the lignum vitæ, and the logwood. The areas of lower rainfall, notably the tableland of Brazil and the plains of the Orinoco, are savana lands, with long, coarse grasses and scattered trees which tend to form clumps along watercourses and other favourable spots. Many parts of the Brazilian tableland have thorn forest of an open kind, as in the caatinga area of the east.

Towards the south the tropical savana and forest give way to subtropical hardwood forests of a kind peculiar to South America. In the Paraná valley the chief species of tree is the Paraná pine, while in the Chaco region south of the Pilcomayo grows the extremely hard *quebracho*. In these forests the density is nowhere as great as in the equatorial and tropical counterparts, and the woods may be penetrated more or less easily. The next belt of vegetation towards the south is one of temperate grass, made up of short species which make excellent pasture. But between the Andes and the hills of Córdoba the Salinas Grandes interpose a small area of salt-encrusted marsh. The temperate grass becomes scantier towards Patagonia, where there is a large stretch of true steppe extending right into Tierra del Fuego.

On the west coast the abundant rains of South Chile favour the growth of temperate rain forests of conifers, among whose species the araucaria is prominent. In Tierra del Fuego the masses of moss and decaying trunks which form a thick layer on the ground make passage through the forest difficult. Northwards the woods are thinner and finally shade away into a region of Mediterranean scrub where, though the indigenous species are different, yet the climate welcomes the introduction of plants from the corresponding region of Europe. As in Africa the Mediterranean region merges into the Sahara, so in north Chile

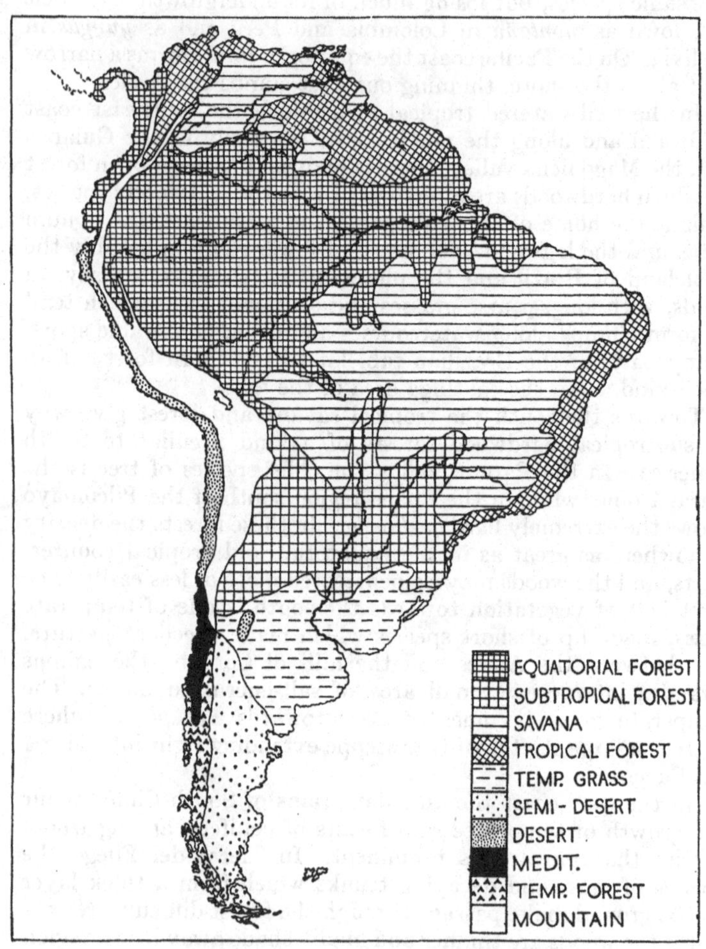

Distribution of natural vegetation

Plate XXI

(*Canadian Pacific Railway*)

FORT WILLIAM, MANITOBA

One of the great Canadian lake ports. The curious construction of cylinders is a grain elevator

Plate XXII

(*Canadian Pacific Railway*)

ONTARIO

Tapping maple trees for the sap which provides syrup and sugar

increasing conditions of drought gradually bring on a state of infertility which reaches its maximum in the provinces of Atacama and Antofagasta. The desert conditions prevail right through Peru, though rivers which carry off the melt-water of the Andine slopes make strips of habitable land at intervals. The transition from desert to equatorial forest is surprisingly sudden and takes place around the Gulf of Guayaquil. The upper slopes of the Andes are clothed with mountain flora not remarkably different from that found in other parts of the world.

Animals. South America is poor in the larger forms of animal life, being in this respect remarkably different from Africa. In the equatorial forest the species are all arboreal or winged, even the large cats (jaguar and puma) taking to the trees, and there are several kinds of tree frogs and snakes. The chief animal is the monkey, nearly all the species of which are prehensile; but there are no large apes. Mosquitoes, beetles, and other insects fill the air, while stinging ants, spiders, scorpions, and centipedes swarm on the trees, bushes, and ground. In the swamps and rivers live alligators, peccaries, water snakes, turtles, and various kinds of frogs. Bird life is plentiful and gaudy, the multicoloured parrot being one of the commonest species. Much the same types are found in the tropical forests, but to them are added the raccoon, the aguti, and the opossum.

The grasslands contain some species of deer and numerous burrowing animals. In the drier part of the temperate region of the south the rhea is also found. But here the most notable form of animal life is the locust which is a recurrent pest both to the pastoral and agricultural farmer. The Andine regions have their own peculiar animals, notably the armadillo and the llama (pronounced *lyama*), which latter, with its cousins the vicuña and guanaco, is the most useful animal in the continent, since it yields wool and is used as a beast of burden. In the colder areas lives the chinchilla, valuable for its fur; and on the crags nests the huge condor, a bird of prey famous for its long sight and powerful wings.

Natural Regions. The various combinations of natural conditions—relief, climate, and plant and animal life—enable the continent to be divided into twelve areas for closer study. They are as follows: (1) the lowlands of the Amazon and the Pacific coast of Colombia and Ecuador, (2) the equatorial highlands,

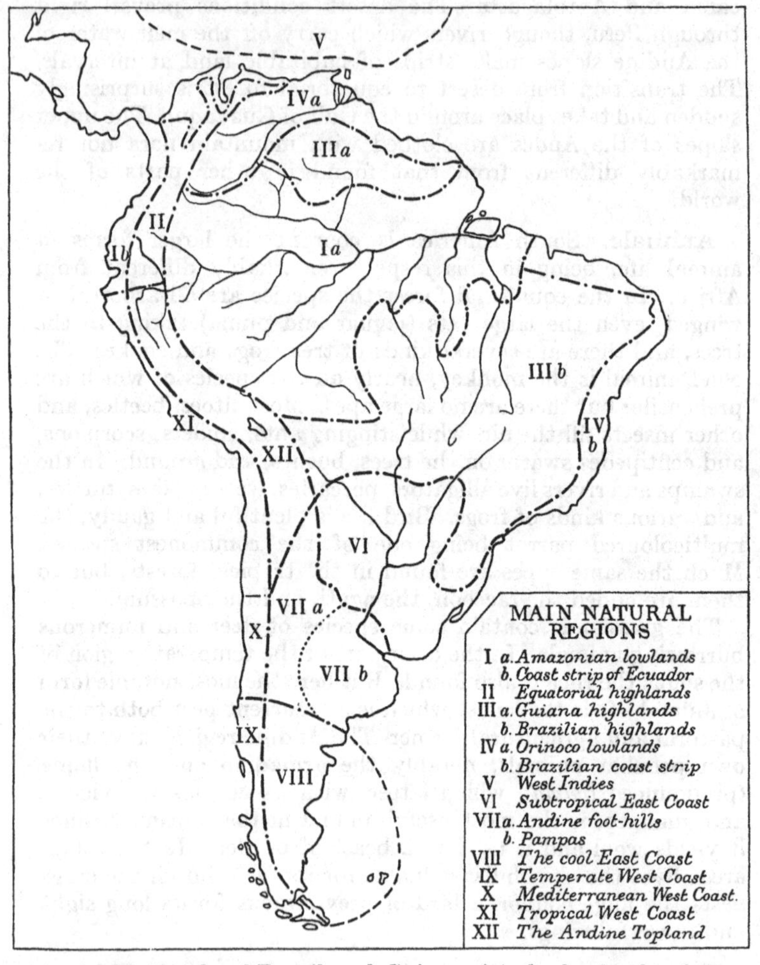

(3) the highlands of Brazil and Guiana, (4) the lowlands of the north coast and the coast strip of southeast Brazil, (5) the West Indies, (6) the subtropical east coast, (7) the Andine foot-hills

and the Pampas, (8) the cool east coast, (9) the temperate west coast, (10) the Mediterranean region, (11) the tropical west coast, and (12) the Andine toplands. The map on page 354 indicates their positions.

THE REGIONS

The Equatorial Lowlands. These include the lower parts of the huge basin of the River Amazon and the Pacific coast strip between the Gulfs of Guayaquil and Panama. The former is the largest area of equatorial lowland in the world, with a maximum length from east to west of two thousand miles and a breadth from north to south of one thousand miles. This greatest breadth is across its middle, the mouth of the valley being constricted by the approach of outlying ridges of the Highlands of Brazil and Guiana. It is this constriction which forces into a common channel the many large streams which make the Amazon the largest river in the world. The biggest of these feeders are the Rio Negro and the Madeira. The floor of the valley is formed of sediment laid down by the rivers and at one time the area so built up extended eastwards beyond its present limits. During that period the Tocantins, which now merely empties into the estuary of the main river, was a feeder of the Amazon. Subsidence has caused the sea to drown the mouth of the valley, but a delta is rapidly being built up around the nucleus of the island of Marajó.

All the feeders of the Amazon which descend from the Highlands of Brazil and Guiana are broken by rapids or falls as they enter the lowlands, a fact which partly accounts for the largely unexplored character of the Highlands. But for such interruptions, river communication would be possible from the Amazon to the Orinoco, since the latter sends off a navigable branch known as the Cassiquiare Canal to join the Rio Negro. The Amazon itself rises in one of the longitudinal valleys of the Andes and, after flowing north for some four hundred miles, cuts through the Central Cordillera and enters the lowlands under the name of Marañón. Rapids and falls interrupt navigation at intervals for the first thousand miles of the river's course, but from Iquitos downwards there is an easy passage for river steamers for a thousand miles to Manaos. At this point the river is joined by the Rio Negro and becomes navigable for ocean-going steamers for the thousand miles which separate Manaos from the North Atlantic. So gentle is the final gradient of the river (1 in 25,000) that the tide runs upstream for five hundred miles, and wide strips on either bank are inundated when the

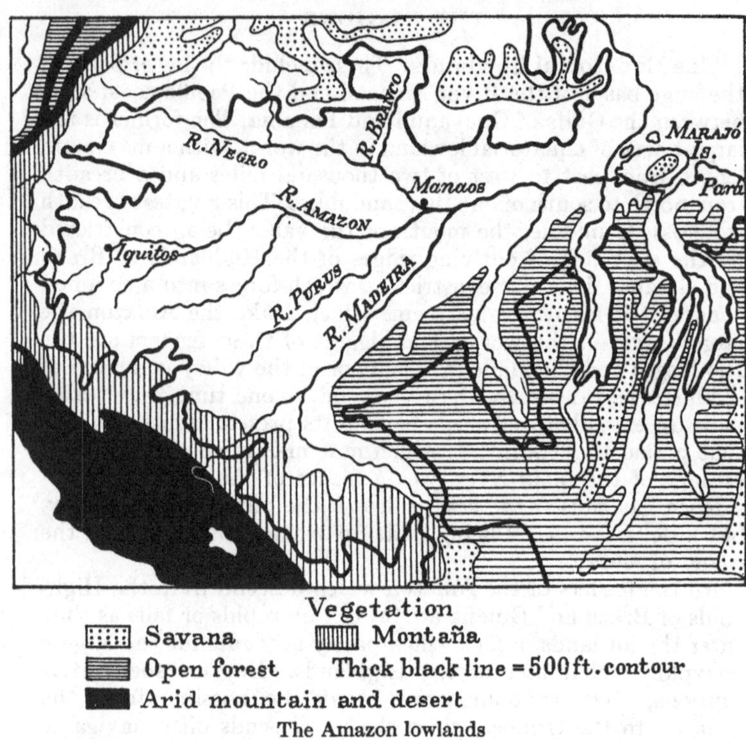

The Amazon lowlands

river is in flood. Hence, Manaos and other settlements are placed where ridges of higher ground approach the banks and afford a dry situation. The volume of the river is kept relatively constant by (1) the well-distributed rainfall, (2) the melt-water from the Andes in spring and autumn, and (3) the maximum tropical rainfall in January and July in the Highlands of Brazil and Guiana respectively; but none the less there is a difference of

40 feet between the highest and lowest conditions of the river at Manaos.

The climate is marked by the usual equatorial features of a steady thermometer, a rainfall fairly evenly distributed among the seasons, and a high degree of cloudiness. There are few stations in the region at which records have been kept, but these confirm the impressions of travellers. At Manaos the coolest month averages 80° F. and the hottest 83° F. Pará (Belem) is slightly cooler, but has the same small range. The nights are scarcely cooler than the days, since the blanket of clouds, which prevents the excessive midday heat of the more clear-skied Tropics, checks radiation after sunset; and the thermometer seldom falls below 70° F. The constant high temperature causes a heavy rainfall mainly due to convection. Manaos has a mean of 66 inches and Pará 86. Slight maxima occur at about the time when the midday sun is overhead. Rain falls nearly every day in the early afternoon and is often accompanied by thunder-storms. So great is the humidity of the air that the slight reduction of temperature which occurs at night causes an extremely heavy dew.

The vegetation is of three types. On the swamps and riparian areas which are often flooded stands a forest of trees adapted to growing in water, while the banks are fringed with mangrove wherever they are reached by the tide. Above the level of the floods is the true equatorial forest, where the vegetation becomes almost a solid, tangled mass. This type varies from place to place with local conditions of temperature, rainfall, and soil. It contains timber woods, of which the Spanish cedar is the best known, and the best of the various species of rubber-yielding trees, notably the *hevea*. Thirdly, on the lower slopes of the Andes there is a peculiar open type of equatorial forest whose rank luxuriance is reduced by the fall in temperature due to altitude. The trees in general are of the same species as on the lowlands, but some peculiar ones occur, among which are the cinchona tree and the coca shrub, both well known for the drugs they yield. In modern times the 'balsa' wood, of extreme lightness and durability, has become important in the building of aircraft. In some areas the forest thins out or even disappears, owing to local conditions which reduce the rainfall. The largest of these districts are the savanas of the island of Marajó and a more extensive tract around the headwaters of the Rio Branco.

The exploitation of the region has not gone very far at the moment, and European settlement is confined to a relatively few backward posts along the Amazon and its feeders. Iquitos (pop. 10,000), Manaos (pop. 83,000 in 1928), and Pará (or Belem, pop. 275,000 in 1928) are by far the largest towns and owe their size to their character of collecting centres for the produce of the lower, middle, and upper parts of the river basin. The state of Pará, which contains rather more than a third of the million inhabitants of the region, alone has any considerable rural settlement. Production is mostly limited to the collection of the yield of the forest, though cattle-raising has become of some importance on the savanas of Marajó and the Rio Branco.

The development of the region is hindered chiefly by bad communications, swamp belts along the Amazon, lack of native labour, and a climate unfavourable to Europeans. Although the rivers form excellent main highways, the swamps and rank forest growth prevent the opening of a network of minor routes necessary to systematic exploitation; and the river system checks entrance to the region except through the mouth of the Amazon. The river banks, which would normally be the early footholds of settlement, are uninhabitable except where ridges of higher ground reach the water, and miles of forested swamp must usually be penetrated to reach *terra firma.* Added to this difficulty is the fact that the well-drained areas are most densely covered with luxuriant vegetation. These drawbacks have caused the region to be thinly populated even by natives, whose habit of life as well as their fewness in numbers makes them unsuitable as a source of supply of labour. In fact, the full development of the region seems impossible without the help of imported coloured workmen. Besides all this, the climate tends to sap the energy of the European and to dull his brain.

Rubber has been the chief product of the region in the past, but the quantity exported has declined of late owing to the better organised plantations of Malaya. In 1932 the amount exported was 17,294 metric tons, valued at £475,000. More important now is the brazil nut, the total quantity collected in 1932 being 23,000 metric tons, valued at £525,000. Timber, cinchona, and coca are also produced, though in small quantities. The rising cattle industry is not yet a serious competitor in world markets, exports taking mainly the form of hides. In the state of Pará there is a certain amount of cultivation of cocoa, cassava, sugarcane, and tropical fruits for local needs, and cotton growing is being tried in the Ucayali valley in the western part of the basin. This western area has its outlet at Iquitos, which is in Peruvian territory and is connected with Lima by an air service maintained by the Peruvian navy.

Most of the Amazon valley is within the borders of Brazil. The central portion forms the state of Amazonas, whose capital is at Manaos; and most of the state of Pará and portions of Maranhão and Matto Grosso are included in the region. Outside Brazilian territory are relatively large areas on the west belonging to Bolivia, Peru, and Colombia, and smaller areas attached to Ecuador and Venezuela.

The Pacific coast of Ecuador and Colombia is a very narrow region consisting of a lowland belt and a higher strip with a general altitude of 2000 feet. The former is true equatorial forest land, but the latter is savana. The climate is peculiar, being largely influenced by local conditions of wind and ocean currents. The following table shows the temperature at Ancón on the Santa Elena Peninsula, north of Guayaquil:

Ancón (1932)	Jan. 79·8	Feb. 80·6	March 81·3	April 81·3	May 79·1	June 76·1	
	July 73·9	Aug. 70·9	Sept. 71·6	Oct. 71·1	Nov. 72·4	Dec. 75·2	Range 10·4° F.

North of Ancón the range increases, and Buenaventura has an altogether equatorial régime. During the season from May to January the prevailing winds are from the south and southwest and are an extension of the Southeast Trades. This is the dry season. But between July and September the land is shrouded in dense wet mist, known locally as the *garua*, caused by the meeting of the sea breezes with air currents over the land. In the wet season (which, though the warm period of the year, is called *el invierno* or 'winter') an equatorial counter-current known as *El Niño* flows south from the Gulf of Panama. Winds from this warm current give abundant rains all along the coast. In the south, however, the rainfall is fitful, the amount received at Ancón varying from 40 inches in 1925 to 3 inches in 1928.

The soil of the region is fertile and produces such tropical products as rice, sugar, cocoa, bananas, and oranges. The cultivation of 'balsa' wood is increasing. Some years ago there was a large export of cocoa, but this has diminished partly owing to competition with producers in the West Indies and the Gold Coast and partly owing to the depredations of the 'witch-broom' disease. It still furnishes 30 per cent. of the exports. The first place is now taken by mineral oil which is worked mainly on the Anglo-Ecuatorian Co.'s fields in the Santa Elena Peninsula. Other agricultural products besides cocoa are coffee and rice. An important export is the panama hat made from local *toquilla* straw.

The Equatorial Highlands. Between the two areas of equatorial lowlands there is a wide belt of highland in which altitude modifies the climate. It consists of the ranges of the Andes northward from the Loja 'knot' and the upper valleys of the Magdalena and Cauca. Except for a group of lofty volcanic cones, of which Chimborazo (20,702 feet) and Cotopaxi (19,498 feet) are the highest, the Andes become gradually lower towards the north, scarcely reaching 2000 feet in the Coast Range of Venezuela. The Eastern Cordillera broadens out into a series of tablelands.

The climate is marked by the same even temperature throughout the year as has been described in the lowlands, but the thermometer readings are lower. Thus:

Quito (9350 ft.)	Jan. 54·5	Feb. 55·0	March 54·5	April 54·5	May 54·7	June 55·0	
	July 54·9	Aug. 54·9	Sept. 55·0	Oct. 54·7	Nov. 54·3	Dec. 54·7	Range 0·7° F.
Bogotá (8730 ft.)	Jan. 57·6	Feb. 57·9	March 58·6	April 58·6	May 58·5	June 58·1	
	July 57·2	Aug. 57·0	Sept. 57·0	Oct. 57·4	Nov. 58·3	Dec. 58·1	Range 1·6° F.
Caracas (3020 ft.)	Jan. 68·5	Feb. 68·9	March 69·3	April 72·5	May 73·9	June 73·0	
	July 72·0	Aug. 72·7	Sept. 72·5	Oct. 71·4	Nov. 71·2	Dec. 68·9	Range 5·4° F.

Owing to the rareness of the air radiation is active, and there is often a diurnal range of 20° F. Morning frosts are common. The wind system is complicated. Between Loja and Pasto no regularity can be noticed, and the abundant rainfall (44 inches at Quito) is due to convection, thunderstorms with heavy showers being the rule in the afternoons. On the western slopes of the Andes rain is mostly due to the condensation of moisture in the southwest winds which prevail from May to January. The rest of the region is in the régime of the Northeast Trades which penetrate up the valleys, giving a rainfall which varies with the altitude. Thus, Caracas has 40 inches, Bogotá 64.

Local usage divides the northeastern portion of South America into three parts according to the temperature. The *tierra caliente* (='hot land') is outside our region, which however includes the *tierra templada* (='temperate land') and the *tierra fría* (='cold land'). This division depends on altitude, the limits being roughly 3000 and 6000 feet. In the *tierra fría* the weather is subject to fierce storms of rain and wind often accompanied by slaty hail. Otherwise, it is a pleasant country, though its inhabitants are liable to heart and throat trouble and the other illnesses connected with life in a rarefied air.

The *tierra templada* is the main agricultural area, coffee being the chief crop. Colombia almost has the monopoly of the market for mild coffee in the United States. Cotton is also an important crop. Potatoes (which are indigenous to the region) and maize are grown for local consumption. The broad tablelands of the *tierra fría* are used for cattle-rearing, and hides are exported. But the chief wealth of the region lies in its minerals, of which gold is the chief and is found nearly everywhere. The emerald mines at Muyo and Chivor in Colombia are the chief source of the world's supply of that precious stone, the working of the mines and the output being strictly controlled by the Colombian Government. Mineral oil is found in large quantities, especially in Venezuela, and the potential supply in Colombia has hardly been touched as yet. Other minerals are copper, lead, mercury, manganese, platinum, coal, and salt.

In such a mountainous region communications are difficult. The main highway is the River Magdalena which is navigable for nearly 600 miles, but is impeded by sandbanks and liable to considerable variation in volume. River steamers and

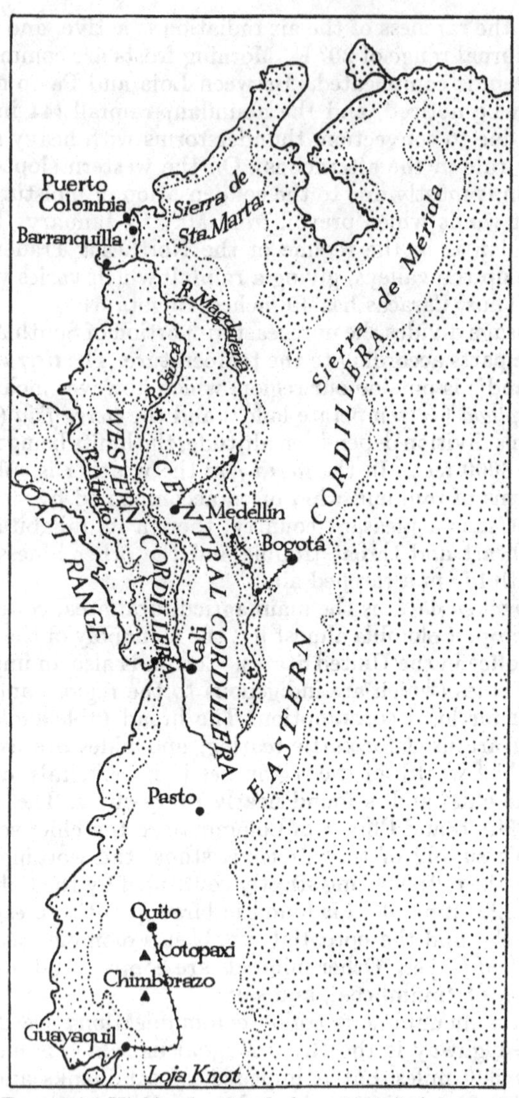

The Equatorial Highlands. Land above 3000 ft. left unstippled

balsas (native rafts) ply up and down stream to Barranquilla and Puerto Colombia at the mouth of the river. Railways, usually short and disconnected, run from important inland centres to the coastal towns or to convenient points on the Magdalena. One of these lines connects Quito with Guayaquil and another joins Cali with Buenaventura. Communication between Bogotá and Puerto Colombia is partly by rail and partly by the Magdalena, the journey taking a week. An air service which has been working for some years accomplishes the journey in a day, connecting with the airmail from Barranquilla to New York (3½ days). An overland telegraph line runs from Buenaventura through Cali to Bogotá, and there are wireless stations at all the larger towns.

As a result of its moderate temperature, this region was among the first settled by the Spanish conquerors in the sixteenth century, and, as the Vice-Royalty of New Granada, the country had its capital at Santa Fe de Bogotá, now known simply as Bogotá. Many churches, monasteries, and public buildings still preserve the peculiar 'colonial' style of architecture of the first century of Spanish settlement. Bogotá, which is situated on a plain said to be the bottom of a former lake, is the largest town in the region and contained 235,000 inhabitants in 1928. Other places of more than 100,000 inhabitants are: Cali (123,000 in 1928), Medellín (120,000 in 1928), and Quito (104,000 in 1932), the capital of Ecuador.

The Eastern Highlands. The Guiana Highlands consist of a mass of old crystalline rocks overlaid by thick horizontal strata of sandstone. The latter have been worn away, except in certain places where they stand up as flat-topped hill-masses, presenting steep or even precipitous faces. The loftiest and biggest of these sandstone elevations is the Roraima *massif*. Generally speaking, the slope of the Highlands is downwards towards the north to within some 25 miles of the coast, where the sudden drop to the lowlands is marked by rapids and falls in the rivers. On the south there is a high, steep escarpment which has been breached in the middle by the Rio Branco. The surface of the Highlands is extremely broken, since the rivers have cut deeply into the rock. The mean height would seem to be about 5000 feet.

Not a great deal is known about the region, which can hardly be said to have been thoroughly explored. So far as is known, its climate is tropical, modified by altitude. No records of temperature have been kept, but there is a cool season in February, when the air becomes distinctly cold at night on the mountains. The Northeast Trade Winds prevail from September to March, but during the rest of the year the region is in the Doldrums belt of calms. At this time there is convection rain, while during the period of the Trades relief rains occur throughout the region. Hence, there is a heavy rainfall which has enabled the rivers to score the surface deeply and has covered the region with dense tropical forest.

Owing to difficulties of penetration and to the unfavourable climate, European influence has hardly touched the region, in spite of the fact that early reports of gold attracted prospectors. It was here that Raleigh hoped to find the Golden City of Manoa. There is gold along the northern edge of the Highlands, but hardly enough to be worked with profit. The real wealth of the region is to be found in its timber. Over a hundred species of trees have been discovered which yield useful wood. The chief are the greenheart, mora, and wallaba, all of which are hardwoods of great strength and durability.

The Brazilian Highlands form a vast triangular mass bounded on the north by the Amazon lowlands, on the west by the valleys of the Madeira and Paraná, and on the east by a narrow strip of coastal lowlands which separates them from the South Atlantic. They rise to a height of about 3000 feet in the southeast by a steep, clearly marked escarpment, whence, broadly speaking, they slope gradually away to the north. Their surface is, however, broken by a zigzag ridge which begins at Cape S. Roque and ends in the foot-hills of the Andes. This is the main watershed of the region, and all the drainage runs off from it either north or south. Parallel with it and farther east is another line of high ground whose central portion is known as the Serra do Espinhaço and which has been cut up by the erosion of the coastal streams into what look like a number of ranges of hills with a northeasterly direction. The largest of these is the Serra do Mar which overlooks Rio de Janeiro and has been cut off from the main tableland by the Parahyba do Sul. Between the two great ridges is a huge trough eroded by the San Fernando and the Paraná. The former, which is a very old river, once entered the sea at Aracaty along the course of the present Juguaryba, but has been captured by one of the streams of the southeast coast. This development is marked by a great elbow and by the Falls of Paulo Alfonso. As a general rule, the rivers of the Highlands are broken by falls, cascades, or rapids, the most notable case being the Falls of Iguassú which lie outside the region near the confluence of the River Iguassú with the Paraná.

The climate is everywhere tropical, with a temperature which rises in summer above that of the equatorial region, but falls below it in the cool season; with a rainfall which shows a distinct summer maximum and a long dry season; and with a vegetation which varies from dense open forest in some places to savana and scrub in others. The temperature régime will be seen from the following table:

São Paulo	Jan. 68·9	Feb. 69·1	March 68·0	April 64·6	May 60·4	June 58·6	
	July 57·9	Aug. 59·0	Sept. 61·5	Oct. 63·0	Nov. 65·5	Dec. 68·0	Range 11·2° F.
Cuyabá	Jan. 79·9	Feb. 80·2	March 80·2	April 79·0	May 77·2	June 73·9	
	July 75·0	Aug. 76·8	Sept. 80·8	Oct. 80·8	Nov. 80·6	Dec. 80·1	Range 6·9° F.

Differences naturally occur over so vast a region, the range increasing towards the south, and the mean temperature decreasing with altitude. In the interior of the tableland the mean temperatures are greatest, and the lowest are found in the São Paulo district where frosts occur not infrequently. In July the Southeast Trades are well developed and blow across the two great ridges, giving abundant rain to the high ground. But the valley of the San Fernando and the northeastern area lie in a rain shadow. In January the Trades weaken and calms or variable winds occur, though along the north and northwest of the tableland there is an indraught of warm air from the equatorial region which gives some rain. In the southern corner of the Highlands the rainfall is fairly evenly distributed throughout the year.

The vegetation on the whole is savana, but numerous local

modifications occur. In the north the forests of the Amazon lowlands invade the lower parts of the Highlands and penetrate long distances up the river valleys. Generally speaking, all the river valleys are wooded, with the exception of that of the San Fernando, which harbours a scrubby vegetation owing to the low rainfall. Towards the southeast coast, where relief rains continuously moisten the hills, the ridges are crowned with open forest.

So large a region cannot easily be described adequately as a whole, since the variations often become too great for generalisation. It will be as well, therefore, to divide it into four subregions, whose divisions are marked on the map on page 367.

Subregion 1. The Western Tableland. This subregion consists of horizontal strata and is flat, except where broad river valleys make undulations. It is crossed by many streams flowing into the Amazon, the chief of which are the Tapajós, the Xingú, and the Tocantins. The extreme west drains partly into the Madeira, and partly into the Paraguay or Paraná. The area is not well known, but Cuyabá may be taken as representing its climatic type. The mean monthly temperatures for this place have been given above and show its tropical régime. The rainfall conforms, there being a wet season lasting from October to April and a dry season during the rest of the year. The rain-bearing winds come from the equatorial region to the north and are very hot. The vegetation consists of gallery forest along the valleys, and savana on the higher ground. The wooded district at the head of the Paraguay is the only place where the ipecacuanha tree grows, and this is therefore the source of the world's supply of that drug.

Apart from the scattered tribes of Indians who still dwell in an uncivilised state in this subregion, population clings to the high ground formed by the great ridge, and it is along this line that the only towns occur. The only industry is cattle-raising, but it is not carried on in a scientific way and is far from flourishing. A certain number of hides are sent down the Paraná for export. The subregion comprises the Brazilian states of Goyaz and Matto Grosso, each with a capital of the same name as its state. Matto Grosso (='great forest') was so named because of the forests at the head of the Paraguay.

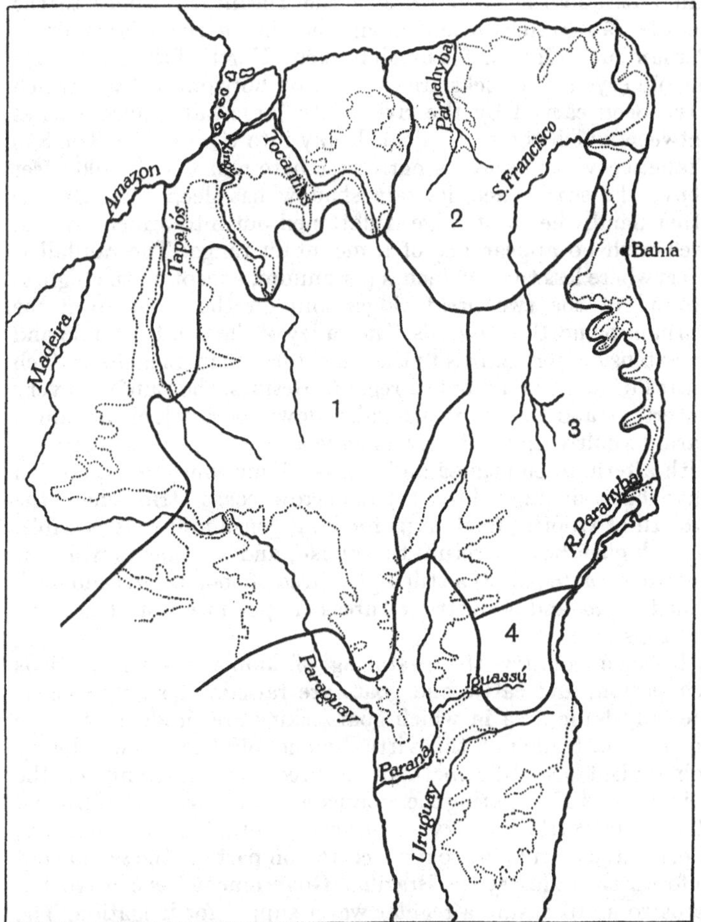

Main natural regions of the Tableland of Brazil

1. The Western Tableland
2. The Dry East
3. The Coffee and Mineral Area
4. The South

Subregion 2. The Dry East. This includes the area between São Luis and the great bend of the Tocantins, thence to the middle São Francisco and along the rim of the Highlands to Pernambuco. Its relief is much broken. North of the great ridge the old crystalline rocks are overlaid by horizontal strata which have been carved by the rivers into large flat-topped masses between which the valleys cut deeply into the surface. The São Francisco valley is at its narrowest here and is only 1000 feet above the sea; hence, its rain shadow has deepened. On the other hand, the coast ridge is flattened out into a mere upland area with no appearance of a mountain range. The rainfall is everywhere less than 40 inches per annum, except on the highest ground of the two great ridges and on the coast from the Parnahyba northwestwards. The valley of the San Fernando and an extension northwards to the sea get less than 25 inches, which means drought in a tropical region. Besides, the rainfall is very uncertain and occurs in irregular downpours which do much harm to cultivation. The rivers have a typically tropical régime, with alternate emptiness and floods. Four zones of vegetation have been distinguished: (1) a narrow coast strip where the growth of tropical food crops for local consumption is possible, (2) a higher belt of scrub, thornbush, and succulents which is known as *caatinga*, (3) a still higher area of poor savana and bush called *sertão*, and (4) restricted areas of open forest on the rainier hill tops.

In such country the pasturing of animals is the obvious occupation, and cattle and goats are raised. This subregion is the only large area in which goats' skins are produced for the purposes of trade and this virtual monopoly keeps the industry going. But scientific methods of breeding, of caring for the animals, and of preparing the produce are unknown to the farmers. Hence, the cattle industry is not very profitable here. Recently, efforts have been made to grow cotton on parts of the *sertão*, and to foster the industry the Brazilian Government has constructed dams so as to ensure a regular water supply for irrigation. The largest of these is the Pedro Reservoir at Quixada.

The chief towns are on the coast and are Pernambuco and São Luis. The former owes much of its importance to its use as a port of call by ships on the route to Rio de Janeiro, Montevideo, and Buenos Aires. Port Natal is to be the South American terminus of the new French trans-Atlantic air service from Dakar on the Senegal.

Subregion 3. The Coffee and Mineral Area. This is the heart of the state of Brazil. It extends from the middle São Francisco valley to just south of São Paulo, and from the rim of the Highlands on the east to the top of the great ridge on the west. It is thus very largely an area of tableland averaging 3000 feet in height and forming the divide between the San Fernando and Paraná. In it the rim of the tableland rises to its

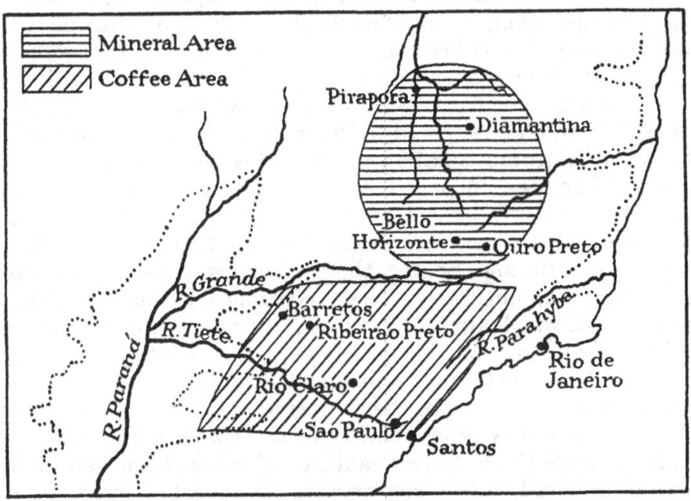

The coffee and mineral districts of Brazil

highest, the culminating point being over 8000 feet above the sea. In the south the feeders of the Paraná give the semblance of an east-to-west grain, but on the east the Parahyba, the Doce, and other coastal streams which flow towards the southwest have sawn through the plateau-edge and given a different direction to the grain along the rim of the tableland.

The temperature is subtropical in degree, though tropical in régime. Frosts often occur in valley bottoms owing to inversion of temperature, and snow has fallen on the highest ground at rare intervals. The Southeast Trades blow steadily from off the sea, but, though they give heavy rains to the eastern slopes of the plateau-edge, they are descending winds on the tableland and yield only a moderate rainfall. Thus, São Paulo has an annual mean of 56 inches. On the higher ground, however, Ouro Preto gets 80 inches. The vegetation is savana, but the tops of the hills are forested as well as the valley bottoms. In the Paraná valley the chief tree is the Paraná pine which yields a valuable hardwood timber. The drier areas of the subregion contain a degenerate savana vegetation in which patches of scrub occur.

Practically the whole of Brazil's coffee exports are derived from this subregion, which thus yields nearly two-thirds of the world's supply. The coffee plant, which will not endure extremes of cold, heat, drought, or wet, seems to find its ideal conditions here at an altitude of between 2000 and 6000 feet. Much below 2000 feet it would be exposed to the frosts in the valley bottoms. The strong iron impregnation of the soil is also held to benefit the crop. The map on page 369 shows the extent of the coffee area. The chief commercial centre is São Paulo, but there are collecting centres at Ribeirao Preto and Rio Claro. Most of the export passes through Santos, though some is shipped from Rio de Janeiro. The output is strictly controlled by the Brazilian Government, which aims at regulating the prices of coffee in the world market. In 1932, an average year, the quantity exported amounted to 770,000 tons, valued at £2¼ million.

Other economic crops in the subregion are cotton, sugar, and tobacco. The last is grown chiefly in the north in the state of Bahía. Besides these, large quantities of maize, beans, and hill rice are produced to feed the workers on the coffee plantations and in the mining areas.

Stock-raising is also an important occupation, though the farmers lag behind the rest of the world in their methods. The number of cattle exceeds eight million, and the export of preserved meat—dried, salted, frozen, and chilled—is second in value to coffee alone. São Paulo is the chief centre, and shipments are made through Santos.

The district around the headwaters of the San Fernando and Doce (see map on page 369) is one of the richest mineral fields in the world, owing to intrusions of igneous rock. Gold is worked on a large scale at Ouro Preto and Bello Horizonte, the production in 1930 amounting to 96,750 oz. The whole of the mineral area is one huge iron field, supposed to be the largest reserve of the metal in the world. It is as yet hardly touched, but efforts are being made by foreign capitalists to exploit mines at Itabira. At present the chief mineral export is manganese, of which 95,550 metric tons were shipped in 1931, mainly to the United States. Diamonds are widely found over the district, the mining being centred around Diamantina. Besides all this, there are great quantities of coal, the vast reserves of which, estimated at 5000 million tons, have scarcely been touched.

The wealth of the district has given rise to manufactures in which local raw products are used, The power employed to drive the mills is hydro-electricity derived from the Falls of the Tiete and Parahyba. Cotton weaving is the most important of these industries, the number of spindles being two millions and the number of looms 60,000. There are also a number of leather works, jute mills, and tobacco refineries. So far these industries have not been able to do more than supply a part of the needs of Brazil, but they are gradually increasing in number and efficiency.

The network of railways is closer in this subregion than in any other part of South America outside the Argentine, but it is not sufficient for the needs of the area. All the agricultural and mining centres are connected up, and the lines aim at three ports: Santos, Rio de Janeiro, and Victoria. The Central Railway is the principal system, running from Diamantina to São Paulo, and finally joining up with the railways of Uruguay, the Argentine, and Paraguay. Its mileage is altogether 2082.

Subregion 4. The South. This is the smallest of the subregions and consists of a low extension of the Highlands, averaging 1000 feet above the sea. The plateau-edge is very near the coast, and the drainage is westwards to the Paraná. The climate is cooler than farther north, but the distinguishing characteristic is the fairly even distribution of rainfall throughout the year. The upper parts of the high ground are savana, but the slopes on both sides are covered with tropical forests whose trees are mostly of hardwood species. The only industry is stock-rearing, but this is not important, since it is overshadowed by its rivals to the north and south. The subregion includes most of the Brazilian states of Paraná and Santa Catharina.

Tropical Lands of the North and East Coasts. These consist of two separate areas: the southeast coast strip of Brazil and the lowlands of the north coast, including the *llanos* of the Orinoco. The coast strip of Brazil comprises the belt between the plateau-edge of the Highlands and the South Atlantic from the neighbourhood of Port Natal to just south of Santos. In width it varies from 60 miles in the north to a mere nothing in the south, and where river valleys cut deeply into the Highlands, as at Bahía, the breadth is increased to as much as 200 miles. South of Rio de Janeiro it becomes negligible. It is crossed by numerous short rivers from the Highlands, whose mouths being

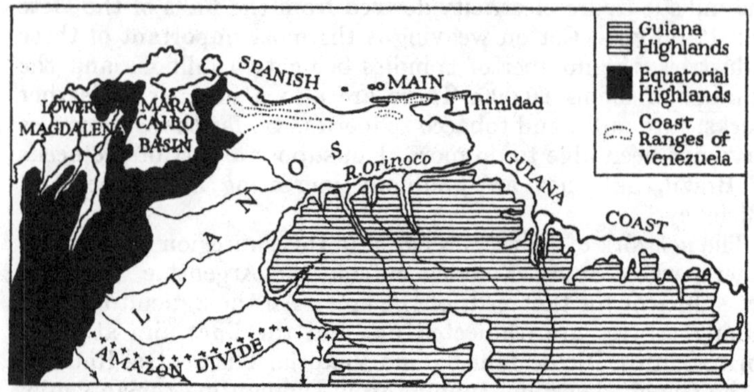

Physical divisions of northern South America

drowned form good harbours. That of Rio de Janeiro is said to be the finest in the world (see photograph facing page 416).

The climate is typically tropical with a marked cool season. Towards the south the range increases and the annual mean decreases by 8° F. Thus, Rio de Janeiro ranges from 78° F. in February to 67·5° F. in July, the yearly mean being 72·5° F. The Southeast Trades blow constantly throughout the year, and the abrupt rise to the Highlands causes a large rainfall. Santos has an annual mean of 82 inches in all, with a range of from 12 inches in February to 2·4 in June. Thus, there is a distinct summer maximum.

Except in the south, where the narrowness of the strip makes it negligible, the damp heat of the climate is not favourable to Europeans. Hence, only the seaports are of any importance.

These are Bahía (or San Salvador, pop. 330,000 in 1930), Rio de Janeiro (pop. 1,470,000 in 1930), the second largest city in South America, and Santos. The last is merely a port, the residential parts being in São Paulo, which is reached by an electric railway. The relation between these two towns emphasises the fact that the ports on this coast strip are dependent on the backland on the Highlands. The population, mostly of negroes and half-breeds, is agricultural, producing such tropical crops as rice, cane-sugar, cocoa, maize, and beans. The produce is consumed locally.

The north coast is equally tropical in climate. The high ground near the sea is forested, but areas protected either by distance from the ocean or by hills lying athwart the Northeast Trades are savana. The region is divided, mainly by relief, into five subregions: the Orinoco lowlands, the Guiana Coast, the Spanish Main, the Maracaibo basin, and the lower Magdalena valley. The Orinoco lowlands are a wide alluvial plain enclosed between the Guiana Highlands and the Andes. They have a constricted outlet to the Atlantic in the northeast and are separated from the Amazon valley by a low, broad saddle in the southwest. The main line of drainage is the Orinoco River which hugs the foot of the Guiana Highlands. Consequently, the plain is crossed from west to east by numerous tributaries from the Andes. The main river empties through a delta whose channels are too shallow to admit ocean-going vessels. Although the Orinoco itself is navigable for some 600 miles as far as the rapids of Maipure, its feeders on the plain are almost useless owing to their tropical régime which causes them to have very little water in the dry season. Nevertheless, communication is mainly by river in this roadless country. The eastern portion of the plain is grass-covered, but towards the west the vegetation degenerates into bush and scrub. The name locally given to this area is *llanos* (Spanish='plains'; pronounced *lyanos*). Towards the south forest begins, and the region shades into the Amazon basin. Little use has been made of the plains, though cattle and horses of an inferior type are reared. The people are mostly half-breeds and live on horseback. The chief town is Ciudad Bolívar, a little river port of 16,000 inhabitants.

The Guiana Coast is a lowland shelf supposed to have been formed of sand and clays brought down by the Amazon and deposited on the shore by ocean currents. Its width increases from east to west and has an average of some 20 miles. Rivers from the Guiana Highlands cross it, and form deltas on its outer edge. The full heat of the tropical climate is lessened by the Northeast Trades, but, with the rainy season which lasts from October to April, Europeans find life difficult. Hence, settlements are limited to the sea coast, where sugarcane is cultivated. The delta of the Demerara River has given its name to a type of sugar with large, clear crystals. Farther inland there is dense tropical forest which provides some timber for export. Communications are mainly by river, the Essequibo and Demerara being navigable for small boats for about 60 miles. The British,

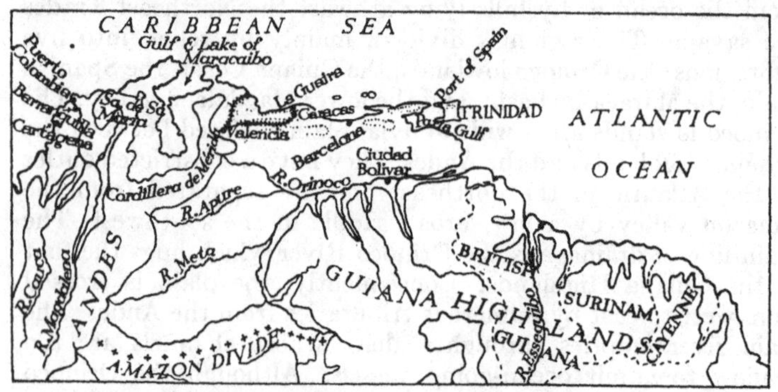

Northern South America. Names mentioned in the text. Broken lines show the political divisions of Guiana

Dutch, and French each have a share of the area, the portions belonging to the latter, Surinam and Cayenne respectively, being of little importance. The chief settlement in British Guiana is Georgetown on the Demerara River.

The Spanish Main consists of a strip of the north coast from the Cordillera de Mérida to the island of Trinidad. Averaging about 80 miles in width, it consists of two parallel ridges of the Coast Range of Venezuela, with an intervening valley. The range has been breached by the sea at Barcelona and again in the Gulf of Paria. The remains of a third ridge form the islands of the Spanish Main. The coast is mangrove-fringed and tropical forest climbs the slopes beyond; but on the tops of the two ridges and in the intervening valley there is forest only in the eastern half, the parts farther west being grass and bush land. This subregion is the heart of Venezuelan life and the wealthy British colony of Trinidad. The forests yield mahogany, cedar, lignum vitæ, and chicle (the basis of chewing gum), while there are large plantations of cocoa and sugarcane. Valuable deposits of mineral oil are found at the foot of the hills, and the well-known 'pitch lake' of Trinidad yields quantities of asphalt and petroleum. A railway runs along the intermontane valley from Valencia to Caracas, with three branches to ports on the coast. One of these is an electric railway which connects Caracas (pop. 135,000 in 1926), the capital of Venezuela, with its little port at La Guaira. The chief town in Trinidad is Port-of-Spain (pop. 70,000 in 1930).

The Maracaibo basin is an area some 300 by 200 miles enclosed between the two northern arms of the Andes. Its centre is occupied by a gulf whose narrow mouth is obstructed by a sand-bar which admits ships drawing less than 20 feet. The subregion is covered with tropical forest from which are obtained mahogany, cedar, and other hardwoods. The most important product is mineral oil, large deposits of which have been discovered around the Gulf. Foreign enterprise extracts the oil and ships it to the United States or Great Britain.

The lower Magdalena valley is a swampy area receiving the waters of the Magdalena and Cauca. The northern part is in the rain shadow of the Sierra de Santa Marta and is covered with grass or swamp plants, but the rest of the subregion is clad in tropical forest. Hence, there are forest products of the same kinds as in the Maracaibo basin. In addition, there are increasing plantations of bananas. But any importance the area has is mainly due to its containing the outlets for the produce of the equatorial highland region. Barranquilla, though not on the sea, receives the products of the backland and sends them on either by rail to Cartagena or by canal to Puerto Colombia.

The West Indies. The north coast of South America is fringed with a chain of islands, known as the West Indies, which enclose the Caribbean Sea. It was on one of these islands, Guanahani or San Salvador in the Bahamas, that Columbus first landed after his historic voyage across the Atlantic in 1492. As the new discoveries were thought to be a part of the southeast coast of Asia, they were named the 'Indies' and for distinction from the lands reached by the eastern route they were called the 'West Indies'.

The islands fall into three distinct parts: the Greater and the Lesser Antilles and the Bahamas. The last group consists of a swarm of coral islands built on a base of submerged highland. They are low, attenuated, and of no great economic importance owing to poor soil and comparative lack of rain. The Greater Antilles are the unsubmerged portions of a mountain system which begins in south Mexico and forms the backbone of Central

Trend lines of the Antillean ranges

America. These Antillean ranges had an east-to-west trend and were not connected with the Pacific Mountains of North America or the Andes of South America. The map above shows the main surviving lines of fold. It will be noticed that the coastal outlines indicate an east-to-west grain.

The Lesser Antilles consist of the tops of a single line of fold mountains and are a continuation of either the Antillean or the Andine Ranges. Outside the line and to the east is the island of Barbados, which is composed partly of volcanic rock and partly of deep-sea rock formations. The volcanic activity connected with the subsidence of this eastern chain is still marked by active and dormant volcanoes, fumaroles, sulphur springs, and earthquakes. In May, 1902, the Souffrière of St Vincent and Mont Pelée in Martinique erupted after a long dormant period, causing much damage. The town of Saint Pierre, which lay at the foot of Mont Pelée, was overwhelmed by a stream of incandescent mud, and its 40,000 inhabitants killed.

The North Equatorial Current of the Atlantic, after running along the coast of South America for some distance, is partially turned by the West Indies and flows northwestwards outside the

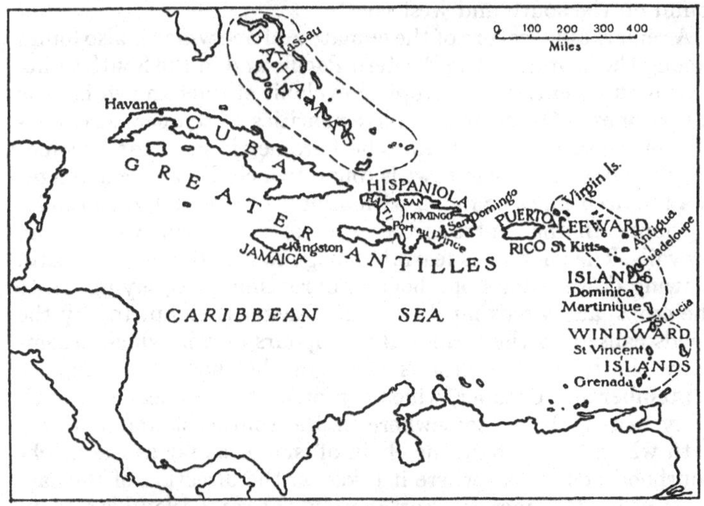

Islands and groups of the West Indies

group. But portions sweep between the southern islands of the Lesser Antilles into the Caribbean, whence they pass into the Gulf of Mexico to issue from Florida Strait as the famous Gulf Stream. North of the Bahamas this current is joined by the main body of the North Equatorial Current.

The climate may be described as maritime tropical, since it has the features of a tropical régime moderated by sea influence. The mean temperature decreases towards the northwest, where the northernmost of the Bahamas is outside the Tropics; but even in the more southerly islands the mean for the hottest month seldom exceeds 96° F. There is a markedly cooler season in December–February. The Northeast Trades blow steadily throughout the year, but are stronger in the cool season than at other times. Since even the smaller islands rise to two, three, or even four thousand feet above the sea, the Trades give relief rain at all times of the year, but in the hot season the added force of convection increases the rainfall, causing a maximum at that period. The leeward sides of the islands are in the rain shadow of the hills and are therefore less wet. Owing to this and to the calmness of the sea under the lee of the land, the chief towns are found on the south and west coasts.

A remarkable feature of the climate and one which is also found among the islands of the Western Pacific and in the South China Seas is the occurrence of tropical cyclonic storms known here as 'hurricanes'. Owing to local irregularities of temperature, convection currents are set up. The barometric gradient becomes very steep, and the wind swirls round in a circle, whose diameter is a hundred miles or more, with a force of from 80 to 120 miles an hour. In the centre is the 'eye' of the storm, where calm prevails. The wind does fearful damage to the islands in its path, blowing down even stone houses, uprooting trees, laying waste the crops, and wrecking ships. The disaster is completed by the floods caused by the torrential downpours of rain which accompany the wind. The storms occur in the months of August, September and October, which comprise 'the hurricane season'. They usually begin somewhere in the southeast and follow a path which curves with the chain of islands as far as the neighbourhood of Florida, where it takes on the direction of the east coast of North America. Fortunately, the same island is seldom affected oftener than once in twenty or thirty years.

The islands are clad in forests whose fresh green appearance largely contributes to the beauty for which the West Indies are famous. The trees yield many excellent kinds of hardwood, the best known being mahogany and cedar. The logwood, which is cultivated especially in Hispaniola, gives a fast black dye. Many of the trees bear eatable fruits: thus, the avocado pear, the sapodilla, and the guava. The breadfruit tree, which was brought from the Pacific, and the mango tree from India also grow well. These fruits, together with the banana, figure largely in the diet of the poorer folk. A large number of beautiful shrubs, tree ferns, feathery bamboos, and flowering plants and creepers grow wild.

In some of the islands dangerous snakes, tarantula spiders, centipedes, and scorpions occur, though rarely. Wild animals, which were never very numerous, have been killed off, a few raccoons and opossums and swarms of the imported mongoose being the only survivors. There are many beautiful birds.

The Spaniards found the Bahamas and the Greater Antilles inhabited by the gentle race of Arawak Indians, who were soon killed off by the harsh treatment of the invaders. The Lesser Antilles were peopled by the fiercer Caribs, who kept out the

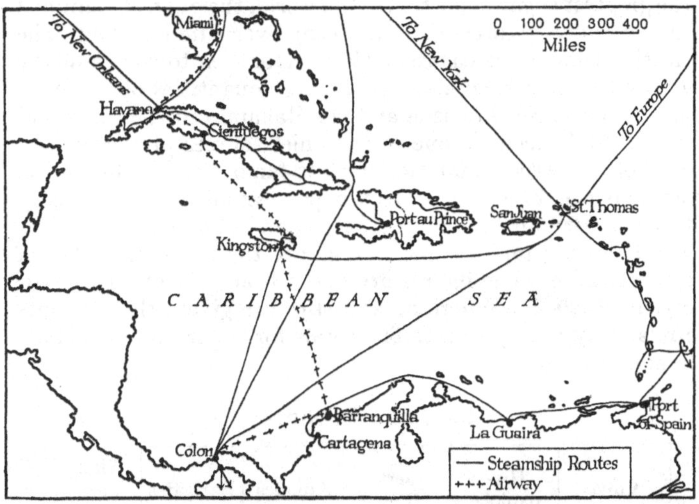

Communications in the West Indies

Spaniards, but were subdued over a century later by the English and French settlers. A few score survivors remain in the islands of St Vincent and Dominica. The Caribs being useless as labourers, negro slaves were imported from Africa, and the descendants of these people to-day form the overwhelming majority of the population. During the wars between England and France in the eighteenth century, the former captured many islands belonging to the latter. Hence, a French patois is still spoken by the negroes in St Lucia, Grenada, and other islands. To-day the French hold only Martinique, Guadeloupe, and a few less important islets. It is an interesting fact that after the conquest of Canada in 1763, the English Government proposed to exchange Canada for Guadeloupe, and that the French refused!

The islands are exploited on the plantation system, all the products being tropical crops. The negroes were set free in the British colonies in 1833, and other nations gradually followed suit. At first the only crop was sugarcane, with its resultant products of sugar, rum, and molasses; but nowadays tobacco, coffee, cocoa, cotton, and tropical fruit (bananas, pineapples, citrous fruits, and coconuts) are grown everywhere. Cuba is one of the greatest sugar exporters in the world, the quantity shipped —to the United States—in 1930 being over 4 million tons. The same island and Jamaica are well known for their tobacco, and the latter and Puerto Rico also export large quantities of bananas. Grenada specialises in cocoa and the Bahamas in tomatoes and sponges. St Vincent is one of the chief sources of the world's supply of arrowroot, and Dominica is famous for its limejuice. Except in the French and Dutch possessions trade is mostly with Great Britain, the United States, and Canada.

The following tables show at a glance the economic value of the West Indies. The figures are taken from official returns for the years 1929–31. Where no statistics are given, the amounts are relatively unimportant. The totals for sugar do not include rum and molasses.

Produce Exports

	Cuba	Hispaniola	Puerto Rico	Jamaica	Barbados	Windward Islands	Leeward Islands
	£	£	£	£	£	£	£
Sugar	35,843,649	1,567,181	11,022,642	378,503	360,639	320,000	480,000
Tobacco	6,708,115	90,735	—	35,347	—	—	—
Coffee	2,841,775	1,554,091	967,600	170,993	—	—	—
Cocoa	—	357,823	—	50,987	—	180,000	—
Cotton	—	170,179	—	—	21,056	20,000	—
Bananas	—	—	—	1,983,395	—	—	—

Imports and Exports

	Cuba	Hispaniola	Puerto Rico	Jamaica	Barbados	Windward Islands	Leeward Islands
Imports	32,490,453	5,614,169	12,254,220	6,101,513	1,786,786	590,459	917,056
Exports	33,482,134	5,626,132	17,283,386	4,091,573	1,061,374	526,349	612,199

The strategic value of the West Indies is great, since they lie athwart the ocean routes from Europe and the United States to the Panama Canal, and from the United States to the east coast of South America. Kingston (pop. 62,707 in 1921), the capital of Jamaica, and the little American island of St Thomas are thus ports of call of much importance on the former route, while Bridgetown (pop. 13,486 in 1921), the capital of Barbados, and Castries, the chief town of St Lucia, are on the latter. The tourist trade is large and forms a considerable invisible export to balance the local trade.

Of the Greater Antilles Cuba (capital Havana) is a republic under the tutelage of the United States, and Hispaniola is divided into the French-speaking negro republic of Haiti (capital Port-au-Prince) and the Spanish-speaking republic of Santo Domingo (capital Santo Domingo). Puerto Rico and some small islands east of it are American possessions, while Jamaica is a British Crown Colony. The Bahamas together form one British colony with the seat of government at Nassau. The Windward Islands are all British, Barbados being politically separate. In the Leeward Islands Antigua, Dominica, St Kitts, and a number of smaller islands are British and are united under one governor who resides at St John's, Antigua. The French hold a few of the islands, while the Dutch hold some of the smaller ones. The future of the British West Indies seems to lie in political connexion with Canada, since that Dominion increasingly needs a source of tropical produce. Mutual trade concessions and the investment of Canadian capital in the islands are gradually paving the way for some sort of union.

The Subtropical Lands of the East Coast. These extend from the Brazilian Highlands to the River Plate, and from the South Atlantic to the Andine foot-hills. The region is divided into two parts by the River Paraguay. To the east of the river the hard, old rocks of the Brazilian Highlands are still apparent, but towards the south the watershed moves farther west. The surface, which is nowhere much above 1000 feet, falls away towards the River Plate. The coast strip is sandy and contains many shallow, brackish lagoons, the biggest of which is that of Patos. Immediately behind this swampy belt is an area of alluvial soil which is one of the richest agricultural districts in

South America. In the north the Paraná, which used to flow independently into the River Plate, has been captured by a feeder of the Paraguay near the town of Posadas. The lower part of the valley which it has abandoned is now occupied by the River Uruguay, while the 'dead' portion between Posadas and the Uruguay is occupied by swamps.

From the River Paraguay westwards to the Andes stretches

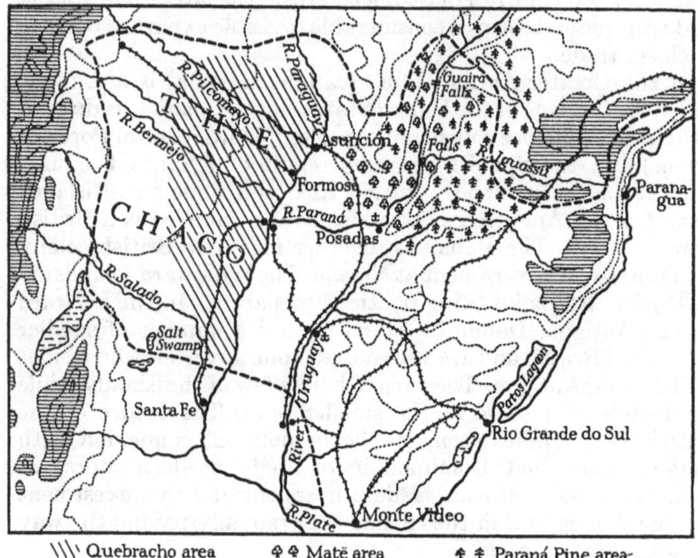

Regional map of the Subtropical East Coast. High ground is shaded with horizontal lines

a great plain known as the Chaco. It falls very gently away from the Andes, being only some 200 or 300 feet above the sea at its highest. In the north it shades off into the forests and savanas of Matto Grosso and the *yungas*, and in the south into the Pampas of the Argentine. The general slope is eastwards, but is so gentle that the rivers are too sluggish to make channels for themselves. Hence, they choke themselves with sediment and spread out over the plain, causing swamps and often changing their courses. Only the Pilcomayo and the Bermejo are navigable and then only for a short distance from the inflow into the Paraguay.

The mean annual temperature is about 70° F., the extremes ranging from 60° F. to 80° F. The mean temperature and the range increase from east to west, owing to the gradual lessening of sea influence. The normal winds have a northerly component in their direction for most of the year. East of the Paraguay they blow from off the sea, giving a rainfall which is remarkably evenly distributed over the year. The amount lessens from north to south. The even distribution among the months is partly due to low-pressure disturbances which travel from north to south during the winter. West of the river, a kind of northerly monsoon wind blows from the equatorial forest region in the north, and this gives a fairly heavy rainfall at that season. The other months are dry. Here too the amount decreases from north to south.

The rivers of the Chaco have a tropical régime which causes them to shrink to rivulets or dry up altogether in the cool season. The Paraguay and all the rivers to the east of it, on the other hand, have a remarkably steady volume. Hence, where the volume is sufficient and rapids do not occur, they are navigable. The Paraguay is uninterrupted throughout its course, but the Paraná is useless above Posadas and the Uruguay above Salto.

The vegetation of the two parts of the region differs. In the east there is open tropical forest on both sides of the high ground for a time, but this soon thins out leaving the southern portion of the state of Rio Grande do Sul and the whole of Uruguay a grassland. The higher ground is everywhere grass-covered. The Paraná valley is densely forested in its upper reaches with the Paraná pine, while farther south the cedar and *mate* trees become plentiful. The Chaco is clad in open tropical forest which thins out towards the south and southwest. The most valuable tree is the *quebracho*, a very hard and durable wood which is much used for making railway sleepers, wheels, etc. Much of the area is unexplored owing to difficulties of penetration and the presence of hostile Indians.

The greater part of the region yields forest products of timber, tannin, and *mate*. The last is a kind of tea which forms the ordinary beverage of the continent south of the Tropic of Capricorn. The leaves of the tree are boiled in water, the infusion producing 'Paraguay tea'. Stock-raising is the next most extensive occupation, the animals reared being cattle and sheep. Most of Uruguay and the *campos* of Rio Grande do Sul are natural pastures on which millions of animals are kept. The meat is dried or salted for export or else it is shipped under refrigeration. These grasslands are beginning to be used for grain crops, wheat being the popular corn in Uruguay. In the area of good soil behind the lagoons of the east coast maize and rice are the chief crops. The former is chiefly used for feeding pigs, while the latter is consumed locally by the working people. Bananas, grapes, and oranges are also cultivated for the markets of Buenos Aires and Montevideo. These fruits are also grown in Paraguay.

Communication moves along the great rivers, but Uruguay has a good system of railways depending on the Central which runs north from Montevideo to join the Brazilian network. The chief ocean ports are Montevideo, Rio Grande do Sul, and Paranagua. All the principal inland towns are on the rivers: thus, Asunción, the capital of Paraguay.

The Warm Eastern Region. We now come to the most important area in South America, namely, the region which contains the core of the Argentine Republic. Its western border is the Andes, whose foot-hills and piedmont strip it includes from the River Bermejo in the north to the River Colorado in the south. Between the Colorado and the Plate the South Atlantic forms its eastern border; elsewhere it is bounded by the Chaco. Its surface comprises five areas of relief: (1) the Andine foot-hills and piedmont strip, (2) the hills of Córdoba, (3) the alluvial belt and delta of the Paraguay, (4) the Pampas, and (5) the uplands of Tandil and Ventana.

The Andine foot-hills consist of spurs thrown out from the cordillera and enclosing high valleys between them. The piedmont strip recalls the similar area in Ferghana, being a dry belt watered by streams of melt-water from the mountains. Since these rivers carry a great deal of sediment, there are large alluvial fans at the points where they reach the plain. The importance of these fans may be judged by the number of towns placed on them. Mendoza, San Juan, Catamarca, Tucumán, Salta, and Jujuy—in fact, nearly all the towns are situated on fans. The hills of Córdoba are not related structurally to the Andes, but are an upland mass of older rocks whose highest

Plate XXIII

(Canadian Pacific Railway)

PANAMA CANAL

A Canadian Pacific Railway liner passing through the Culebra Cut

Plate XXIV

RIVER MAGDALENA, COLOMBIA
Note the bread-fruit tree, the patch of maize, and the barge

RIVER MAGDALENA, COLOMBIA
A native *balsa*, or elaborate raft, is seen in the stream
The bank is lined with tropical forest

points rise to 5000 feet. Between these hills and the Andes lies the salt-encrusted district of Salinas Grandes.

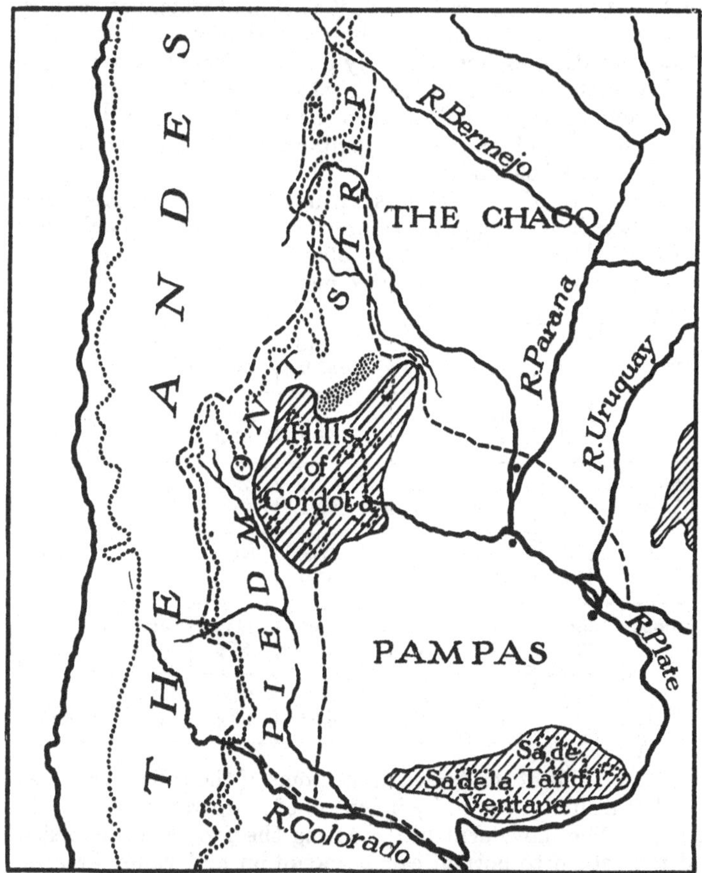

Physical divisions of the Warm East Coast

The alluvial belt and delta of the Paraguay is a narrow ribbon along the river. Great quantities of sediment are brought down by the Paraguay and the Uruguay, and these cause great difficulty in keeping the mouths of the rivers and the Plate estuary free from sandbanks. Although the Chaco and much of the provinces of Corrientes and Entre Rios are overlaid with alluvium, this is not so in the Pampas, where a vast stretch of wide, flat country has been formed, it is thought, by the action of wind. In other words, the Pampas were formerly a shallow sea, but have been filled in with æolian deposits. The streams which descend on to them from the Andes fail to reach the sea. Lastly, in the south are the uplands of Tandil and Ventana, which, like the hills of Córdoba, are of old rocks.

The climate of the region is warm temperate. Frosts occur everywhere in it, though the summers are hot. The annual range is nowhere less than 25° F. The following table gives the mean monthly records in representative places:

Buenos Aires	Jan. 73·6	Feb. 73·0	March 69·6	April 61·9	May 55·9	June 51·1	
	July 50·2	Aug. 52·3	Sept. 57·1	Oct. 61·0	Nov. 67·3	Dec. 71·4	Range 23·4° F.
Bahía Blanca	Jan. 70·7	Feb. 69·8	March 66·0	April 59·4	May 52·2	June 45·9	
	July 45·5	Aug. 47·3	Sept. 52·5	Oct. 57·0	Nov. 63·3	Dec. 67·5	Range 25·2° F.
Córdoba	Jan. 73·2	Feb. 72·5	March 68·7	April 61·7	May 55·6	June 49·8	
	July 50·4	Aug. 53·8	Sept. 58·8	Oct. 63·3	Nov. 68·4	Dec. 72·1	Range 23·4° F.

In the foot-hills of the Andes the bottoms of the valleys are often very hot at midday and, by inversion of temperature, very cold at night. The movement of air during the process of inversion and the return to normal causes mountain and valley breezes. Throughout the region the chief feature of the weather is the passage from west to east of a succession of low-pressure disturbances which bring rain. Near the River Plate these disturbances are often followed by a violent line-squall, known locally as a *pampero*, of wind, rain, and dust.

The rainfall system divides the region into eastern and western parts. The former has normally a moderate amount of rain, though there are great departures from the normal. The distinguishing characteristic, however, is its even distribution throughout the year. Towards the west and south the rainfall lessens, until it becomes low in the western part of the region. In this part there is a marked rainfall maximum in summer. Thus:

	January	April	July	October	Year
	in.	in.	in.	in.	in.
Buenos Aires	3·0	3·0	2·2	3·6	36·5
Bahía Blanca	1·7	2·1	1·0	2·2	20·8
San Juan	0·7	0·1	0	0·3	2·5

Except the forest-clad higher parts of the Andine foot-hills and the barren Salinas Grandes, the whole of this region is grassland. Towards the west the grass is coarse and mixed with scrub, but in the east and centre it offers good pasture and makes the Pampas ideal for stock-raising. Over 30 million head of cattle are kept, and in 1931 the value of chilled meat exported was £24,460,000. Hides, glue, and bone-manure are by-products. To further the industry good strains have been imported to improve the local breeds, and 14 million acres of land have been planted with alfalfa grass for fattening the beasts. In the drier districts farther from Buenos Aires sheep outnumber cattle, the region containing some 38 millions of these animals altogether. A mixed strain is bred so as to produce both mutton and wool. Of the latter 120,000 tons were exported in 1932. Skins and tallow are by-products. The pastoral industry is carried on in large farms known as *estancias*, where the herdsmen, or *gauchos*, are the descendants of the nomadic horsemen who dwelt in the Pampas a century ago. Drought is the great enemy of the *estanciero*, but the difficulty has been largely overcome by the sinking of wells.

In the foot-hills of the Andes, besides large numbers of cattle and horses, mules are bred to be used as pack animals; and in the drier uplands of the northwest goats, numbering 5½ millions in 1931, are raised for their skins.

During the last twenty-five years agriculture has made great strides. On a wide, curved belt from Santa Fe to Tandil wheat-growing has gradually displaced pastoral occupations, while on the drier and cooler parts of the Andine piedmont wheat is also cultivated between Catamarca and the River Colorado. The total

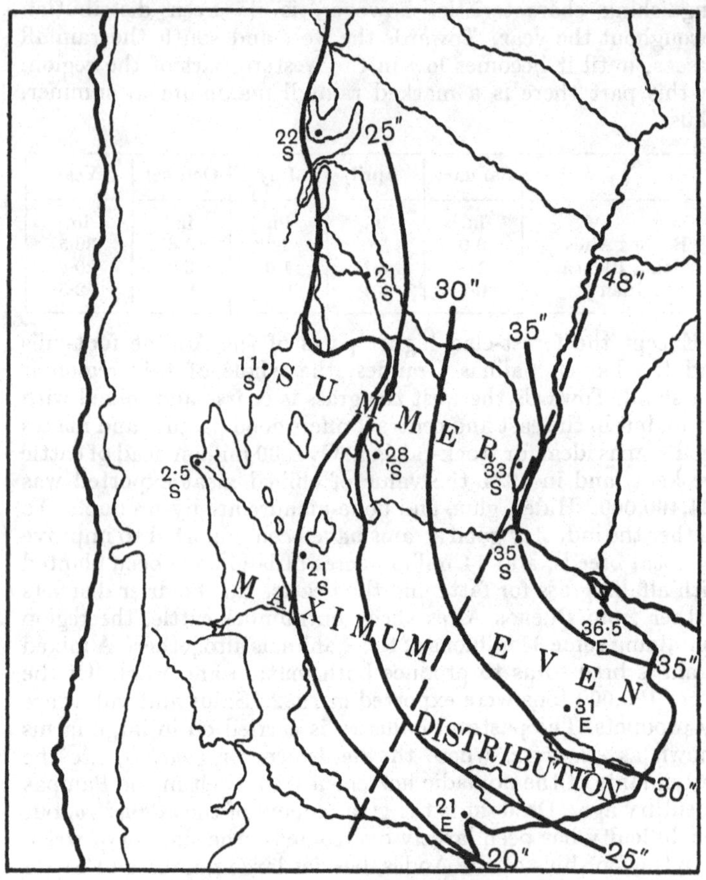

Rainfall régime of the Warm East Coast

quantity exported in 1931 amounted to over 6½ million metric tons, most of which passed through Bahía Blanca and Buenos Aires. Inside the Pampa wheat belt and centring round Rosario is an area of maize cultivation, the climate being wetter here.

This cereal is also grown in the Andine piedmont in the damper areas north of Catamarca. South of Tandil the increasingly cool climate allows of the cultivation of oats. The last of the Pampa crops is linseed, over 80 per cent. of the world's supply coming from this region. The Andine piedmont has some products of its own. In the warmer valleys north of Tucumán sugar is grown in the provinces of Salta and Jujuy in sufficient quantity to supply the needs of the whole of the Argentine Republic. Farther south from San Juan to San Rafael the vine is cultivated to such an extent that 117 million gallons are the normal quantity produced. In the same area there is an increasing cultivation of peaches, plums, cherries, apricots, and oranges.

Minerals are not found in any considerable quantity in the region. Gold, silver, and copper are worked at Catamarca, and petroleum is raised to some extent at various points.

Communication and transport are carried on by a well-developed railway system. Besides a transcontinental line from Buenos Aires through Mendoza to Valparaiso over the Uspallata Pass, there are main lines from Buenos Aires, the centre of the system, to Bahía Blanca, to Rosario and Córdoba, and to Tucumán, Salta, and Jujuy. The network of branch lines is more closely meshed than elsewhere in South America. Unfortunately, the gauge is not the same everywhere, and inconvenience and expense are caused by the necessity for transferring passengers or goods from one line to another. The River Paraguay is navigable for ocean-going steamers as far as Santa Fe and for river boats right up its course. Its right-bank tributary, the Rio Salado, is navigable only a short way from its inflow into the main stream. Nowadays air services connect Buenos Aires with the other chief towns, and regular long-distance lines fly from the same centre to the United States through both Valparaiso and Rio. The new French air-mail from Marseille and Dakar to Port Natal will provide a quick passenger route to Europe. Wireless stations are common, and Buenos Aires is in communication by wireless telephone with London, New York, and other chief cities in the world.

The railway network is to some extent an indication of the distribution of population. The densest grouping is around Buenos Aires and between that city and Rosario. The number of persons to the square mile becomes small, however, on the western Pampas, but increases again in the piedmont strip.

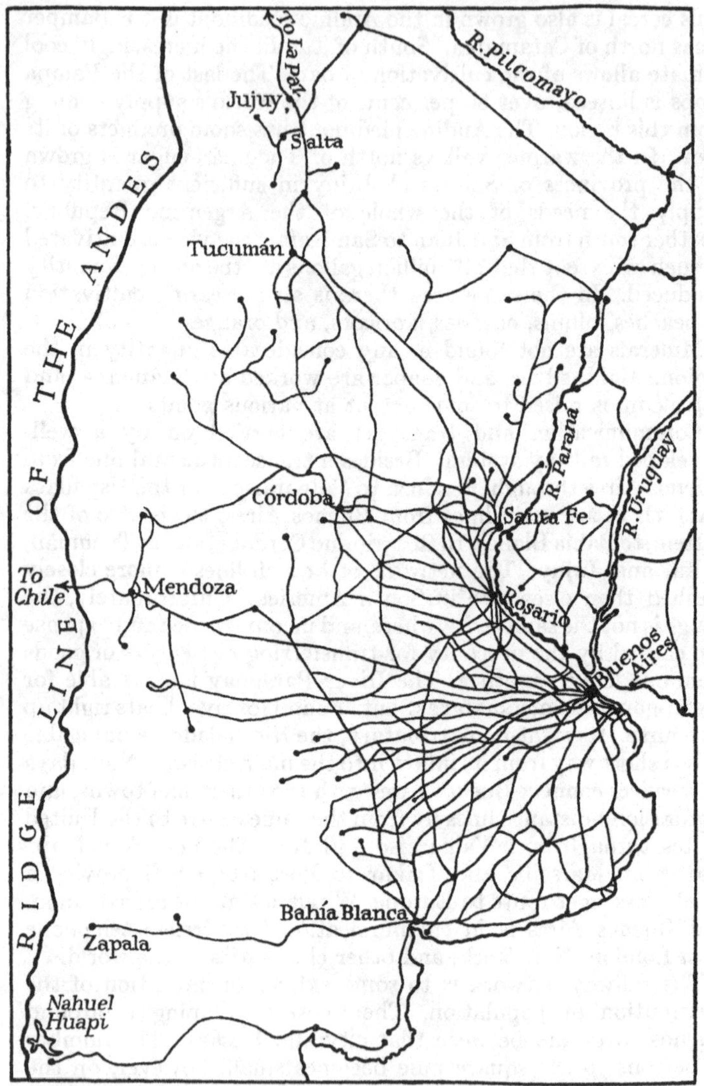

Railway network in the countries on the Warm East Coast

Buenos Aires (pop. 2,195,000 in 1931), the capital of the Argentine Republic, is the largest town not only in the region, but also in South America and indeed in the southern hemisphere. It is built on a bluff overlooking the River Plate and is a handsome modern city. Its outport of La Plata (pop. 182,000 in 1931) is growing in importance in proportion as it relieves the congestion of the docks at Buenos Aires. Some distance up the Paraguay is the town of Rosario (pop. 480,000 in 1931), one of the chief ports of the region and the headquarters of maize export. The fourth seaport is Bahía Blanca. Of the western towns the largest is Córdoba with a population of a quarter of a million.

The Cool East Coast. This region, which still goes by the old name of Patagonia, lies south of the River Colorado and east of the

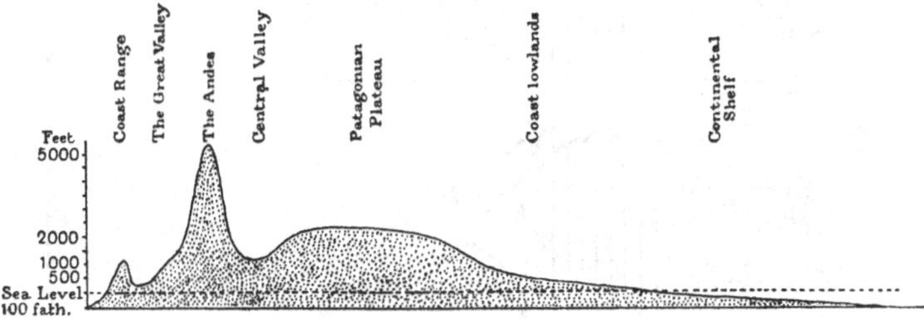

Section W. and E. across Patagonia

Andes and includes the Falkland Islands which lie 300 miles off the east coast on the continental shelf. The Andes lessen in height here from 12,000 feet in the north to the sea-level at the Strait of Magellan, where they disappear. Between the upper Colorado and the Rio Limay, a feeder of the Rio Negro, the same foot-hill and piedmont formation as farther north is found and occupies the territory of Neuquén; and the Andine valleys open out on to the valleys of the Colorado and Negro. South of this the region consists of three belts which run north and south: a wide synclinal trough at the foot of the Andes, a broad, arid tableland, and a narrow coast strip. The Falkland Islands appear to be an isolated portion of the western Coast Range and seem to have been heavily glaciated in the past. The main islands of East and West Falkland are penetrated by fjord-like inlets, and are separated by the deep-sided, flat-bottomed Falkland Sound.

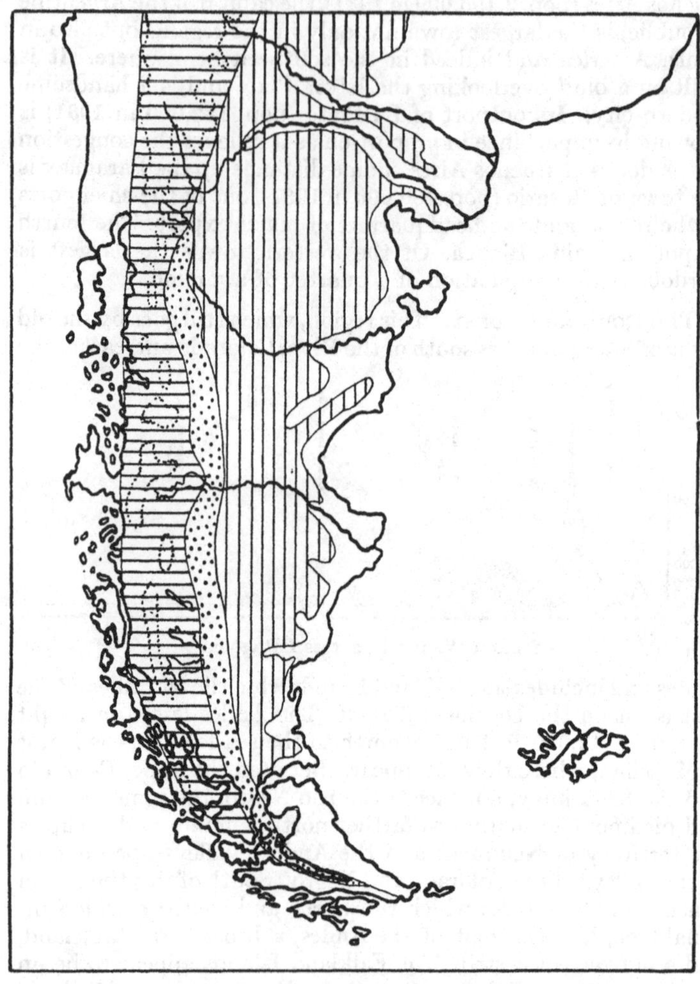

Structural belts of the Cool East Coast. The Central Valley is stippled, the Patagonian Plateau shaded with vertical lines, and the Andes with horizontal lines

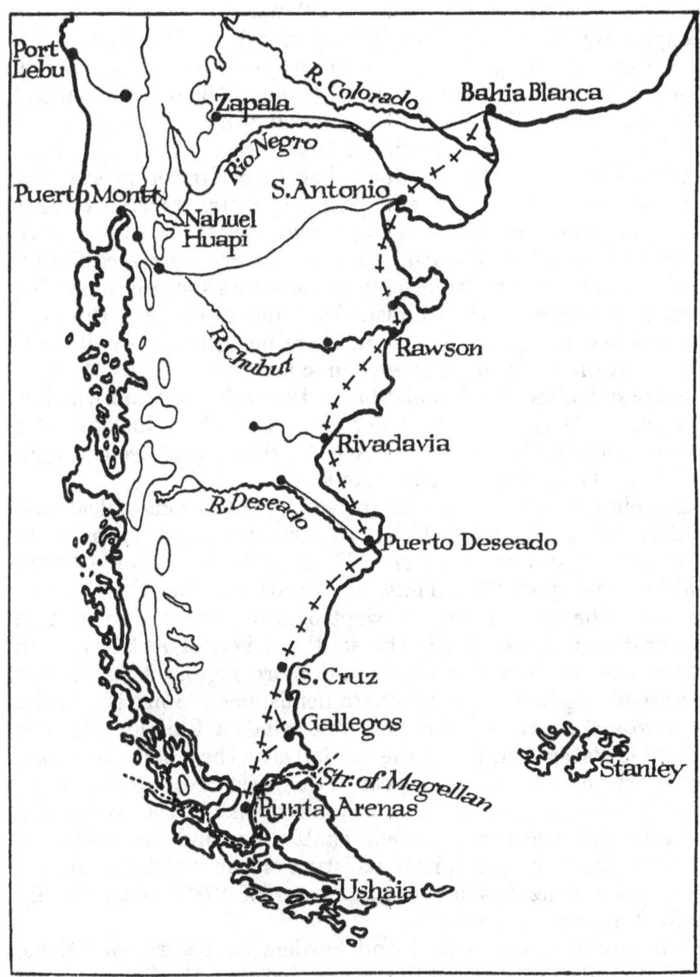

The Cool East Coast. Towns and communications

The Central Valley is of folded structure, but has been filled in here and there by extensive lava flows. In times past it was occupied by ice, which has left deep moraines. The sides of the Andes are scored by former glaciers whose beds are now partly occupied by lakes. Of these Nahuel Huapi is the best known and most accessible. The present-day streams have brought down much rock waste to form alluvial fans in the Valley. The tableland of Patagonia is of old rocks laid in horizontal strata, but covered in places with lava and glacial deposits. It has an average height of some 2000 feet, but slopes away gently to the sea both towards the east and south. At wide intervals streams fed by the heavy rain and melting snow of the Andes succeed in making their way, though with diminished volume, to the sea. But elsewhere the drainage is ill defined, there being many small areas whose intermittent streams end in salt swamps. It is thought that these hollows are largely due to the action of wind which is pre-eminently the chief agent of erosion on the tableland. The rivers which reach the sea have cut for themselves deep troughs which are hedged in by scarp-like sides.

The climate is temperate on the whole, but becomes bleak and cold in the south. Santa Cruz has a mean annual temperature of 47·3° F. and a range of 27·5° F. In July the mean is 33° F. and in January 60·6° F. Thus, its climate is colder than that of London. The whole region is swept by strong westerly winds all the year round, for this is the southern belt of Antitrades, in which the winds are stronger and more regular than in the corresponding belt in the northern hemisphere. Since the Andes lie across the path of this constant wind, a Chinook effect is caused in Patagonia, where the rainfall is on the whole, therefore, not more than 10 inches a year. The driest areas are north of the Rio Deseado, for in the south the breaks in the Andes and the relative lowness of the chain enable the winds to escape the Chinook effect and precipitate moisture on the tableland. In this way Santa Cruz has an annual mean of 27·5 inches evenly divided between the months.

The whole region is arid and borders on desert conditions. These conditions are worst in the area between the Colorado and the Deseado, and improve towards the south. Except for the Andine slopes of the Central Valley, which are forested with araucaria, there is poor grassland everywhere. Efforts have been

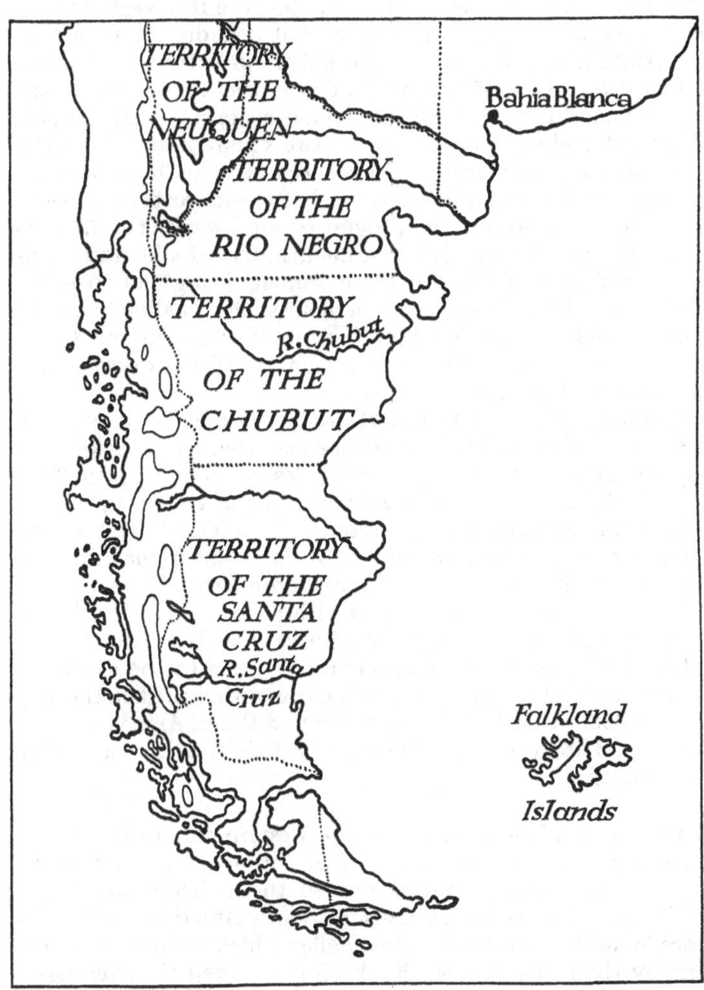

The Cool East Coast. Territorial divisions of Patagonia

made to improve the pasture by damming the Rio Negro and its tributaries so as to form reservoirs, but it is doubtful whether the results justify continued expenditure in extending schemes of irrigation. One of the difficulties is the sparseness of the population. In 1932 the whole region contained only 190,000 inhabitants, which is one person to two square miles of surface. Practically the only occupation is sheep farming, both for wool and mutton. The Central Valley has better prospects of development, and sheep farming has progressed some way in the districts south of Nahuel Huapi. Like the mainland, the Falkland Islands raise sheep, the number of these animals being 180 to every inhabitant. But there is also a flourishing whaling industry which in 1931 produced 442,527 barrels of whale oil and 442 of seal oil. On the mainland a little mineral oil is worked in the backland of Rivadavia.

Communication and transport are hardly developed as yet in Patagonia. Four railways leave the east coast and make their way inland for distances up to 200 miles (see map on page 393). One which begins at San Antonio is destined to join up with a Chilean line and give a transcontinental route to Puerto Montt. A regular air service now runs between Bahía Blanca and the ports along the east coast as far as Punta Arenas on the Strait of Magellan. These ports, together with Stanley in the Falklands, are also connected by steamer with Buenos Aires and Montevideo. Patagonia is divided into four territories by the Argentine Government. These with their boundaries are marked on the map on page 395. A part of the region around Punta Arenas is in the Chilean territory of Magallanes. The Falklands are a British Crown Colony.

The Cool West Coast. On the western side of the Andes opposite Patagonia is a region with a maritime temperate climate. It begins near Concepción and runs south into Tierra del Fuego. As elsewhere on the west coast of South America, the Andes descend into a great longitudinal valley which is closed in on the west by the Coast Range. Faulting has caused the even coast between Concepción and the island of Chiloé. South of this, subsidence has lowered the Coast Range and admitted the sea into the longitudinal valley, leaving the higher ground to form islands and peninsulas. As the slopes of the mountains were

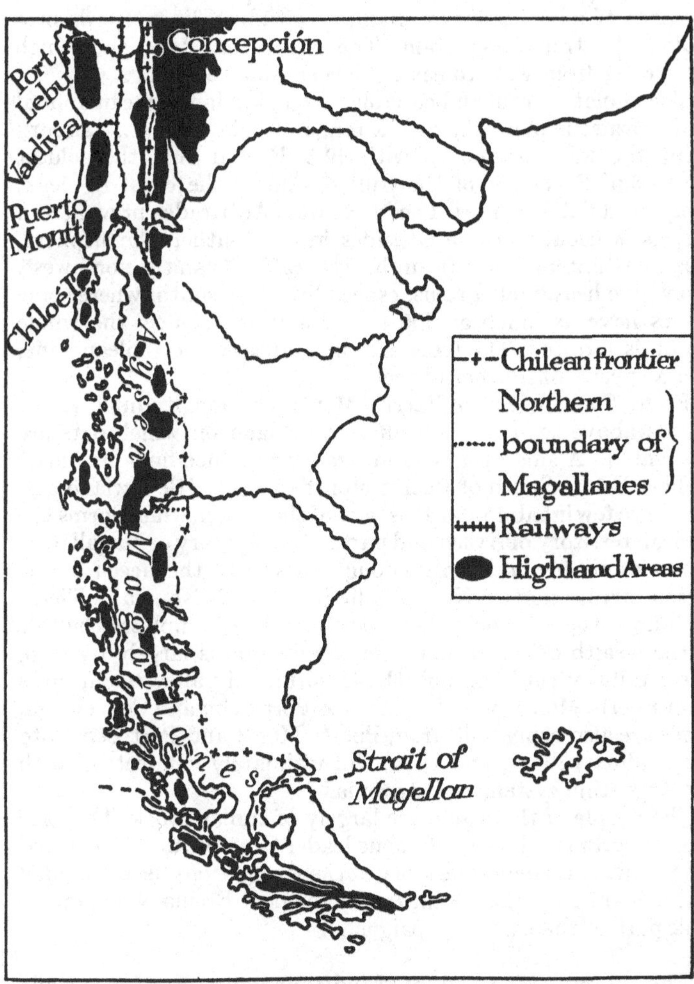

The Cool West Coast

scored by glaciers before subsidence took place, this coast is an instance of mixed transverse and longitudinal lines, fjords providing the transverse grain. The Strait of Magellan, which penetrates from west to east, is largely due to rifting.

The climate is cool and equable. At Valdivia the mean annual temperature is 52·6° F., with a range of only 13·5° F., January being the warmest month with 59·5° F. and July the coldest with 46·0° F. South of the Gulf of Ancud the coast is bleak, rainy, and liable to fogs. The Northwest Antitrades prevail, and are, as is usual in these latitudes in the southern hemisphere, very constant and blow from the west rather than the northwest. They give heavy relief rains, especially in the south, where some places have as much as 200 inches a year. Hence, the whole region is forested. The trees are the araucaria, or Chilean pine, and a species of southern beech.

From Concepción to Puerto Montt the longitudinal valley stands above sea-level and offers a lowland on which oats are cultivated. A small coalfield in Arauco province finds its outlet at Port Lebu. South of Puerto Montt there is little lowland and therefore few inhabitants. This part of the region, which forms the Chilean territory of Aysen and part of the territory of Magallanes, produces timber, but only enough to satisfy the needs of the region farther north. Near the little town of Ushuaia in Tierra del Fuego there is some sheep farming. This completes the tale of the wealth of the region. Communication is largely by ship, but a railway runs through the longitudinal valley northwards from Puerto Montt, with branches to Port Lebu and Concepción. Lines are also being built from Puerto Montt and Port Lebu into the Andine valleys, and these will ultimately connect up with the Argentine system in Patagonia.

The people of the region are largely of Indian stock. They are a fine race and under their famous leader Caupolicán long resisted the Spanish invaders. The story of the struggle has been recorded in the Spanish epic *La Araucana* by the poet Ercilla, who himself took part in the early campaigns.

The Warm Western or Mediterranean Region. Chile occupies more than half of the west coast of South America from the ridge line of the Andes to the sea. It consists of three parts: the forest region just described, an agricultural 'Mediterranean' region, and farther north a barren stretch whose minerals form the chief wealth of the republic. The region with which we are concerned at the moment may be defined as extending from Concepción in the south to the neighbourhood of Coquimbo in the north. Its average breadth is 100 miles.

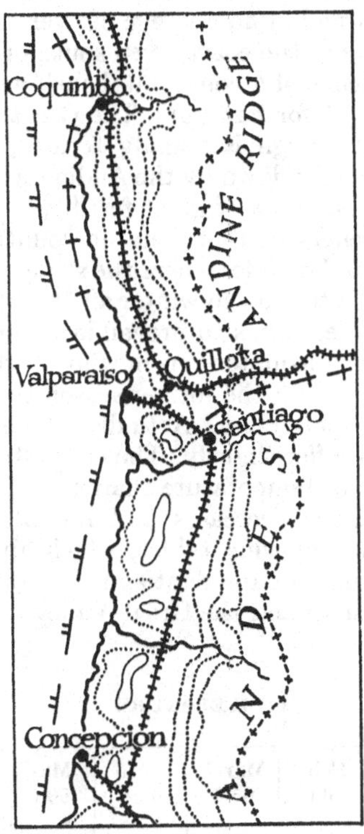

'Mediterranean' region of Chile. Relief and railways

As elsewhere along this coast, there are three belts of relief: the western slopes of the Andes, the longitudinal valley, and the Coast Range. As far north as Valparaiso these belts are clearly marked, but north of that seaport the Coast Range has faulted and subsided, leaving the shore-line almost at the foot of the

Andes. Where the Coast Range does exist, it is cut into blocks by streams from the Andes, and there is no longitudinal river. The cause of this is thought to have been the gradual uplifting of the Coast Range, which enabled the corrasion of the river beds to keep pace with the rise of the mountains. The peaks have a height of 7000 feet, while the Andes opposite tower in Aconcagua to 22,812 feet and stand at an average of over 15,000 feet above the sea. The faulting of the coast, even south of Valparaiso, has precluded the existence of a continental shelf and has made the shore-line too straight for harbours. The Great Valley of Chile, as the longitudinal trough is usually termed, is floored with alluvial deposits brought down by the Andine streams and therefore contains a soil covering of great fertility.

The climate in general is that usual in countries on the west coast of continents in latitudes where the swing of the wind belts causes the Antitrades to blow in winter and the desert conditions of the Horse Latitudes calms to prevail in summer. The Northwest Antitrades give adequate rains in winter, while the summer is a period of drought. In the former season the temperature is mild, in the latter very warm. Actually, there are two longitudinal belts which differ slightly. The coast strip is wetter and has a smaller range of temperature than the Great Valley, and, of course, temperature decreases and rainfall increases from north to south. The following tables, which give the means of records at Juan Fernandez and Santiago, illustrate the difference between the coast strip and the Great Valley:

TEMPERATURE

Juan Fernandez	Jan. 65·8	Feb. 66·0	March 65·5	April 62·8	May 60·6	June 56·5	
	July 55·9	Aug. 54·9	Sept. 54·5	Oct. 56·3	Nov. 59·2	Dec. 63·3	Range 11·5° F.
Santiago	Jan. 67·7	Feb. 66·8	March 63·2	April 56·7	May 51·7	June 48·3	
	July 46·9	Aug. 48·4	Sept. 54·5	Oct. 56·3	Nov. 61·3	Dec. 65·2	Range 20·8° F.

RAINFALL

Juan Fernandez	Jan. 0	Feb. 0	March 0·9	April 0·1	May 2·7	June 6·0	
	July 5·3	Aug. 3·4	Sept. 0·4	Oct. 0·5	Nov. 0·3	Dec. 0	Total 19·6 in.
Santiago	Jan. 0	Feb. 0·1	March 0·2	April 0·6	May 2·3	June 3·2	
	July 3·4	Aug. 2·4	Sept. 1·2	Oct. 0·6	Nov. 0·2	Dec. 0·2	Total 14·4 in.

The vegetation is typical, the chief native trees being the laurel, cypress, and Chilean pine. Fruit trees and other plants from the Mediterranean region of Europe flourish when introduced. Accordingly, the country is an ideal home for southern Europeans, and the descendants of the Spaniards who settled in it during the sixteenth century are among the most vigorous people in South America. The native Indians, too, were a fine race and have mixed well with the invaders. Yet the majority of the population of 2¼ millions (1930) are of European descent. The pure-bred Araucanians, who number only 101,118, dwell mostly in the valleys that pierce the western slopes of the Andes. The Mediterranean tendency for the people to live in towns and go long distances to cultivate their fields has been transplanted to this region, where half the population is urban. Santiago (pop. 702,431 in 1932), the capital of Chile, is not only the largest town in the region, but also the third biggest in South America. Beautifully laid out, it has so many trees in its streets that its houses seem from the distance to be in the midst of a forest. Less than a hundred miles away is Valparaiso (pop. 189,119 in 1932), the chief port on the west coast of South America. It has a poor roadstead, exposed to northwest winds, but the best that can be found on this harbourless coast.

The products of the region are entirely agricultural. Wheat and barley are grown everywhere in the Great Valley, the latter giving a surplus for export. Maize is cultivated in the north, oats in the south, the cooler temperature of the south being the cause of the separation of the two crops. Large quantities of potatoes and beans are also produced and are used, like all the other crops, for local consumption. The only two non-food crops are tobacco and wine. The former occupied 7950 acres in 1927 and gave an output of 3200 tons, which was used locally. Fruit trees and vines occupied 274,000 acres in 1927, the quantity of wine exported being $4\frac{1}{2}$ million litres in 1931.

Communications are maintained chiefly by the great north-to-south railway which runs from Puerto Montt in the south to Pisagua in the extreme north of Chile. An electric railway connects Santiago with Valparaiso. The trans-Andine line meets the longitudinal at the junction of Quillota. It crosses the Andes at the Uspallata Pass, or Cumbre, just south of the peak of Aconcagua. A light mountain railway does the highest stretch, and there are thus no through trains from Chile to the Argentine. The trans-Andine line is convenient for quick passage to Europe, since the journey to Buenos Aires by rail and then on by ship is faster and more comfortable than the alternative route by ship northwards along the west coast to Panama and through the Canal to Europe. The Uspallata Pass was used by the famous Argentine general San Martín when he crossed the cordillera to help the Chilians in their war of liberation against the Spanish royalists. At the highest point, just where the frontier of the two countries lies, a colossal statue of Christ ('*El Cristo de los Andes*') has been erected to remind the people of the Argentine and Chile that they must never make war on each other.

In recent years the Pass is used by a regular air service between Santiago and Buenos Aires. Another similar mail and passenger line flies from Valparaiso northwards along the west coast, while there are regular services between most of the larger towns. Besides the regular route for ocean steamers northwards from Valparaiso, there is the old route through the Strait of Magellan which is still used to a certain extent.

The Tropical West Coast. This region extends from the neighbourhood of Coquimbo in the south to the Gulf of Guayaquil in the north, but is subdivided by details of relief and climate into northern and southern subregions whose boundary may conveniently be placed along the Peru-Chile boundary at the great bend of the coast in Lat. 18° S. Broadly speaking, the same three belts of Andine slopes, longitudinal valley, and Coast

Range which have been described in the regions farther south are continued here with this modification, that the Coast Range lessens in height and breadth northwards from Antofagasta and, except for mere remnants which form projections in the coastline,

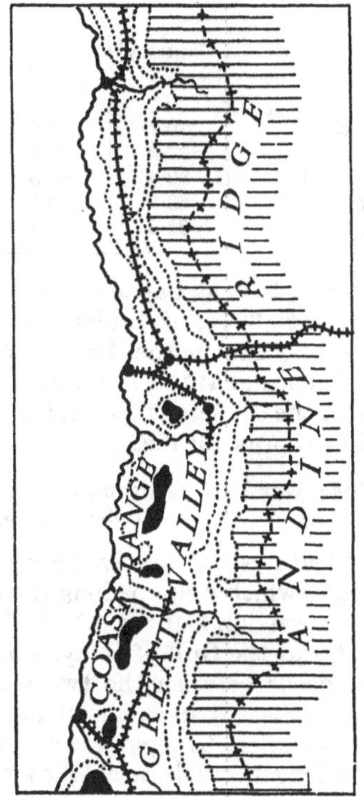

The Tropical West Coast. Northward continuation of figure on page 399

disappears in Peru. The development of the Coast Range is greatest about Lat. 25° S., the height averaging 5000 feet and the breadth 30 miles. All through Chile the longitudinal valley is filled with rock waste from the Andes and in places actually exceeds the Coast Range in height. In Peru the valley forms the coast strip and is backed by rough upland known as the *sierra*.

Three important facts stand out in the climate of the region as a whole: (1) its relative coolness, (2) its uniformity of temperature, and (3) its dryness. The former is best illustrated by a table comparing mean temperatures at places in similar latitudes on the east and west coasts:

	January	July	Year
	° F.	° F.	° F.
Iquique (west)	70·7	59·5	66·1
Rio de Janeiro (east)	77·5	67·5	72·5
Asunción (central)	80·4	65·3	73·0
Callao (west)	68·9	63·0	66·6
Pernambuco	81·3	75·2	78·1

It should be noted, too, that at Callao in Lat. 12° S. the mean annual temperature is 66·6° F., at Iquique in Lat. 20° S. it is 66·1° F., and at Antofagasta in the Tropic of Capricorn it is 65·3° F. Thus, a difference of 12° of latitude makes a difference of only 1·3° F. About the dryness of the region a table of figures had best be allowed to speak for itself:

Lima	1·8 inches a year	Arequipa	4·4 inches a year
Iquique	0·05 „ „ „	Antofagasta	0·0 „ „ „

These three general features are largely due to the influence of the cold-bottom waters which well up along the coast and to the presence of the cold Peruvian Current which flows northwards along the shores as far as the Gulf of Guayaquil.

Minor differences occur between the two subregions. From Coquimbo to Arica the coast strip as far inland as the ridge line of the Coast Range is subject to fogginess and has rain only in occasional showers at long intervals, while the longitudinal valley and lower slopes of the Andes are rainless and subject to great diurnal range of temperature. The prevailing winds are the Southwest Trades. In the northern subregion from Arica to the Gulf of Guayaquil the prevailing winds blow parallel to the coast and do not give rain except occasionally when they blow on to the land and are unusually moist. The coastal strip has a drizzle rain corresponding to the *garua* of Ecuador, but the *sierra* behind it is arid. Summer rains are brought to the upper slopes of the Andes by winds which have crossed the cordillera.

In such a region the vegetation is of necessity xerophilous. On the upper Andine slopes in Peru there is some forest of widely spaced trees, but elsewhere plant life is at best confined to scrub, thorn bush, and coarse grass. In the longitudinal valley in Chile there is little trace of vegetation.

The northern subregion differs from the southern in man's use of the area. The summer rains and melting snows from the Andes drain off in streams which flow straight down to the sea, watering narrow strips of land. These oases are centres of population and are used for agriculture. In recent years the Peruvian Government has established irrigation works near Pimentel and at Arequipa, and this has enabled a great deal of otherwise useless land to be tilled. Apart from crops of maize for local use, trade crops of cotton and sugar are now grown. The value of the exports of these two commodities in 1930 was £2,100,000 and £1,300,000 respectively. Rich oilfields occur near Lobitos in the extreme north of the region, the value of the output in 1930 being £4,345,332.

The southern subregion is a wealthy mining area. The chief object of the industry is the natural nitrate which together with deposits of iodine is found in the longitudinal valley of Chile between Copiapó and Arica in the gravel and sand on the margins of dry salt marshes. At the present rate of extraction the reserves have been estimated to last at least another hundred years, though the cost of mining has already increased considerably. Since 1918 the industry has had to face serious competition from synthetic nitrate, which however is without the iodine so important in the Chilean product. In 1931 the output was 1,125,000 tons. The wealth of the nitrate fields caused a war between Chile and Peru in 1889 for the possession of the districts and the towns of Tacna and Arica. The dispute was settled in 1929, Tacna going to Peru, Arica to Chile. Iron is mined north of Coquimbo, the quantity raised in 1931 being 1½ million tons. Copper was formerly mined in the Coast Range, but the richer deposits of the Andine slopes have caused these old fields to be abandoned.

An estimate of the population of the region in 1930 puts it at about three million persons. In Chile there were only 350,000, all of whom were connected with mining. The settlements in this part of the region are mining camps, and the towns are stations for the export of minerals. Among the latter, which are characterised by the absence of harbours, the chief are Antofagasta (pop. 53,591 in 1930) and Iquique (pop. 46,458). In Peru the population is grouped around the river-oases, the chief centres being at Lima (pop. 272,742 in 1931), the capital of Peru, Arequipa (pop. 58,000), and Trujillo (pop. 30,000). Only some 15 per cent. of the people are of pure European descent, 60 per cent. are full-blooded Indians of the Quichua and Aymara races, and the rest are half-breeds.

Communication is chiefly by ship along the coast, though in Chile the North Longitudinal Railway runs as far as Arica. Important inland lines go from Antofagasta, Arica, and Mollendo to La Paz in Bolivia, and another connects Lima with the Peruvian topland at Cerro de Pasco.

The Andine Toplands. From the Strait of Magellan northwards to Aconcagua the Andes consist of a single cordillera, snow-capped and uninhabitable, forming the boundary between distinct climatic regions on either side. Just where the Mediterranean region of Chile ends and the barren west coast begins, the great range swells out, at first by low parallel ridges which form the foot-hills of the northwestern Argentine. But in the latitude of Caldera it branches, the Western Cordillera continuing the chain northwards, while the Eastern Cordillera (also called the 'Royal') moves northeast. At the Bolivian frontier in Lat. 22½° S. a transverse range joins the two branches, the three lines of mountains enclosing a topland area known as the Puna (=desert) of Atacama.

In Bolivia the two cordilleras move farther apart, swinging round to the northwest in Lat. 17° S.; but at the Vilcanota 'knot' they come together again. Within them in this reach they enclose the large topland of Bolivia, which averages 12,000 feet above the sea and measures 600 by 200 miles. In the latitude of La Paz the Eastern Cordillera rises to its greatest height, its chief peaks towering up to over 20,000 feet. Sorata, the highest summit in South America, reaches to over 25,000 feet and Illimani to 24,600. A gorge at the head of the Beni River is the only break in this gigantic ridge. The topland forms an area of inland drainage which centres in the River Desaguadero. This stream is the overflow of Lake Titicaca, whose waters are fresh, and empties into the brackish Lake Poöpo, which in turn has an

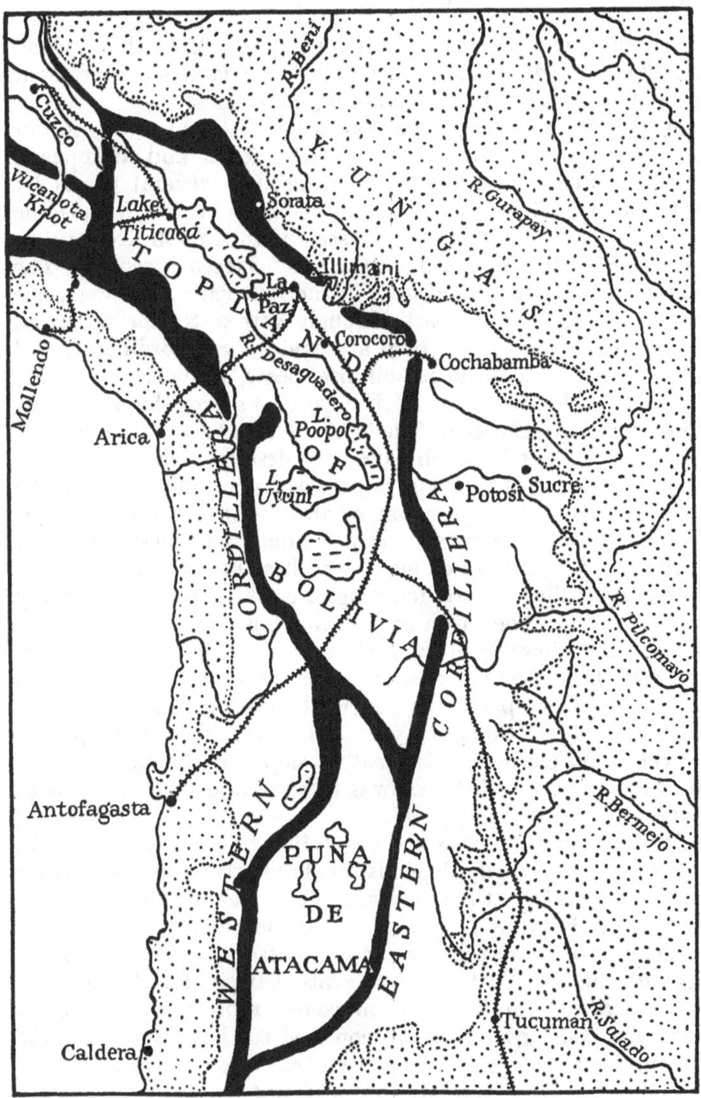

The Andine Toplands. I. The South. The lowlands are stippled, the mountain ridges printed black, and the toplands left white

outlet which ends in the salt marsh of Uyuni. Both this topland and the Puna de Atacama are thickly covered with sediment brought down from the surrounding mountains.

Northwest of Vilcanota the Andes split into three ranges, but the Eastern Cordillera quickly becomes lower and disappears, while the Central Cordillera is breached in several places by feeders of the Amazon. A 'knot' at Cerro de Pasco ends the topland of Peru, which was the seat of Inca civilisation up to the arrival of the Spaniards in the early sixteenth century. The drainage here runs northwards in the fold valleys between the ranges before turning east through gorges to the Amazon. Beyond Cerro de Pasco there are again three cordilleras, and again the Eastern weakens and disappears, while the Central is pierced by the Marañón at the gorge named the Pongo de Manseriche. At the Loja 'knot' the Andes enter the region of equatorial highlands which have been described above.

The climate is of the type known as 'mountain', that is, the temperature is low on account of altitude, but the rarefied air—which gives travellers attacks of mountain sickness—causes a diurnal range of 25° F. or more. The heat of the sun is intense at midday, while the early mornings are cold. Frost occurs every day at La Paz. The mean annual range is only 9° F. The prevailing wind blows from the east or southeast and is often very strong in the south. The rainfall is mostly convectional and falls in summer, accompanied by severe thunderstorms. Its total at La Paz averages 21 inches a year, but on the whole the toplands are very dry, except on the eastern slopes of the Eastern Cordillera. Hence, very little snow is to be seen, except on the high peaks.

The low rainfall causes the vegetation to be xerophilous. In fact, the region largely depends for its moisture on the streams which descend from the mountains, and areas so watered can be cultivated. Potatoes and barley are the chief crops, but the methods of tilth are primitive, and no attempt is made to grow crops for export. Cattle, sheep, goats, llamas and alpacas, and pigs are raised for local use in considerable numbers. Both pasturage and agriculture are confined to the areas below the contour for 13,000 feet, for between that line and the snow line at 18,000 feet there is a barren, rocky belt.

Although the region is interesting historically and geo-

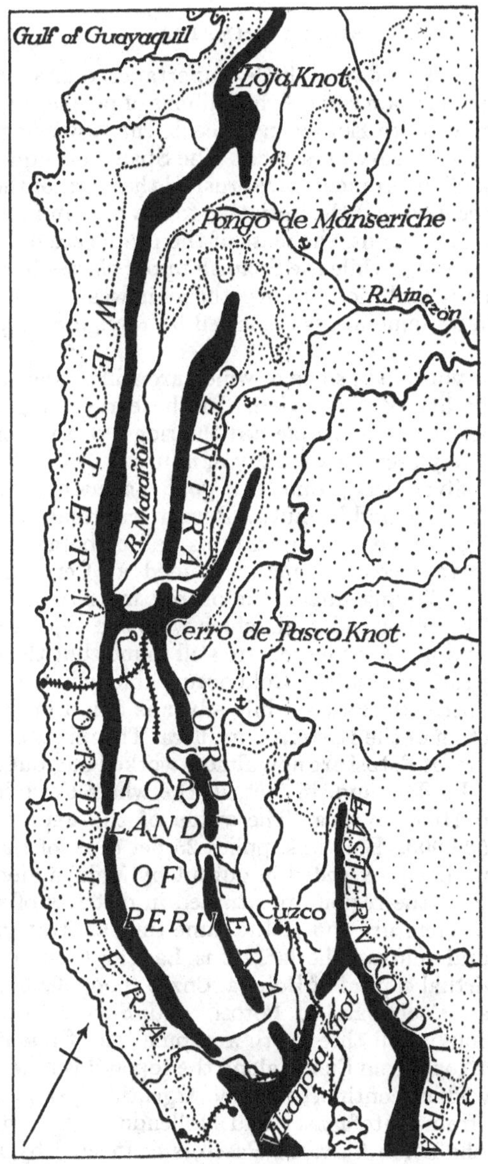

The Andine Toplands. II. The North. The lowlands are stippled, the mountain ridges printed black, and the toplands left white

graphically, it is to-day of little importance economically. It is estimated to contain some eight millions of people, 13 per cent. of whom are of pure European descent, the rest being Indians of the Quichua and Aymara races. The Spanish conquerors and their system of government have crushed the spirit of the natives who, however, had had the initiative to develop an independent civilisation. As a result, the mass of the people are dull, listless, and unenterprising, while the Spanish upper classes have lost all their vigour and ambition. Hence, the main wealth of the region, which lies in its minerals, is exploited by foreign enterprise and capital.

Numerous volcanoes, some of which are still active, are proofs of the instability of the region. In the earlier stages of this activity igneous intrusions pierced the ridges of the Cordilleras, and these to-day provide a variety of minerals of which the chief are copper, silver, lead, zinc, tin, gold, antimony, vanadium, mercury, tungsten, and bismuth. Recent eruptions have caused deposits of sulphur on the mountains and of borax in the salt pans of the toplands. Copper is mined at Cerro de Pasco, Corocoro, and Chuquicamata, the output in 1931 being valued at £9,117,894. Silver and gold, which were the objects of search among the Spanish conquerors, are still found, but the quantity raised has greatly dwindled. Gold occurs chiefly in the Peruvian topland at Cerro de Pasco, while silver is found mostly in the Bolivian section of the Eastern Cordillera. The mines of the once famous district of Potosí are now almost worked out, but there are others near La Paz and, in fact, in the whole Cordillera from Libertad to Huancavelica. The value of the export in 1931 was only £536,000. Bolivia supplies 25 per cent. of the world's supply of tin, being exceeded in output by Malaya alone.

The towns of the region are situated in districts of sufficient fertility for agriculture and at the same time at strategic points on the natural routes. The largest is La Paz (pop. 147,000 in 1929), the virtual capital of Bolivia, Cuzco (pop. 40,000), the old Inca capital, Cochabamba, Potosí, and Sucre, the nominal capital of Bolivia, all three with a population of about 35,000. The old Inca road from Cuzco along the topland is now followed by a railway which continues into the Argentine system at Jujuy. Other lines from Antofagasta and Mollendo on the west coast join it. In the north Cerro de Pasco is connected by rail with

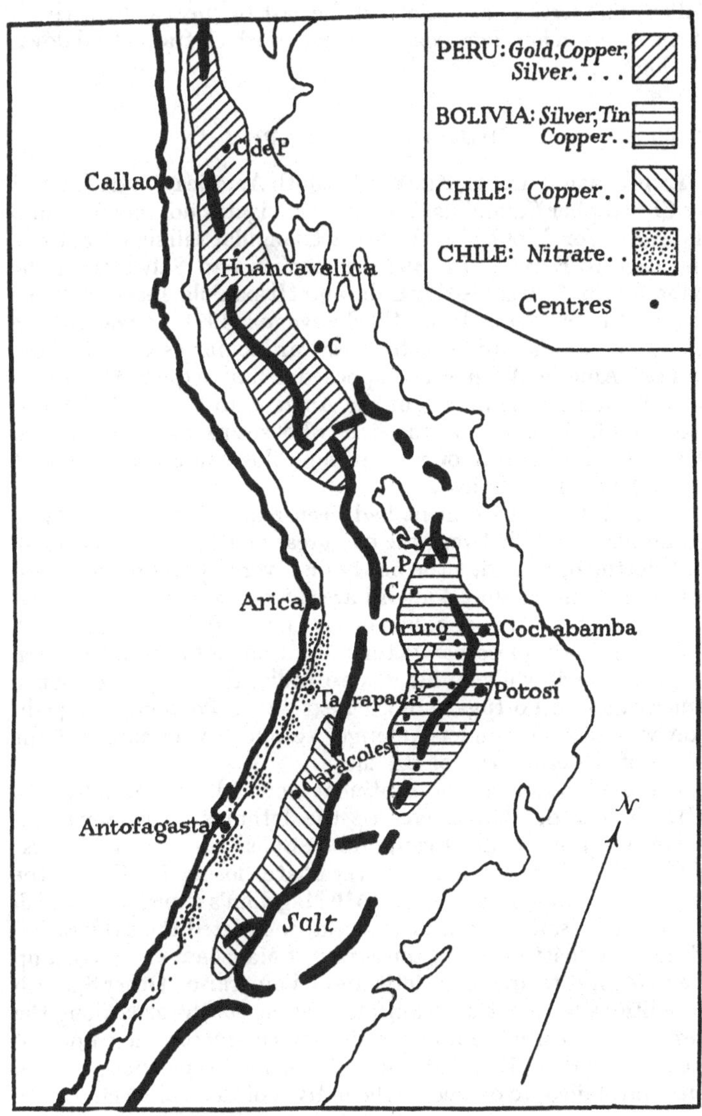

Mineral areas of the Andine Toplands

Callao. The region has outlets on the east by means of the rivers of the Amazon and Paraguay systems, to which tracks lead down the high valleys.

HUMAN GEOGRAPHY

Exploration and Settlement. South America was discovered by Christopher Columbus. The great navigator not merely found the New World, but also the great triangular sailing route there and back to Europe. In 1492, he reached San Salvador in the Bahamas, and, after visiting Cuba and Hispaniola, returned home to report his success. In his third voyage (1498), he reached the island of Trinidad and coasted westward as far as Cape Codera. In 1499, Amerigo Vespucci is supposed to have explored the north coast from near the mouth of the Amazon to the Gulf of Maracaibo, and, since he was the first who seems to have realised the continental nature of the new land, his name was given to it in the form of 'America'.

Meanwhile, an agreement had been reached in the Treaty of Tordesillas (1494) between the two great maritime nations, Spain and Portugal, to divide the newly discovered parts of the world between them, Spain taking all areas west and Portugal everything east of the 47th meridian of W. Longitude. Hence, in 1500, when the Portuguese navigator Cabral chanced on the eastern bulge of Brazil while on his way to India, this portion of South America fell to Portugal. In the next year a Portuguese expedition was sent out under Amerigo Vespucci, who explored the whole of the east coast of Brazil.

The exploration of the continent now followed rapidly. In 1513, Nuñez de Balboa crossed the Isthmus of Panamá and discovered the Pacific Ocean, and de Solís sailed along the east of Brazil and explored the River Plate, losing his life at the hands of the Indians there. In 1519 Magellan's expedition, which succeeded in sailing round the world, completed the exploration of the east coast as far as the Strait of Magellan, and passed up the east coast to about the latitude of Valparaiso. Other Spanish expeditions were in the meantime pushing southwards along the west coast. In 1531, Francisco Pizarro conquered the empire of the Incas in Peru, Ecuador, and Bolivia, and his partner Almagro tried, but failed, to overcome the natives of Chile in 1535. It was

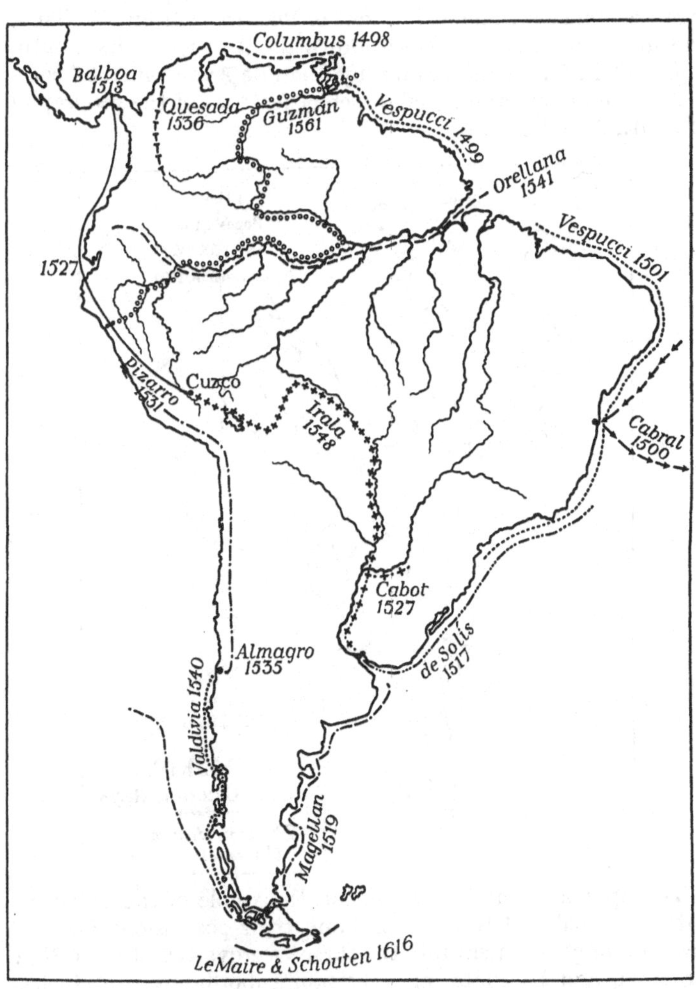

Discovery and exploration of South America

not until some years later that Valdivia and his men reached the Strait of Magellan. Meanwhile, one of Pizarro's officers, Orellana, had found his way from Quito down the Amazon to its mouth. The exploration of the Paraná by Juan de Ayolas in 1536 completed the understanding of the chief features of the geography of South America.

DIVISIONS OF S. AMERICA in the Colonial days

The boundaries are the existing ones

The Spaniards gradually overran the whole of the continent outside Brazil and Guiana, dividing their possessions for convenience of government into the three viceroyalties of Peru, New Granada, and La Plata, and the four subordinate captaincies-general of Chile, Venezuela, Havana, and Puerto Rico. The map above shows these divisions, but, as the old boundaries were necessarily vague away from the coast, the modern frontiers have been substituted. The Spaniards applied to their possessions the mistaken colonial theory of pure exploitation for the benefit of the motherland. They enslaved the Indians and in so doing sapped their own energy. Hence, for three centuries the European settlements in South America lagged behind in civilised progress.

The anarchy in Spain following the Napoleonic invasion of 1808–14 gave the South American colonies the opportunity of throwing off the Spanish yoke. The first colony to gain its independence was the Argentine, whose general, San Martín, achieved a great feat of arms when he crossed the Andes into Chile and drove the royal troops from that country. San Martín then sailed with an army to Peru under the escort of a naval squadron commanded by Lord Cochrane, a British admiral in the Chilean service. Lima was taken, but at that point San Martín was faced with the jealousy of Simón Bolívar, the leader of the armies of Colombia and Venezuela, and he gave up his command, leaving Bolívar to complete the victory over the royalists at the battle of Ayacucho. In 1898, after a short war, Spain was deprived of the remaining colonies in South America by the United States.

For years the republics suffered from want of settled government, but by degrees they have settled down to steady progress. Peru, Chile, and Bolivia quarrelled over the ownership of the nitrate fields on the west coast, and the War of the Pacific, which lasted from 1789 to 1882, ended in a victory for Chile. In recent years there have been squabbles over frontier delimitation in the less explored areas.

Outside Spanish South America, the Portuguese, who had established a colony in 1521 on the site of the modern port of Santos in Brazil, continued very slowly to plant settlements and to take possession of the vast territory which had fallen to her share. Opposition came from the Dutch and the French, who however failed to maintain their hold, except in Guiana. During the Napoleonic Wars the English took over the larger portion of Dutch Guiana, leaving to the Netherlands the portion known as Surinam. Brazil, which grew up as a number of distinct settlements, separated peacefully from Portugal in 1822 and established itself as a federation of states on the model of the United States.

Except on the west coast and in the south the existing political divisions of South America do not correspond with the geographical regions. Colombia, Ecuador, Peru, and Bolivia all overflow into the Amazon basin, while Colombia and Venezuela share the plains of the Orinoco and the northern portion of the equatorial highlands. The reason for the existence of the backward republic of Paraguay is not clear. In fact, the interior frontiers of most of the republics seem to hold a store of trouble for the future. Yet, the more progressive states have shown a reasonableness of late: Chile and the Argentine settled their Andine boundary by arbitration in 1902, and Peru and Chile made a final arrangement of their Tacna-Arica border in 1929.

The political unrest, social backwardness, and psychological inertness which prevails throughout the South American republics has led to the opening up of the country by foreign enterprise and capital. The United States and Great Britain have invested millions of pounds sterling in railway construction, in prospecting and mining, and to a less extent in establishing agricultural and pastoral industries. Large British 'colonies' exist in the Argentine, where the sheep farming in the newly opened territories of Patagonia and Tierra del Fuego are to a very large extent carried on by Englishmen. The congenial climate of the Argentine makes it suitable for the permanent habitation of northern Europeans. Hence, this republic is destined to take its place among the great powers of the world. Besides British, there are large numbers of Italian immigrants, some of whom used to be seasonal, who usually settle down to maize and wheat planting in the province of Buenos Aires. Chile also has a considerable element of British and American population.

Since the war of 1914–18, the governments of the republics have grown restive over the exploitation of their mineral and agricultural wealth by foreigners. This exploitation enriches the foreign companies, but it opens up the country, increases its facilities for communication and transport, and altogether adds enormously to its wealth. The republican governments have, however, begun to grudge the return due to enterprise and have recently placed an embargo on the transfer of money abroad. The immediate effect of this step has been that foreign investors have been unable to get payments due to them, and the ultimate

Plate XXV

(*Mondiale*)

RIO DE JANEIRO

View taken from the Corcovado. The harbour and the famous Sugar Loaf can be seen

(*Mondiale*)

THE UPPER AMAZON

A break in the equatorial forest

Plate XXVI

(*Mondiale*)

IGUASÚ FALLS, RIVER PARANÁ

result will be that a check will be put on the development of the continent.

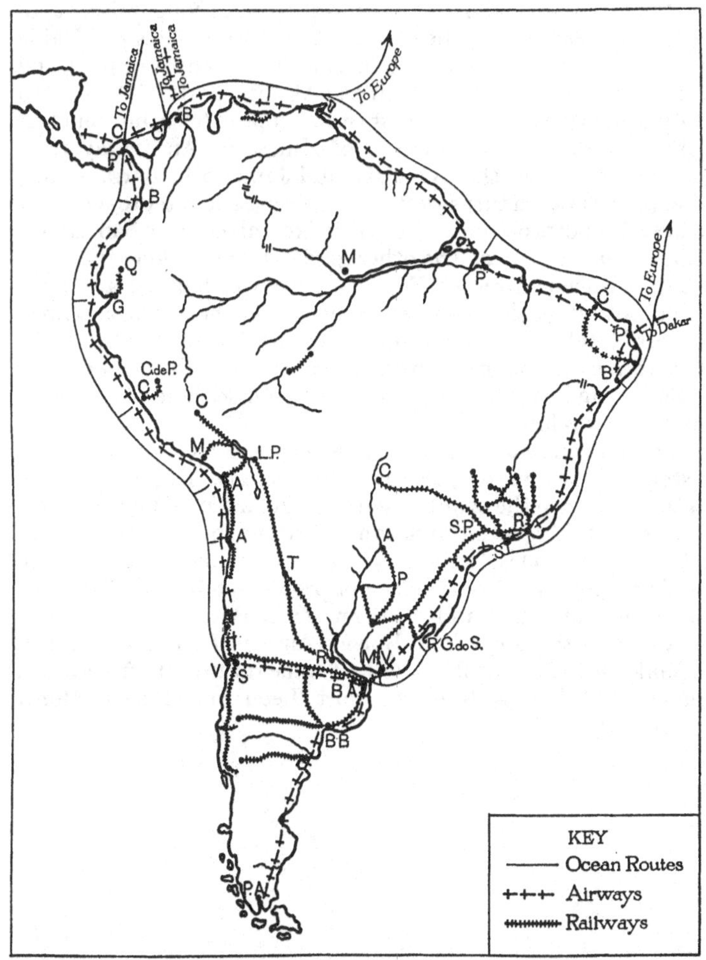

Main lines of communication

Communications. The external communications are carried on by ocean steamer, aircraft, and various forms of telegraphy.

The main lines all aim at the east coast of the United States or the west coast of Europe. European ocean routes run (1) to Kingston, the Panama Canal, and along the west coast to Valparaiso, (2) to Pernambuco and along the east coast to Buenos Aires and Bahía Blanca, (3) to Kingston, Cartagena, La Guaira, Port-of-Spain, and Barbados. Routes from the Gulf and Atlantic ports of the United States eventually follow the first two European lines, but touch at Cuba and Puerto Rico instead of Kingston. An airway from Florida now passes through Cuba and Jamaica to Barranquilla, whence one branch turns westward to Colon and after touching at the various ports on the west coast strikes inland from Santiago in Chile to Buenos Aires. The other goes eastwards along the coast as far as Buenos Aires. A French service which now starts from Toulouse and reaches Dakar in Senegal will shortly be extended to Port Natal in Brazil and ultimately to Buenos Aires. All the large towns are connected by undersea cable to the chief centres of Europe and the United States and of course there is communication by wireless.

Internal communication is by railway and river steamer. The systems of the Amazon, Paraná, Orinoco, and Magdalena penetrate deeply into the country and are regularly used for transport. Except in the Argentine and, to a less extent, in Brazil the railway network is not closely meshed, the lines being short and connecting a port with some centre of production inland. There are two transcontinental railways: one from Buenos Aires to Valparaiso, and the other from Buenos Aires to La Paz and thence to Mollendo. Others will be built at some future date from Bahía Blanca and San Antonio to Port Lebu and Puerto Montt respectively.

PART VI

AUSTRALIA, NEW ZEALAND, AND THE PACIFIC ISLANDS

BROADLY speaking, the world is divided into two parts: the larger part which looks out on to the Atlantic Ocean, and the smaller which faces the Pacific. In the Americas the Pacific lands form a narrow strip from the ridge of the backbone of mountains to the Ocean, and in Asia there are the wide areas of Japan, eastern Siberia, Manchuria, China, Indo-China, and Malaya with its associated East Indian Islands. But in the great Ocean itself is a vast swarm of land fragments ranging in size from the merest islet to the veritable continent of Australia.

These Pacific lands, whether they are part of a continent or are washed on all sides by the Ocean, form a world apart from the civilisation of western Europe. This is not through any general likeness among themselves, but just through unlikeness to the rest of the world. In modern times the Pacific world has been invaded by the West, and the results are interesting, though sometimes appalling, for we have seen the substitution of a Western people for primitive Pacific folk in Australia, the introduction of industrialism into Japan, and the forcing on the vast multitudes of China of the ideals of Western civilisation. Hitherto, the Pacific lands have been the world's back garden, but the rise of Japan to the position of a Great Power, the awakening of China, and the progress of the west coast strips of America, together with the increasing facilities and speed of transport and communication, have enhanced the importance of the area. The evolution of a 'Pacific civilisation' does not seem possible, since the very improvements in communication which might create such a separate culture will prevent its growth by maintaining a connexion with the West which will ensure a constant mingling of ideas.

The axis of the Pacific lands is the 180th degree of Longitude, that mysterious line along which one day changes into the next and whose place for all practical purposes has been taken in modern times by the arbitrary International Date Line. The Pacific coasts of the Americas and Asia are dealt with elsewhere in this book: here it is proposed to treat of only the lands lying within the southwestern quadrant of the Ocean, together with the purely oceanic islands in whatever quadrant. That is to say, our subject is Australia with New Guinea and the islands immediately connected with them, New Zealand, and the oceanic islands of the Pacific.

Build. These lands are of very varied origin. The western half of Australia is a remnant of that ancient land mass known to geologists as Gondwana Land, parts of which still survive to form peninsular India, most of Africa, and large portions of South America (see map on page 195). The eastern parts are of more recent origin and owe their present form to the forces of earth movement whose seat is in the mid-Pacific. The radial pressure of the ocean floor has caused the uprising of mighty ridges against the old resistant portions of crustal rock, and hence the ring of mountains which encircle the Pacific. These recent folds are as yet far from stable and contain a majority of the world's active volcanoes, the 'fiery girdle of the Pacific'. The accompanying map of the Western Pacific shows a series of four arcs of pressure ridges, one within the other, formed against the old mass of Gondwana Land. The outer arc meets at the Caroline Islands with one of the Asiatic arcs curving southwards from Japan. The ridges of the arcs do not everywhere rise above sea-level even in Australia, but as a rule break the surface here and there to form islands. In the outer arc the ridge is wholly under the sea, though the constructions raised by the coral polyps form clusters of atolls and similar islets. It is only outside these arcs that true oceanic islands are found, but the distance of groups like Fiji from any land of considerable size makes them biologically, if not geologically, free from continental influence. The true oceanic islands, like Easter Island and Hawaii, are for the most part volcanic in origin and rise like gigantic boils on the Earth's skin.

In Mesozoic times Australia was joined to Asia, but at the end of that period separation took place through the subsidence of the area which now forms the East Indies. The line of deep water which passes between the islands of Bali and Lombok, through the Celébes Sea and Macassar Strait to the Pacific, and which goes by the name of the Wallace Line, marks the boundary between Asia and Australasia. The number of active volcanoes in Java and other islands in the neighbourhood and the frequency of earthquakes proves that the Earth's crust is in an unstable

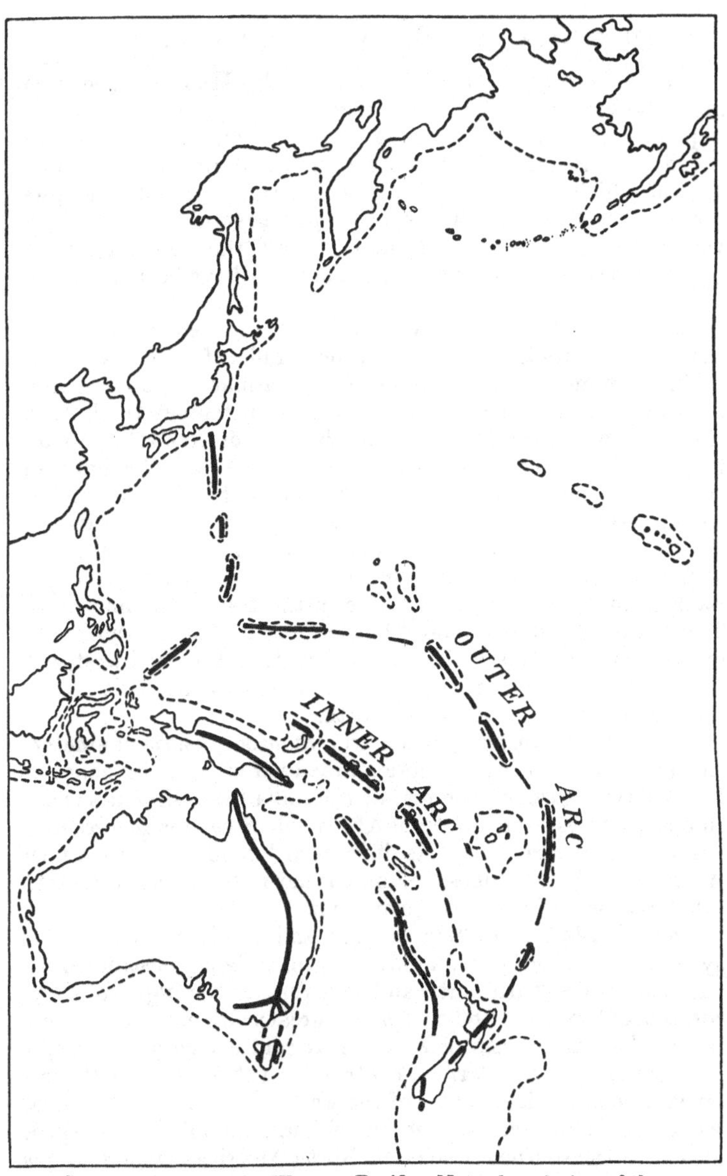

Structural arcs of the Western Pacific. Note the relation of the arcs to the ancient rock mass of Western and Central Australia

condition here still. The folds of the Alps-Himalayan system meet those of the Australasian arcs in this region, and the consequent stress and strain have produced large hollows in the broad continental shelves of Asia and Australia, thus forming seas of which that of Celébes is an instance. Similar hollows occur in the Pacific itself, especially just beyond the Outer Arc, where are found the Tuscarora Deep and the Tonga Deep, in which some of the greatest ocean soundings have been obtained.

Climate. All the lands with which we are concerned here lie between the Tropic of Cancer and the parallel of 48° S. Lat., that is, they are mostly intertropical. Only Tasmania, New Zealand, and a few oceanic islands actually have a temperate climate, though a wide belt along the south coast of Australia is subtropical. New Guinea and the islands to the east of it have an equatorial climate—of an unusually equable type in the smaller islands, where the influence of the sea is great. Over the Pacific the thermal equator swings in February and March southwards as far as the main Fiji group in Lat. 16° S., so that Fiji, Tonga, Samoa, and the New Hebrides lie in the belt of the Southeast Trades from April to January, but in the belt of northerly Trades in February and March. The latter blow from the northwest, not the northeast, since on crossing the Line they are deflected to the left by the rotation of the Earth.

The north of Australia has a true monsoon whose characteristics are no doubt all the more marked through the influence of the Asiatic monsoon. The rest of Australia has a similar succession of climates to the north of Africa, where savana merges into desert and desert into 'Mediterranean' lands. Southeastern Australia with its 'Chinese' climate is naturally the most densely inhabited part of the island-continent.

This distribution of temperature and winds gives rise to a system of equatorial ocean currents which, being obstructed by the East Indies, turn north and south under the impulse of the rotation of the Earth and complete a great swirl, moving equatorwards along the coasts of America as cold currents. The Peruvian, or Humboldt, Current contributes not a little to the relatively cool, rainless nature of the west coast of South America, while the California Current has similar, though less marked, effects on the southwest coast of North America. The numerous

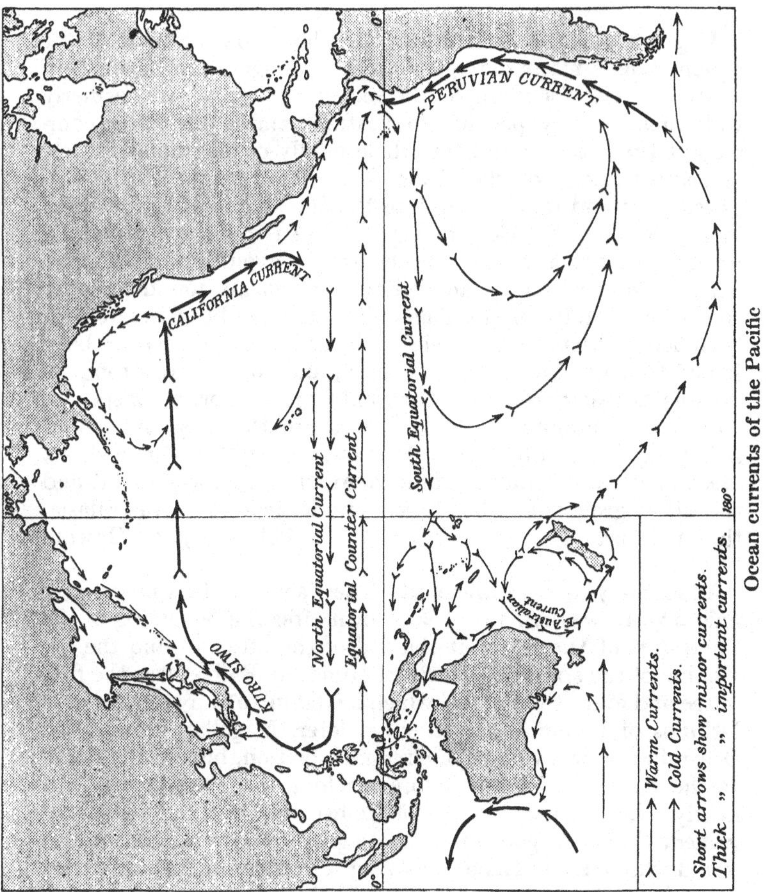

Ocean currents of the Pacific

island groups of the southwestern Pacific break up the South Equatorial Current after it turns south and give it an abnormal course as far as Chile.

Vegetation and Animals. The plants and animals of the region include many old types which no longer exist or are very rare elsewhere; for example, the eucalyptus, which was formerly widespread, is now peculiar to Australia and is one of the commonest types there; and the whole family of marsupials is now almost restricted to Australasia. It has been supposed that this 'fossil flora and fauna' is due to the age-long isolation of these lands; hence, the period during which Australia was separated from Asia must have occurred before the evolution of the more efficient modern plants and animals which are found in other continents. Many of the Pacific islands have been isolated for even longer than Australasia, if indeed they have ever been joined to a neighbouring land mass; and they show strong individual peculiarities. Their flora and fauna comprise few species, and birds, amphibians, and reptiles are the only indigenous animal life. Broadly speaking, it is possible to distinguish six areas of flora and fauna, namely, Australia, New Zealand and the island groups near by, Papua and its fellows in Australasia, the 'high' islands of the western Pacific, Polynesia, and Hawaii.

Discovery and Settlement. The Pacific and its lands were unknown to western Europeans until after the beginning of the Great Age of Discovery at the end of the fifteenth and the beginning of the sixteenth century. Nuñez de Balboa was the first to see and enter the Pacific, which he did in 1513 by crossing the Isthmus of Panamá. Seven years later, Magellan crossed the Ocean in the first voyage of circumnavigation, but touched only at the Ladrones Islands before reaching the Philippines. For nearly ninety years no useful knowledge of the Pacific was gained, though a good many voyages were made and various islands discovered. Even the voyage of Torres in 1606 produced no real result. In 1642, however, Tasman sailed from Batavia westwards round Australia, discovered New Zealand, and returned home via Tonga and Fiji; but though the voyage gave much information to geographers, it led to no settlement or further European contact. In 1768, Cook began the first of his great voyages, during which the famous navigator explored and mapped the coasts of New Zealand, followed the eastern shores of Australia from Cape Howe to Cape York, and visited so many of the island groups that he left very little for his successors to discover.

Cook had been despatched by the British Admiralty to take possession of the southern continent of Tasman in order to fore-

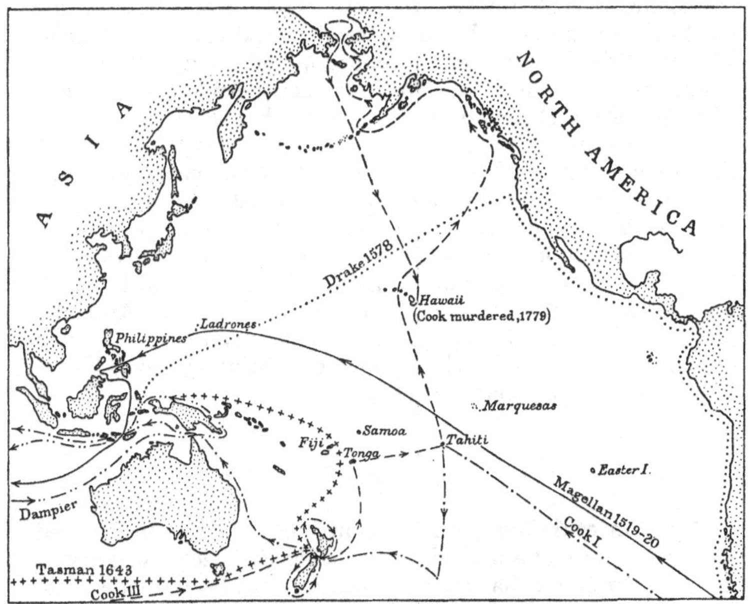

Voyages of discovery in the Pacific

stall anticipated French settlement, and as a result of his report a colony was sent out in 1788 under Governor Phillip. The colonists consisted of convicts and, as time went on, retired soldiers who wished to remain in Australia, together with a gradually growing number of 'free' settlers. Later, a similar settlement was made in Van Diemen's Land, as Tasmania was then called, and also at Brisbane. By degrees the exploration and settlement of the country extended outwards from Sydney until Victoria and Queensland were sufficiently densely peopled to be made into separate colonies. Meanwhile, a convict settlement had been planted at Perth and a free colony established at Adelaide, so that the island-continent was already falling into the six parts which in 1901 were destined to become the states of the Commonwealth.

The various other portions of our area came under European influence through the settlement of sailors, the enterprise of missionaries, or political expediency. New Zealand had become the haunt of so many disorderly seamen and runaways that in 1840 the British Government stepped in and instituted a formal administration. There was much friction between the settlers and the native Maoris over land rights, but after two bloody wars the trouble was removed. In 1874 Fiji was taken over at the request of its native 'king'. After this, the various nations of Europe began to annex all unappropriated territory, and by 1914 all the Pacific islands were in the hands of Europeans. The German islands in Micronesia were seized by the Japanese at the outbreak of the war of 1914–18 and are still held by them under mandate from the League of Nations. Outside Australia and New Zealand, European settlement is sporadic, the only considerable clusters of white population being at Suva, Honolulu, Numea, Apia, and Papeete.

AUSTRALIA

Position and Size. Australia lies between the Indian and Pacific Oceans and is in the latitude of North Africa. As it covers 40° of longitude, it has for convenience' sake been divided into three belts of standard time: (1) the eastern belt, 10 hours ahead of Greenwich, (2) the central belt, 9½ hours ahead, and (3) the western belt, 8 hours ahead. It has an area of about three million square miles and is the smallest of the continents. This and its isolation have sometimes caused it to be spoken of as the island-continent. It shares with the southern extremities of South America and South Africa the distinction of being one of the three considerable land masses in the southern hemisphere, and in consequence it has many points of likeness to those two countries. It is, however, unlike them in having its greatest length from west to east, while they taper towards the south.

Build and Relief. The nucleus of Australia is the great crustal block which forms about half the continent. Likenesses in the nature and arrangement of the rocks have led to the belief that this part of the continent was once joined to India, Africa, and the Highlands of Brazil and Guiana by vast stretches of land which formed the mighty land mass of Gondwana Land, but which have since foundered beneath the sea. It forms a tableland with an average height above sea-level of some 1200 feet. The southern

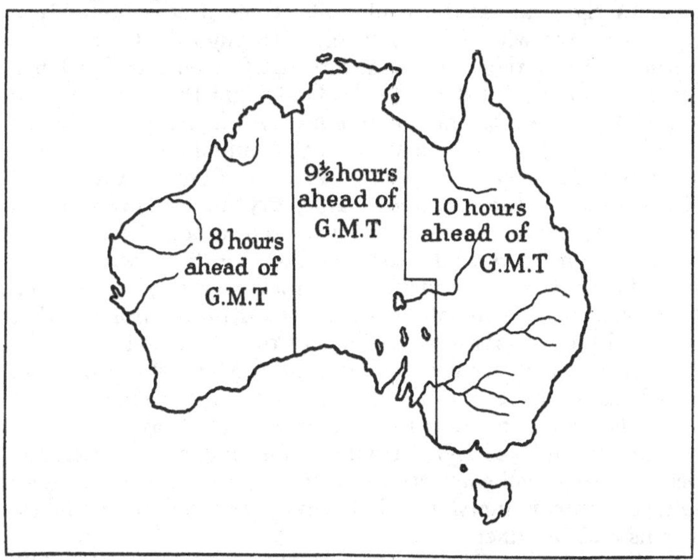

Belts of standard time in Australia

edge falls away sharply to the sea in cliffs 200 feet high, but on the west the drop is more gradual and there is a narrow strip of coastal lowland. The top of the tableland is roughened by lines of worn hills, of which the Macdonnell and the Musgrave Ranges are the most important. Much of the surface drains inland into salt lakes ranging in size up to Lake Amadeus. The Nullarbor (='treeless') plains of the south, which are made of limestone, are irretrievably barren, while farther north patches of bare rock are intermingled with poor steppe dotted with tufts of spinifex or clumps of mulga scrub. In the northwest there is an area of true sand desert, but it is not as large as it is usually said to be. The south coast of this part of the continent is unbroken and offers no shelter from the storms of the Great Australian Bight; but on the west and northwest an occasional river mouth forms a harbour, while not a few barren islands fringe the northwestern coast. It was the forbidding nature of this part of the Australian coast that repelled Dutch settlement in the sixteenth century.

To the east of the great tableland is an area of lowland which stretches right across the continent. Once it was probably a large inland sea with a broad entrance through the Gulf of Carpentaria. A low ridge divides it into the Carpentaria Lowlands—which slope gently down to the Gulf—and the Eyre Depression, a huge area of inland drainage which, lying astride the Tropic of Capricorn, gets a very scanty rainfall with a summer maximum. Its streams, of which Barcoo, or Cooper's Creek, and Diamantina are comparatively large, dry up for three-fourths of the year, leaving a line of water holes or a bed of damp mud to mark their courses. In summer they are filled with turgid water bearing a large quantity of rock waste in suspension. Many of them end in salt pans, being restrained from further progress by the increasing pastiness of the water and mud. The larger ones empty into vast salt swamps, the biggest of which are named Lakes Eyre and Frome. In periods of drought these 'lakes' become expanses of salt over which a heavy lorry has been driven on Lake Eyre. Owing to faulting, a depression has been formed which connects the Eyre Depression with Spencer Gulf. Its floor is marshy and contains two shallow salt lakes, Torrens and Gairdner.

To the east of the Eyre Depression, the ground rises into wide, rolling plains which form the basins of the Darling and Murray Rivers. The low Grey and Selwyn Ranges separate the plains from the Eyre Depression. Eastwards, the ground gradually rises to the Eastern Highlands, while the basin of the Murray falls away westwards to Lake Alexandrina, a lagoon into which the river empties. Owing to the scanty rainfall in its lower course, the Murray has cut a shallow canyon for itself through the plain.

The Eastern Highlands are one of the arcs of fold mountains which are found on the western side of the Pacific. At one time their nucleus of granite probably formed three islands off the east coast of Gondwana Land, but the push from the Pacific raised them, together with the surrounding rock, into what is called a monoclinal, or one-slope, range. The slope drops steeply to the east, while on the west it is gentle and the upland merges gradually into the plains. The Highlands do not form a single range, but a complex mountain system composed of several masses of which the most important are the Australian Alps, the Blue Mountains, and the New England Range. These are

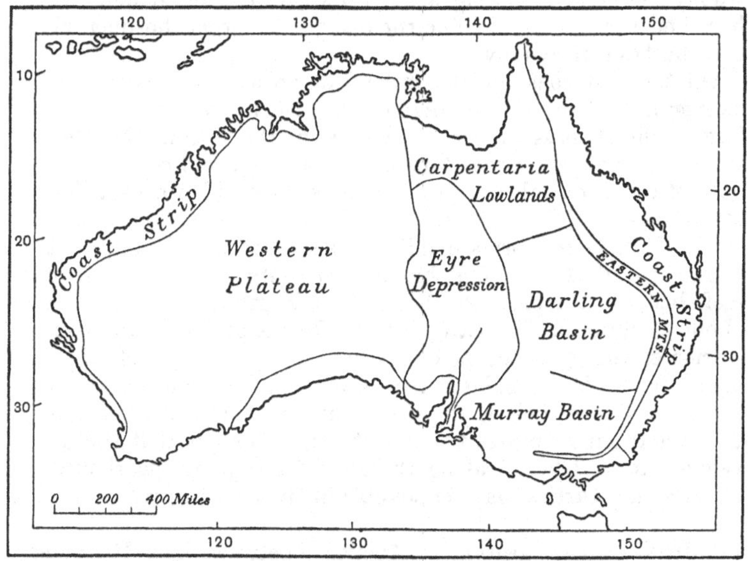

Physical divisions of Australia

separated by gaps which played an important part in the opening up of the country and which to-day carry the principal routes through the Eastern Highlands. The Cassilis Gate, which is connected with the River Hunter and leads inland from Newcastle, the Goulburn Gap at the head of the Murrumbidgee, and the Kilmore Gate behind Melbourne are the chief. In Queensland the mountains come right down to the sea, but in New South Wales there is a narrow coast plain. In Victoria the Highlands send a branch westwards to end in the West Victorian Highlands. South of these is the Great Valley of Victoria, one of the richest agricultural areas in the continent. Closing this valley on the south is a low range which is broken into two sections, the Gippsland and Otway Hills, by the drowning of Port Phillip and which forms excellent pasture land for both sheep and cattle. Tasmania, which is one of the hill masses of the Eastern Highlands, has lost its connexion with the rest through subsidence. From Queensland to Tasmania the coast plain has been almost wholly submerged, leaving a narrow strip of lowland indented here and there with rias like the one which forms the magnificent harbour of Sydney.

Off the east coast of Queensland at an average distance of twenty miles from the mainland is the Great Barrier Reef, the largest single mass of coral structure in the world. The Reef stands on the former coastline of the continent and consists not only of coral, but also of projections of rock which are outliers of the Eastern Highlands and stand up as islands. The section of the Reef from Torres Strait to just north of Hinchinbrook Island is fairly continuous, but farther south it is just a line of reef-clusters, islands, and shoals. The length of the whole is about 1500 miles. The sea within the Reef is protected from the storms of the open ocean, but it has its own terrors, of which uncharted reefs, irregular currents, and a great tidal range are the chief. Nevertheless, the steamship route from Australia to China and Japan passes through it. It is thought that the development of fisheries along the Reef for trepang, pearl-shell, trochus, and turtles may be possible in the near future.

Climate. The position of Australia between Lat. 10° and 40° S. gives the continent a range of climate that varies from warm temperate to tropical. About half the country actually lies within the Tropics. A glance at the map on page 431 shows that Australia is approximately in the same latitude as the Sahara, Tasmania corresponding to the north of Spain, Victoria to the south of the same country, while South Australia, Western Australia, and parts of New South Wales correspond to the Atlas region of North Africa. New Guinea, which is cut by the Equator, lies in the same latitude as the Gold Coast.

Apart from latitude, there are two special factors which influence the climate of Australia, namely, compactness and length from east to west. Unlike South America and South Africa, with which it is usual to compare Australia because these three are the only considerable land masses in the southern hemi-

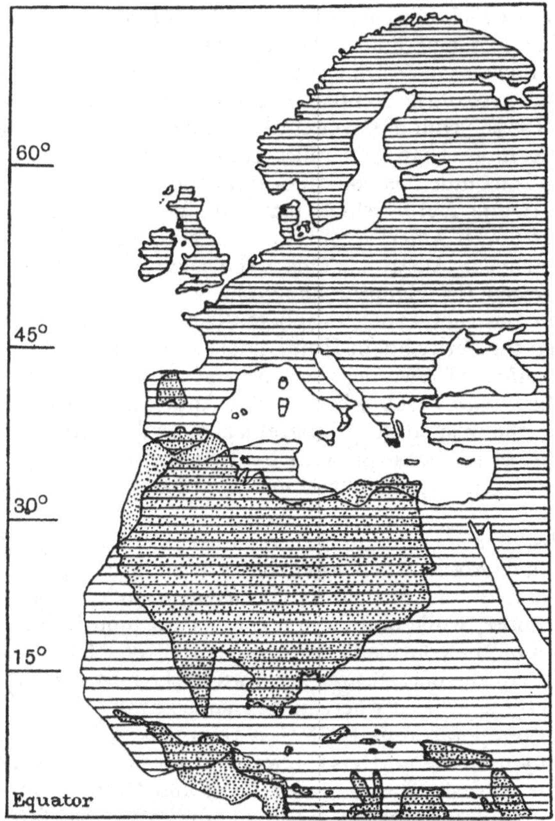

Comparison in latitude between Australia and North Africa

sphere, the greatest length of the island-continent runs from east to west, while the other two taper towards the south. If Australia were turned round so as to have its greatest length from north to south, as all the other continents have, there would be a far greater range of climate. As it is, Australia has the greatest uniformity of climate of all the continents. Its relatively huge longitudinal span—together with the protecting ring of highlands on the east and southeast—causes vast areas of semi-desert which have far smaller and less arid counterparts in South America and South Africa. The extremes of heat and aridity which occur within the continent are unmoderated by the sea, because of the compactness of the land mass, whose only large inlet is the Gulf of Carpentaria. While no part of peninsular Europe is more than 300 miles from the sea, the centre of Australia is as much as 600 miles from the coast.

There is one factor which plays an important part in the climate of South America and South Africa, but which has but little influence in Australia. This is the cold ocean current which flows towards the Equator along the west coasts of the continents in the southern hemisphere. The Benguella Current in Africa and the Peruvian (or Humboldt) Current in South America have the effect of lowering the temperature of the coasts along which they pass. But whether the Antarctic Current which flows past the west coast of Australia is more diffused or whether the outline of the coast presents too little land to the current—whatever the cause, the effect on the climate seems to be very slight.

The distribution of temperature is best studied from the set of maps on page 433, where it will be seen that in July, the coolest month, a fifth of the continent has a mean temperature of 70° F. or more, while in January, the hottest month, more than four-fifths of the continent have a mean of over 80° F. In the January map the hottest area is shown in the northwest and the coolest in the southeast where, if Tasmania be taken into account, the mean temperature is 10° F. less than it is in the southwest. In the July map the north is the warmest area and the southeast is still the coolest. The maps for April and October are intermediate. Midday temperatures in summer often rise far higher than the means shown on the maps. Adelaide, for instance, has known a temperature of 116° F., and Melbourne one of 111° F., while 120° F. is not an infrequent figure in parts of Queensland. These extreme temperatures are not, however, as unpleasant as might be expected, because the dryness of the air makes the human body less sensitive to great heat or great cold.

The correspondence between the temperatures of Australia and those of Spain and North Africa is illustrated by the following table:

Place	Spring	Summer	Autumn	Winter
	° F.	° F.	° F.	° F.
Hobart	54·0	59·8	55·4	45·7
Santiago	51·8	64·8	55·6	45·1
Wilcannia	68·2	81·4	65·4	50·0
Marrakesh	66·7	82·0	69·8	51·6
Darwin	85·5	84·0	84·2	77·2
Kayes	94·1	83·7	84·5	77·2

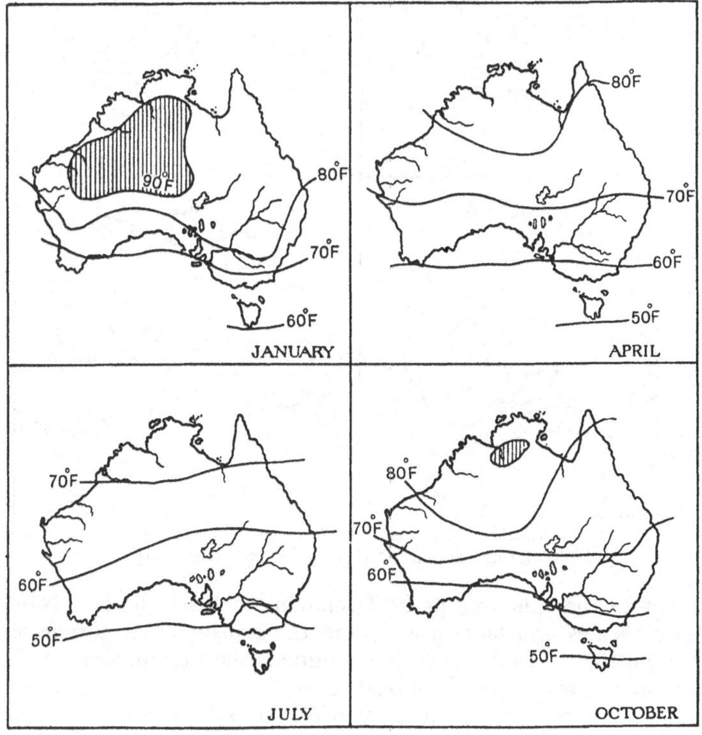

Distribution of temperature in Australia

The maps below show the distribution of pressure and the consequent system of winds in the months of January, April, July, and October. It will be seen that, except in January, Australia is crossed by the high-pressure belt which one expects to find on the poleward side of the Tropic. Offshore winds are

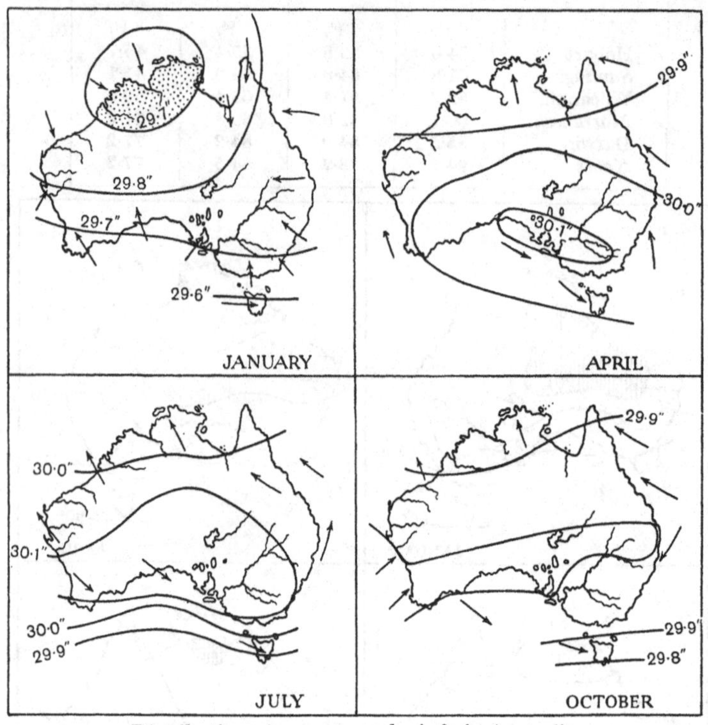

Distribution of pressure and winds in Australia
Note the monsoonal setting in January and July

therefore the rule, except in Tasmania, the east coast of New South Wales and Queensland, and the southwestern corner of the continent. In January (i.e. summer) the high-pressure belt is pushed south and a low-pressure centre forms over the northwest of Australia. Consequently, winds become onshore on every coast, though not with the same effects at all points. In the north this change from southeasterly offshore to northwesterly onshore winds becomes a true monsoon system, though its influence does not penetrate very deeply into the continent.

As a result of this distribution of pressure there are five wind systems in Australia, namely, (1) the monsoon in the north,

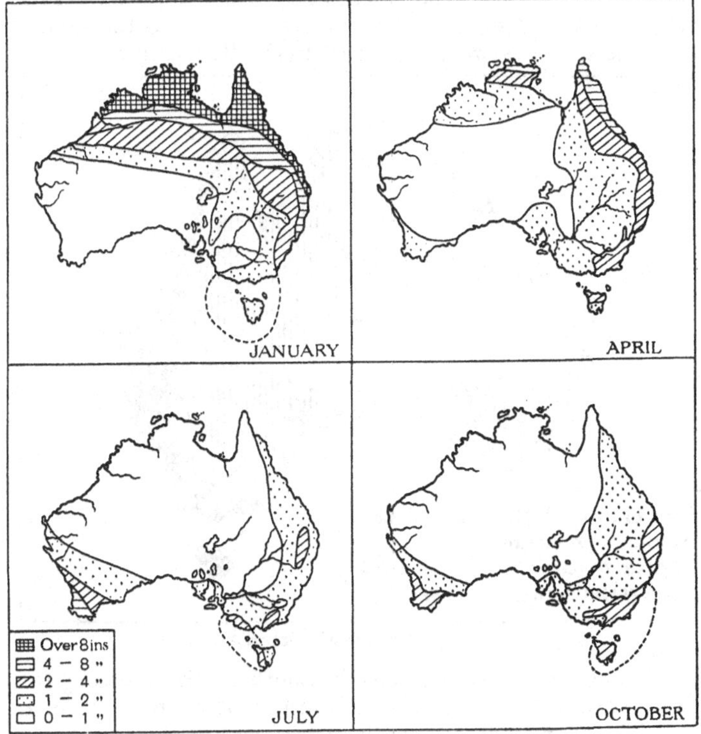

Monthly distribution of rainfall in Australia

(2) the trade wind system of the east coast which extends farther south and is intensified in summer, (3) the south coast system with warm northwesterly winds in winter, but cool southerly breezes in summer, (4) the area of fitful winds in the high-pressure belt of the interior, and (5) the system of Westerlies in Tasmania.

Each of these wind systems has its own rainfall régime. Tasmania gets abundant rain well distributed throughout the year from its prevailing Westerlies. The east coast of the mainland is also well supplied throughout the year, especially in the tropical districts of north Queensland, where there is an average of 100 inches a year. On the western side of the Eastern Highlands the drying effect of descending winds is soon felt, and the rainfall quickly lessens westward from the divide. The monsoon which

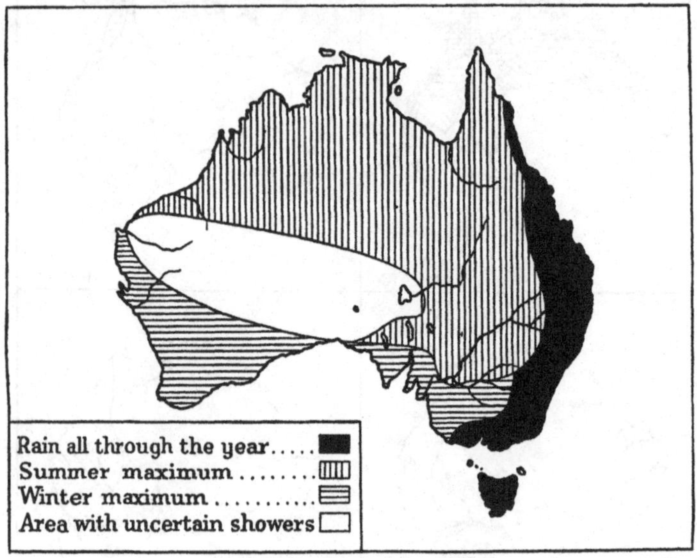

Rainfall regions of Australia

develops in summer strengthens the Southeast Trades along this coast and extends them farther south, thus enabling them to carry their influence farther into the continent and to give summer rains to the plains and the Central Lowlands. In the north the effect of the monsoon is to give an abundant rainfall in summer, with little precipitation in the winter months. The south coast derives its supply of moisture from the northwesterly winds which blow in winter, but gets little rain from the southerly breezes which prevail most of the year, since these are cool winds moving to warmer latitudes. This area has therefore a rainfall of Mediterranean type. The central portions of the continent have less than 10 inches of rain a year and, like all areas of scanty rainfall, suffer from great irregularity of system. They are therefore at best areas of semi-desert, becoming true sand desert over a small space in the northwest between Shark's Bay and the Tropic of Capricorn. The map on page 436 shows the distribution of these five systems.

Drainage. The extent of areas with scanty rainfall makes the river system of Australia a subject of unusually great importance. The majority of streams have a tropical régime, that is, they are flooded during the period of rains in summer and sweep onward bearing a large quantity of rock waste, while in winter, owing to the much decreased rainfall, they are reduced to a mere trickle, to a line of water holes, or even to a discontinuous trail of damp mud marking the course. Such rivers are useless for navigation or irrigation and unreliable for any other purpose. Such, for instance, are all the rivers of north and northwestern Australia, though a few, like the Flinders and the Roper, succeed in being permanent streams, with a much reduced volume in spring.

Of similar régime are the streams of the great area of inland drainage which occupies the centre of the continent. The Barcoo and the Diamantina, which drain into Lake Eyre and are considerable rivers in summer, are the most important; but even these practically disappear in the season of drought. The Eyre basin comprises about one-third of the whole area of inland drainage, the remainder being divided into innumerable small basins of which the largest is that of Lakes Torrens and Gairdner.

The only really important river in Australia is the Murray, whose drainage basin covers the plains of Victoria, New South Wales, and southern Queensland. The main stream rises near the snow-clad peak of Kosciusko, whose melt-water supplies it throughout the year, but gives it a maximum in spring. The district round its headwaters is within the area of rain throughout the year, and so the river tends to keep an even volume with a slight maximum in summer when the rainfall is most abundant. Its two feeders, the Murrumbidgee and the Lachlan, have a similar rain-fed régime, though their summer maximum is more marked. Once on the plains the Murray gets little reinforcement, but flows through an increasingly dry and barren country. About two-thirds of its way to the sea the river is joined by the Darling, which drains the large and almost enclosed basin formed by the plains of northern New South Wales and southern Queensland. The régime of this feeder is typically tropical in accordance with the rainfall system of its basin, and, while in summer the volume exceeds that of the Murray and boats can ascend as far as

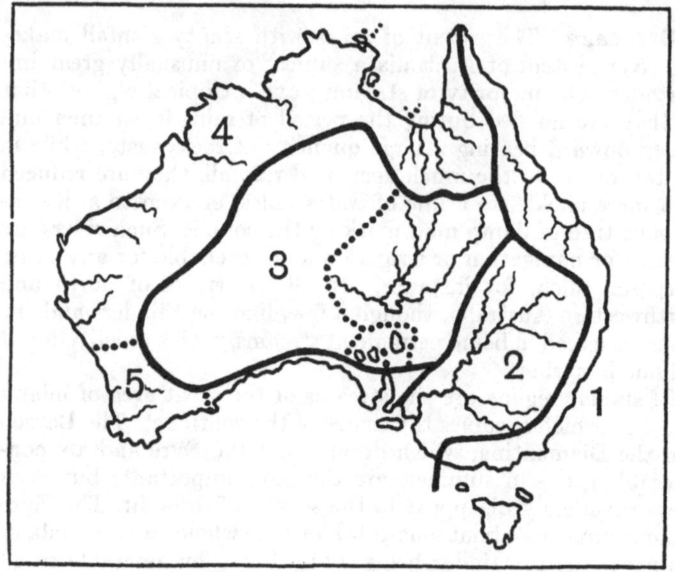

Drainage areas of Australia

1. **Pacific drainage**
2. **The Murray-Darling basin**
3. **The inland drainage systems**
4. **Drainage into the Indian Ocean**
5. **The southwestern area**

Bourke, in winter the flow is at best reduced to that of a puny and unnavigable stream, while at times there is merely a chain of disconnected water holes. After their junction the Murray and the Darling flow on without receiving a single regular feeder and are thus comparable with the Indus, Nile, and Euphrates-Tigris in that these rivers all rise in a region of abundant water supply, but in their lower courses pass through a dry country where they shrink owing to evaporation and seepage. The end of the Murray is an anticlimax, for instead of the broad and noble mouth expected by the early explorers, the river empties into the shallow lagoon of Lake Alexandrina, whence its waters pass to the ocean by percolation through a sandspit.

Apart from the Murray, there are two groups of perennial streams in Australia. The first and least important comprises those of the southwestern corner of the continent where, we have seen, a Mediterranean rainfall system prevails. Here the smaller streams are of the wadi type, flowing amply in winter, but being much reduced in summer; while the larger ones, like the Swan, are truly perennial, though they too show a decided winter maximum. The other group consists of the rivers which flow from the Eastern Highlands into the Pacific, together with those of Tasmania. A glance at the map on page 438 will show that they drain a region in which rain falls throughout the year, and it is clear that they will be perennial. The largest are the Hunter and Hawkesbury in New South Wales, the Burdekin and Fitzroy in Queensland, and the Snowy in Victoria. The last drains the southern slopes of Kosciusko and is therefore snow-fed.

Artesian Basins. As if to compensate in some degree for the scanty rainfall from which large areas of Australia suffer, the continent has extensive artesian basins in which water may be obtained by boring. The map on page 440 shows the distribution of these basins. The water drawn from artesian wells in these areas has two drawbacks: (1) the supply cannot be relied upon indefinitely, and (2) so much salt is held in solution by the water that, though it can be used for stock, it generally impregnates with salt and ruins any field which it has been used to irrigate.

Vegetation. Just as the rainfall of Australia is peripheral, so the luxuriance of the vegetation lessens towards the centre of the continent. Considerable areas of tropical rain forests occur in Queensland, while the Eastern Highlands are everywhere clad in woodland—open tropical forest in the north, warm temperate forest in the south. A scrub forest of Mediterranean type grows in the southwestern corner. To the west of the Eastern Highlands a belt of temperate grassland is found, and here it was that Australia's greatest productive industry grew up, but towards the Central Lowland the vegetation gradually passes into poor steppe and finally into scrubland and semi-desert. A similar, though poorer, belt of temperate grass lies outside the south-western woodland. Across the north of the continent runs a band of savana which becomes progressively poorer towards the south. Finally, the central portions of the continent are occupied by xerophilous vegetation of which the chief plants are the salt-bush, the mallee scrub, and the mulga scrub. The former scrub is a dwarf eucalyptus, the latter a stunted acacia. The very worst

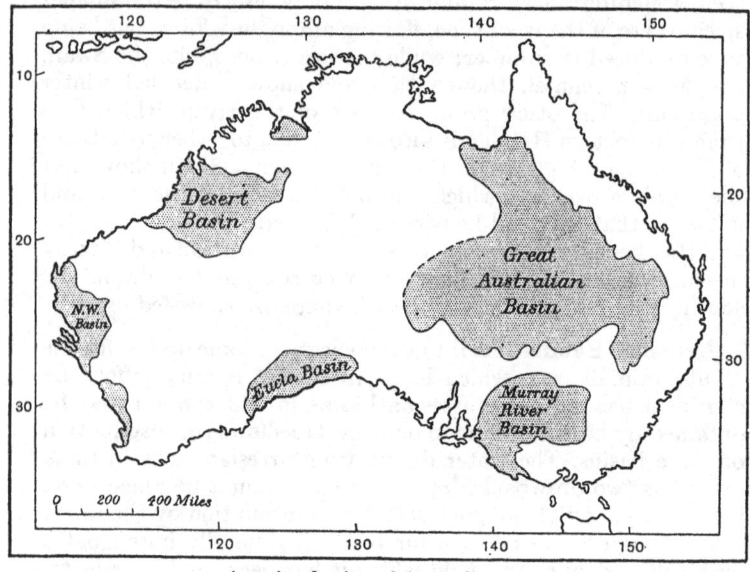

Artesian basins of Australia

areas are without even these drought-loving plants and are characterised by the wiry spinifex which grows in tufts in the sand.

The characteristic type of tree is the eucalyptus, or gum tree, of which many species exist. The majority have the peculiar xerophilous adaptation of hanging their leaves vertically so as to avoid the incidence of the sun's rays. Some yield the familiar oil of our medicine chests, others give hardwood timber. The chief of the latter are the blue gum of the Eastern Highlands and the jarra and karri of the southwest. There is no good softwood timber in Australia, which results in the importation of timber for house building from Canada and New Zealand. The bark of some of the species of eucalyptus is used for tanning, but the best results are produced from the black wattle, a kind of acacia. Apart from these, there are no native plants of any great economic value, all needs having to be satisfied from imported species. Wheat, barley, and rye have been introduced from Europe and maize from North America; the vine and all the various kinds

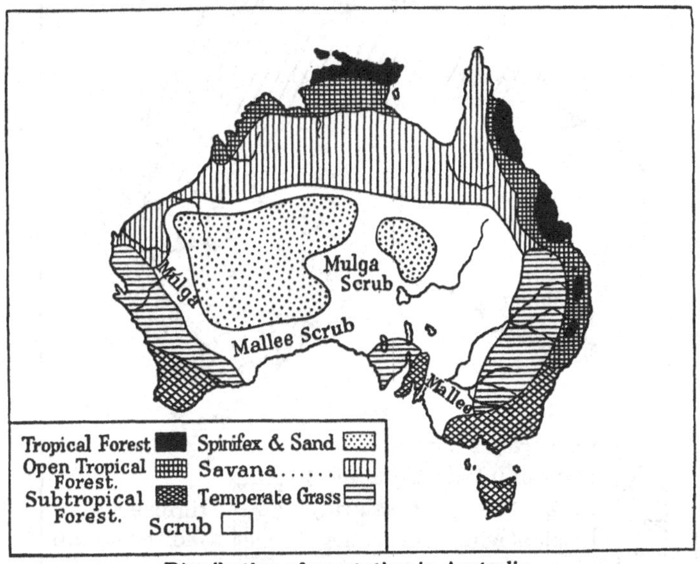

Distribution of vegetation in Australia

of European fruit trees now flourish, especially in Victoria and South Australia. Not all the imported species have proved beneficial, since the native flora and fauna seem incapable of resisting the competition of exotic rivals, and many comparatively harmless plants, like the American prickly pear, have proved to be pests by rapidly overrunning wide areas of country. A similar tendency is noticed among the animals.

Animals. The native animals of Australia are intensely interesting, but are of little importance. The kangaroo, a peculiar kind of grass-eating animal, is the enemy of the sheep farmer, who kills it as readily as he kills the rabbit. The woods have a dwarf bear, the wombat, and various other animals, most of which like the kangaroo are marsupials. River banks are the home of the duck-billed platypus, an animal so strange that the first stuffed specimen to reach England was thought to be a joke.

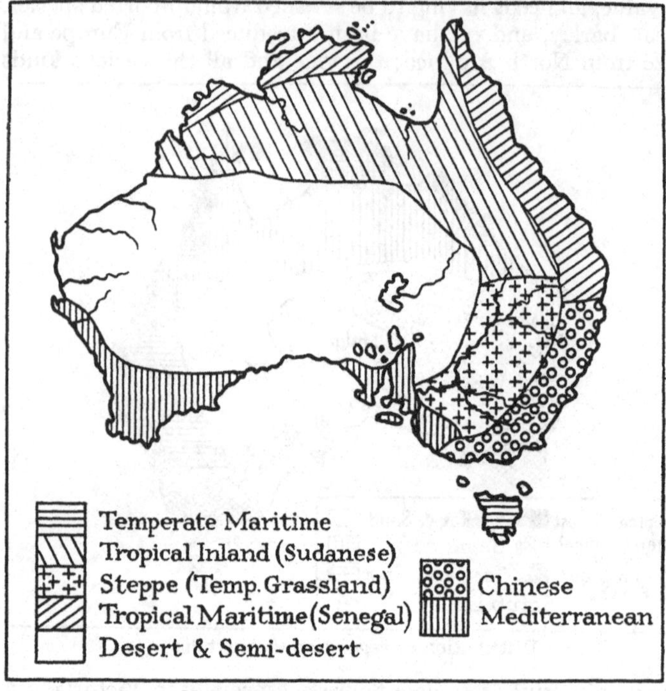

Main natural regions of Australia

There are also the dingo, or wild dog, and the flying fox, a large fruit-eating bat. In the semi-desert are two big ostrich-like birds, the emu and the cassowary.

Natural Regions. The various physical factors which have been discussed in the foregoing pages mark off Australia into seven main geographical regions. The most southerly consists of the island of Tasmania, which, as has been said, is a detached portion of the Eastern Highlands. Its granite core rises on the west to form some rugged country topped by peaks 5000 feet high. The centre and east are lower on the whole and are cut by the trench-like valleys of the Derwent and Macquarie rivers. The climate is temperate maritime in type, with prevailing northwesterly winds and a moderate rainfall evenly distributed throughout the year. The western hills cast a rain shadow over the centre and east, where the rainfall is in some places as low as 20 inches a year. The vegetation of the highlands is woodland in which the chief

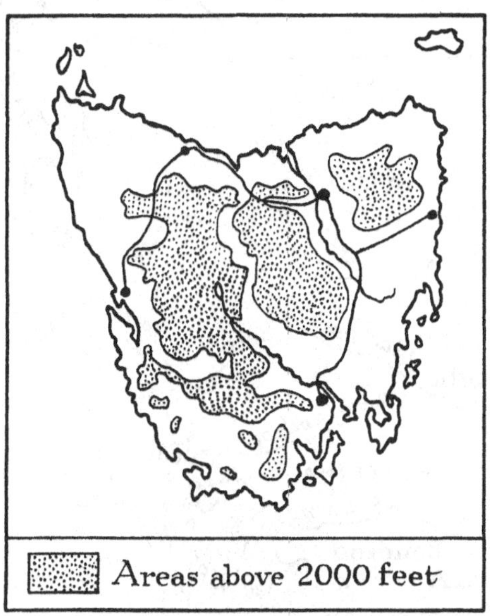

Relief and railways of Tasmania

species are eucalypts, southern beeches, and ferns. A plentiful rainfall and young rivers lead to the production of hydro-electricity for the supply of the whole island. Hobart, the capital, and Launceston stand on rias at opposite ends of the island.

On the coast strip and highlands of the mainland opposite and northeastwards as far as Brisbane lies a region whose climate is of the Chinese type. It has a moderately warm temperature, a fairly abundant rainfall well distributed throughout the year, and a forest vegetation which includes some of the largest trees in the world. The region may be subdivided into (1) the Gippsland and Otway Hills, where dairying is following fast on the clearance of the forest, (2) the Great Valley of Victoria and its continuation

Southeastern Australia

as a coast strip into New South Wales, (3) the Eastern Highlands, which include the West Victorian Highlands, the Australian Alps, the Blue Mountains, and the New England Range. The Great Valley focuses on Melbourne, which has the further advantage of the Kilmore Gate behind it. The town is built on the Yarra, not on Port Phillip itself. Round the shores of the inlet are many pleasure resorts and some considerable towns, including Geelong (pop. 42,000). On the east coast strip Sydney owes its importance to the extra broadness of the plain at this point and to the convenient positions of the Cassilis Gate and the Goulburn Gap relative to the port.

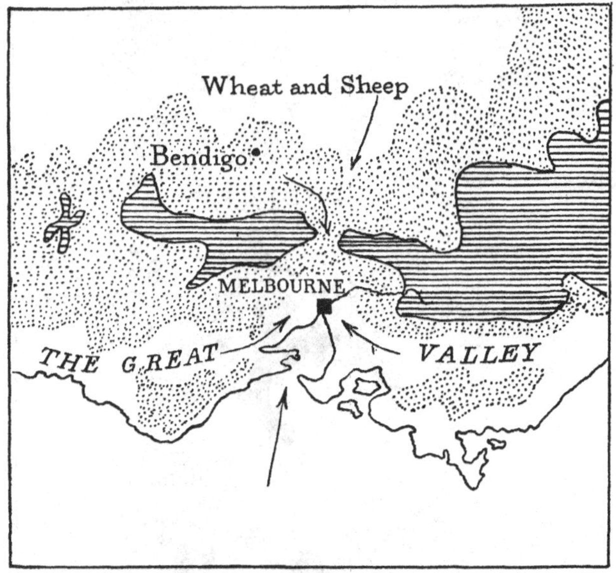

Position of Melbourne

West of the 'Chinese' region lies one whose climate is of the Mediterranean type, but which is broken into two parts by the Great Australian Bight. The eastern portion consists of the South Australian Highlands, the rift valley area, Eyre's Peninsula, and the mallee scrub country southeast of Lake Alexandrina. The mallee scrub lands are being gradually used for wheat, while the flood plain of the Murray is developing under irrigation into an important fruit-growing area. The South Australian Highlands consist of a line of fold hills between two and three thousand feet high. The Flinders Range is the highest portion, and in the south the uplands tail off into the Mount Lofty Range, while Kangaroo Island is a detached section. The slopes are forested, but when cleared produce Mediterranean fruits, especially the grape and the olive. The rift valley, which was formed by the same earth movement as folded the South Australian Highlands, runs from Lake Torrens in the north to Adelaide in the south. Owing to its reliable—albeit not over-

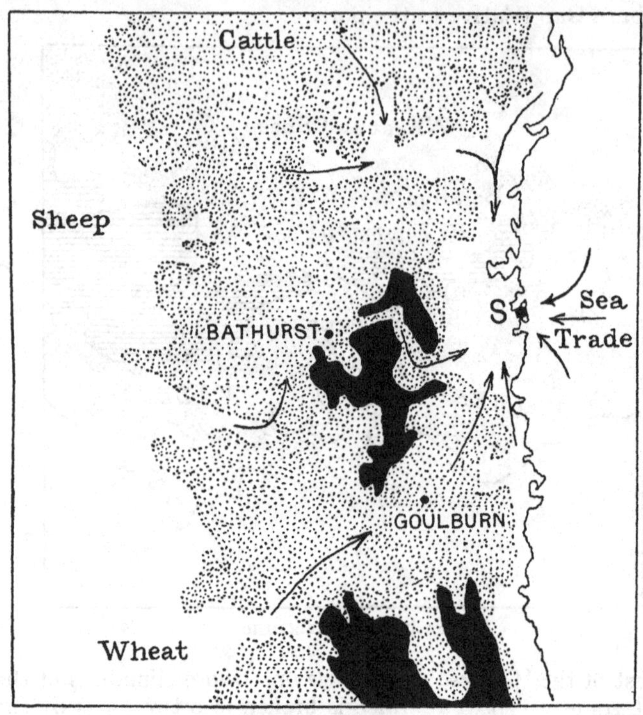

Position of Sydney

plentiful—rainfall, it is the richest part of the region, growing wheat of high quality and providing good pasture for sheep. Port Pirie and Port Augusta have an importance due to matters which are external to the region. Eyre's Peninsula is another area of reliable rainfall and is becoming an increasingly important wheat and sheep country.

The other area which forms part of the 'Mediterranean' region occupies the southwestern corner of the continent. It is

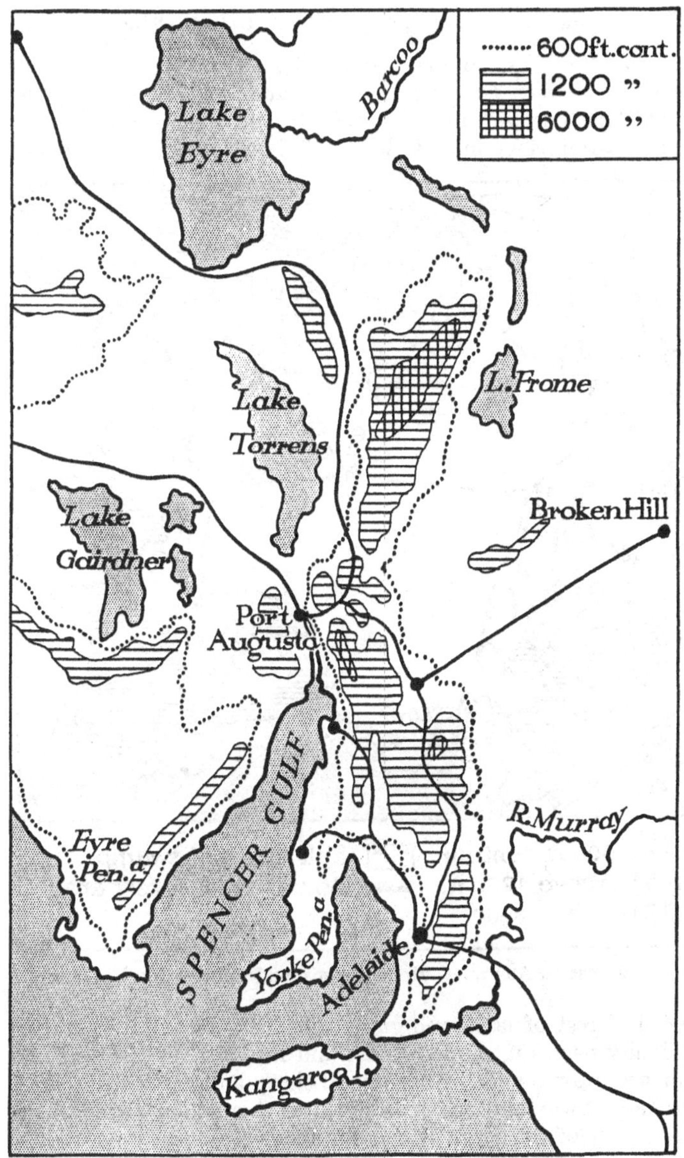

Part of South Australia, showing relief, towns, and railway development

a part of the Western Tableland, with a narrow coast strip on the west between the Darling scarp and the sea. The district about Cape Leeuwin receives 40 inches of rain a year, but the amount diminishes inland. Where it is as much as 25 inches,

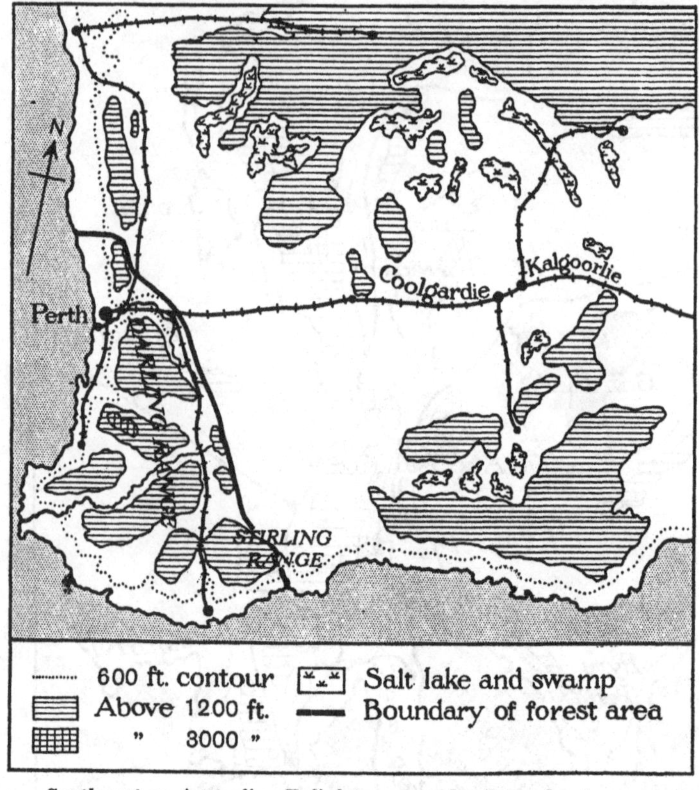

Southwestern Australia. Relief, towns, and railway development

there is forest of jarra, red gum, and karri, but the vegetation gradually passes into open forest and mallee scrub. Wheat and fruit are grown, and farther inland sheep and cattle are reared. The name Swanland has been given to this region owing to the importance of the Swan River as a focus.

Plate XXVII

(*Mondiale*)

AN *ESTANCIA* IN THE ARGENTINE

Plate XXVIII

(Mondiale)

SAVANA IN CENTRAL AUSTRALIA

The inner slopes of the extra-tropical portions of the Eastern Highlands as far as the Darling consist of temperate grasslands. Along the upper slopes are rolling pastures of which the Darling Downs and Liverpool Plains are some of the best parts. Broadly speaking, this eastern portion of the region has the better rainfall and is thus able to grow wheat as well as to raise sheep, while the western parts of New South Wales are too dry for anything except sheep farming. Bourke is the focus for the drier parts of the Darling basin. Between the Murray and the Murrumbidgee is the fertile Riverina which irrigation is gradually turning into one of the great wheat areas of Australia. The Burrinjuck Dam, which was intended to water this plain, is one of the largest constructions of its kind. Farther west between the Murray and the West Victorian Highlands lies the Wimmera, once a waste of mallee scrub, but now a promising area in which, by means of irrigation, wheat crops are being produced. Near the river fruit is grown, the industry centring round Echuca and Mildura.

Within the borders of Queensland this temperate grassland changes gradually into savana, the belt of which sweeps round to the northwest coast. It is a cattle and cotton country which needs only time to become populous and wealthy. The water supply grows more and more uncertain with distance from the sea, and midday temperatures run up to 110° and 120° F. in the shade. Charleville and Normanton in Queensland, Daly Waters in North Australia, and Wyndham in Western Australia are focal settlements.

Round the north and northeast coasts there is a strip where proximity to the sea lowers the temperature and increases the rainfall, thus clothing the land in tropical forest (see map on page 441). In Queensland this region includes the Eastern Highlands, which are here broader and lower than they are farther south. Though broken into three or four blocks, the Highlands here go by the single name of Great Dividing Range. Europeans find the moist heat very trying, and it is doubtful whether they will ever be able to make a permanent home in this region. Sugarcane and tropical fruits, such as the banana and pineapple, are the chief products, and it is characteristic of the region that it grows tropical produce, not to export to Europe, but to sell in the Australian market. There are several mining centres in the

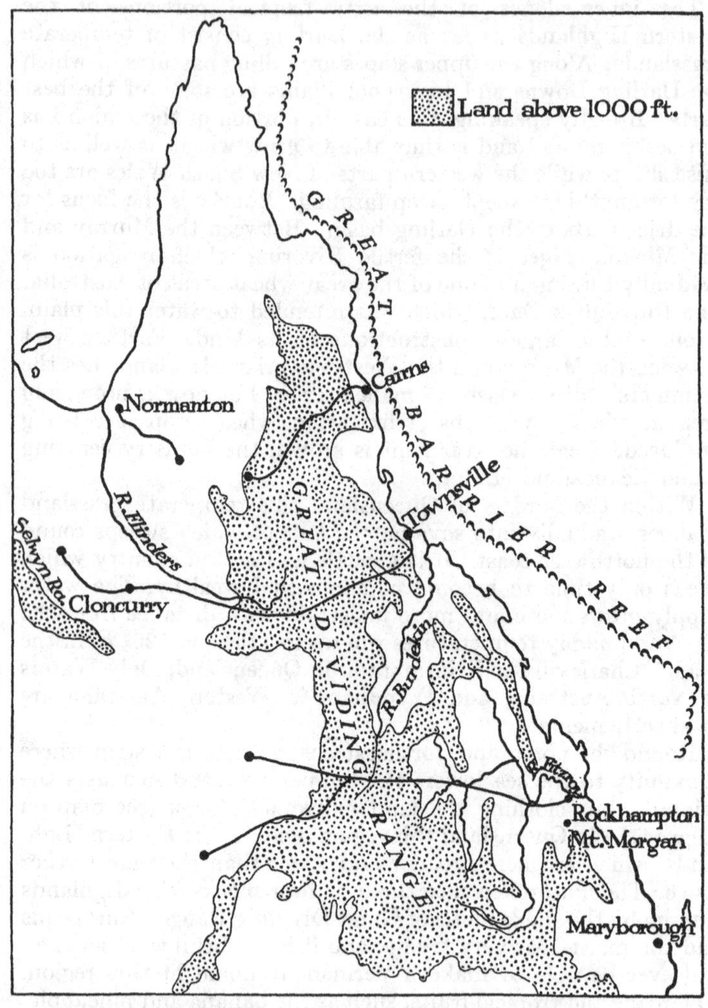

Northeastern Australia. The relief, chief towns, and railway development are shown. The area covers what is often spoken of as 'Tropical Queensland'

Dividing Range, each forming a basis for the importance of one of the four chief ports: Brisbane, Rockhampton, Townsville, and Cairns.

The vast central portions of Australia are given up to desert and semi-desert. The huge salt-impregnated Eyre basin has been well named 'the Dead Heart of Australia', since little except salts can be produced there. In the northwest there is an area of true sand desert with shifting dunes and an uncompromising absence of vegetation, but over the rest of the region are found a xerophilous plant association of mallee scrub in the south and mulga farther north, with saltbush and spinifex everywhere. Hardy crossbred sheep can eke out a living on these plants, but need a wide area over which to range. There is much controversy and difference of opinion as to the capacity of the region for human habitation. Place names like Alice Springs suggest the importance of water in a thirsty land, where the most populous settlements are mining towns like Kalgoorlie, Coolgardie, and Broken Hill.

Human Geography

Exploration. The settlement made at Sydney in 1788 soon spread over the narrow coast strip, and adventurous colonists began to make their way westwards. Fortunately, the aboriginal natives were too few to be much of an obstacle to this expansion; but the crossing of the Eastern Highlands gave some trouble, and it was not until 1813 that the divide was crossed by Gregory Blaxland, who discovered the grass plains of New South Wales. Other explorers followed at once, and in 1824 Hamilton Hume made his way across the Murrumbidgee and over the Victorian Highlands to Port Phillip. It was now supposed that there must be a large inland sea in the interior, since the westward-flowing Murrumbidgee and Lachlan seemed too big to end in salt pans and since no large river mouth had been discovered, though the continent had been circumnavigated by Flinders in 1803. To put this theory to the test, Captain Charles Sturt dragged a boat across the divide in 1829, sailed down the Murrumbidgee into the Murray, discovered the Darling, and finally reached the mouth of the river at Lake Alexandrina. In 1844 another journey of Sturt's and the work of Mitchell in the following year completed the framework of exploration in the southeast.

On the foundation of the colony of South Australia in 1834 a period of vigorous pioneering set in, and many expeditions were made in search of the best farm land. The most important of these journeys was that of John MacDougall Stuart who in 1858–62 crossed the continent from Adelaide along approximately the line of the modern overland telegraph. Burke and

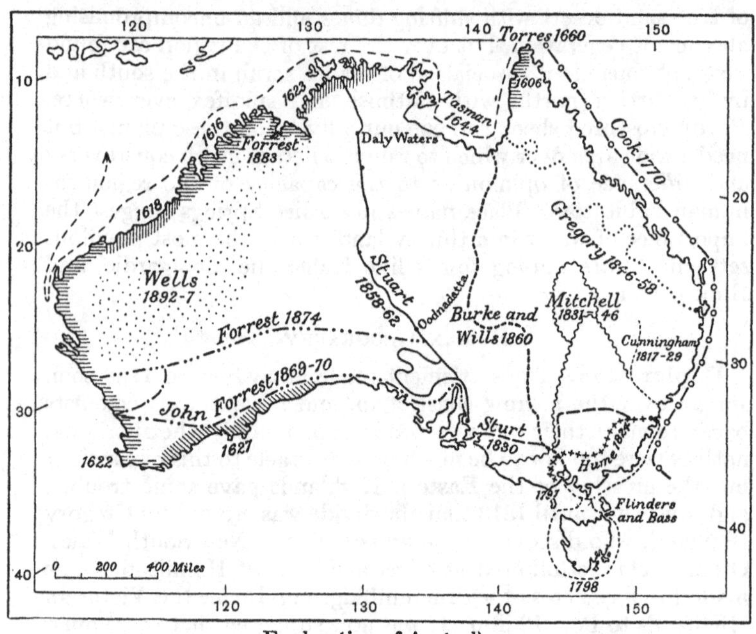

Exploration of Australia

Wills also crossed the continent in 1860, losing their lives on their return journey. The expedition achieved little in itself, but much exploration was carried out by the parties sent to look for Burke and Wills. Meanwhile, Gregory had explored Queensland, and Edward Eyre had added the Lake Torrens district to the map. Since 1870 the great western region has been steadily opened up, the most important names connected with the work being those of Forrest, Giles, and Wells.

Settlement. Three motive forces have been at work in the settlement of Australia: the forestallment of foreigners, the attainment of good pasture land, and the discovery of mineral wealth, especially gold. The first settlers in Sydney were sent there because the British Government was afraid that New South Wales might be occupied by the French. The later settlements at Hobart and Launceston in Tasmania, on the Swan River in Western Australia, and at Brisbane, as well as a small post on Melville Island, were due to the same cause. The search for good pastures and land not subject to drought led to the crossing of the Eastern Highlands and to much of the subsequent exploration, though some of the pioneers, like Burke and Wills, were pure adventurers. After 1850, the discovery of deposits of precious metals became a strong stimulus. Its effect has left its imprint on the map in the many railway lines which run into the country and come to a sudden stop for no apparent reason.

The earliest settlers introduced the ordinary food plants and domestic animals from England and aimed at supplying their own needs as far as possible. The colony was therefore agricultural, though as early as 1796 coal was mined at Newcastle in the Hunter valley. The efforts of John Macarthur led to the rise of sheep farms with the object of exporting wool, and it was the pressure of drought on this industry that drove the settlers across the Eastern Highlands. The plains beyond proved so eminently suitable for sheep farming that a swift and vast development of the industry at once took place. The settlers tended to spread out north and south, occupying on the one hand the eastern basin of the Murray and its feeders as well as the Great Valley of Victoria and on the other the plains of southern Queensland. Those who went north were led by the increasing heat to prefer cattle to sheep. In these early days the sheep farms, or 'stations', were large, unfenced portions of the plain over which the animals roamed at will. The shepherds were mounted men who led a life similar to that of the American cowboy. The attacks of bushrangers (escaped convicts) added a spice of personal danger to the daily routine, while drought and pests were a constant and serious menace to the flocks and herds.

Broadly speaking, the settlements in Tasmania, Western Australia and Queensland had a similar history. Troubles due to bushrangers have more than counteracted the freedom from drought enjoyed by Tasmania, and this island has not progressed as fast as New South Wales and Victoria. Western Australia suffered from its remoteness and the moderate fertility of its land. Queensland, which was separated from New South Wales in 1859, was the first colony to be faced with the problems of a tropical area. At first, natives were imported from the Pacific Islands to work on the plantations, recruiting being carried out on a press gang system known as 'blackbirding'. On the plantations maize took the place of wheat, and the sugarcane began to be cultivated as an economic crop.

In 1834, a settlement was established at Adelaide by Edward Gibbon Wakefield, who had formed a plan for systematic colonisation. Only free colonists were admitted, and the price of land was kept high enough to force new settlers to spend their first few years as farm hands before being able to buy their own land. The system has not been entirely successful, since there is only a relatively small area of favourable climate and soil; but mineral deposits have assisted the growth of population. The discovery of gold at Bathurst and later at Ballarat attracted vast numbers of people to Australia and raised the population from 190,000 in 1840 to 1,145,000 in 1860. An immediate result of the 'gold rush' was the separation of Victoria from New South Wales in 1851.

The increase of population due to the 'gold rush' brought about a complete change in Australian farming. Previously, the great sheep farmers had owned vast areas and were strong enough to see that the colonies were run in their own interests. Now, many smaller farmers began to grow wheat and fruit and to herd cattle in moderately large numbers. Immediately, there was trouble over the tenure of land, various systems being tried in turn, the most successful of which has been that of perpetual leasehold. The result has been a reduction in the size of farms and the greater independence of the smaller leaseholders.

At the present day the population has grown to six millions, 97 per cent. of which are of British descent. Some 50,000 of the aboriginal 'blackfellows' survive—mostly in the northwest—but they are gradually dying out before a civilisation which they seem too primitive to acquire. The total number of persons of foreign descent in Australia is less than 50,000, the majority being Chinese who are engaged in laundry work, market gardening, and baking. The continent has by no means exhausted its capacity for holding population, but the possibilities are a

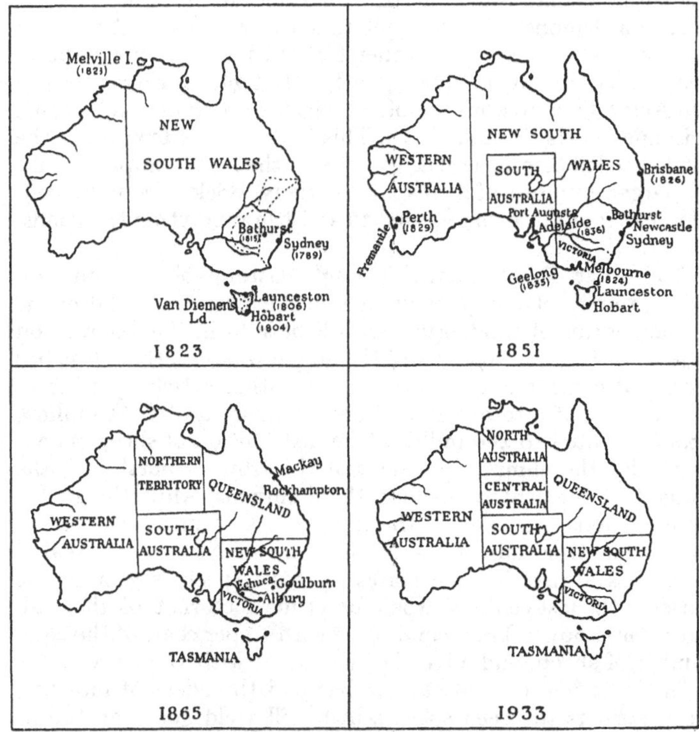

Settlement and political development of Australia. The continent was originally known as New Holland. The area taken possession of by Captain Cook was named New South Wales and out of this have been carved the states of New South Wales, Victoria, Queensland, and South Australia and the Commonwealth territories of North and Central Australia

subject of controversy, the maximum attributed to the mainland and Tasmania together ranging from 15 to 200 million people. The chief limitation to the growth of population is the climate. whose dryness renders unfit for anything but the sparsest settlement quite a third of the continent. The tropical nature of the north repels persons of European descent and has prevented all but a mere sprinkling of population in favoured spots. In consequence, a map showing the distribution of population in Australia demonstrates the fact that the vast majority live in the southeast. It shows another feature, namely, that population clusters round the five big cities. In fact, four out of every five Australians live in one of the cities of Sydney, Melbourne, Brisbane, Adelaide, and Perth. This is due to the fewness of the hands needed to prepare the produce of the land, compared with the larger number of transport workers, dock labourers, and middlemen of all sorts required to assist in marketing the goods.

Economic Production. Though pastoral occupations are losing ground relatively, they are still the most important of the primary forms of production, as will be seen in the diagram on page 463. In fact, wool is still the staple product, though it has decreased considerably compared with other articles produced. In 1927, out of a total value of goods exported of £141 million, wool amounted to £60 million. The distribution of sheep is controlled by the climate, the heat of the tropical north and the drought of the centre repelling these animals, while the southeastern plains are some of the finest areas of sheep country in the world. One half of the animals are in New South Wales, which has always led the states in this respect, except in the sixties of the last century, when for a time Victoria took the lead. Queensland now takes second place with 15 per cent. of the total number of sheep, and Victoria third place with 13 per cent.

The invention of cold storage has had the effect of inducing sheep farmers to breed sheep which will yield good mutton as well as wool of fine quality, and the crossbreds used for this purpose have proved to be more hardy than pure-bred merinos. But severe competition with New Zealand and the Argentine has greatly checked the tendency towards the production of mutton for export. In 1927–8 the value of mutton and lamb preserved by cold process and shipped amounted to £1,188,000.

The old rough and ready methods have disappeared from sheep farming, and every possibility of modern knowledge has been utilised. Scientific breeding, the sinking of artesian wells where possible and desirable, machine shearing, fenced paddocks, and the raising of fodder crops have all improved the industry enormously, and the three chief animal pests—the rabbit, the kangaroo, and the dingo—have been largely removed or held in check. By-products, of which tallow and stearine obtained from the wool for the manufacture of soap and candles, combs and handles made from the horns, glue from the hoofs, and catgut from the intestines, are the chief, have been developed and have become a profitable form of output. The chief customers of

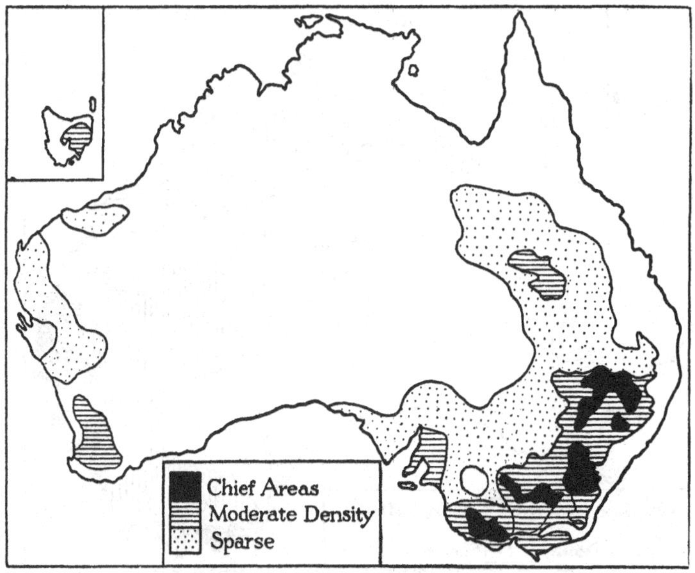

Distribution of sheep in Australia

Australian wool are Great Britain, France, Japan, the United States, Germany, and Belgium.

The range of the ox is greater than that of the sheep, and cattle flourish as well in the tropical north as in the temperate southeast. Nearly half of the total number of animals are in Queensland, where the wide savanas are a natural home of cattle and are capable of holding a far greater number of animals than they do now. The map on page 458 shows that the densest distribution is on the coast strip of southern Queensland, northern

New South Wales, and Victoria, in which districts the animals are largely used for dairying. The cows are kept in the open in paddocks, but are mainly fed on maize. Pigs and poultry are as usual found together with dairying. In 1927–8 the value of the dairy produce exported was £8,600,000, of which butter represented £7 million. In 1927 Australia was the fourth largest supplier of butter to Great Britain, and since the imposition of special duties on products from the Irish Free State, she has

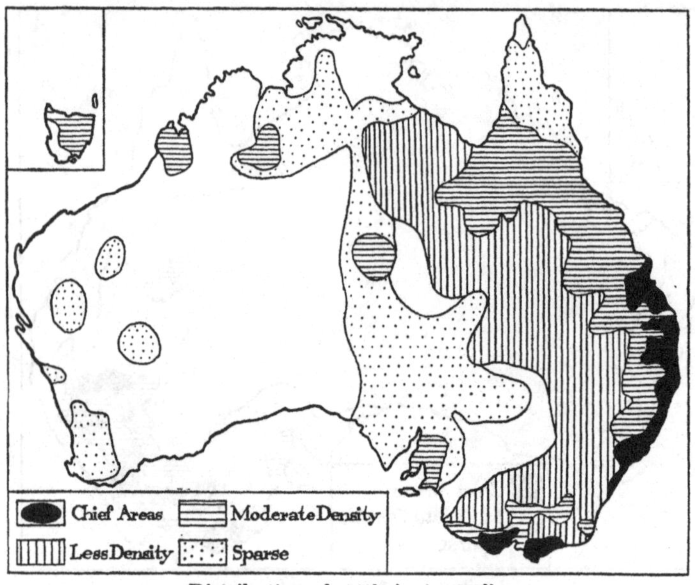

Distribution of cattle in Australia

moved up to third place. Co-operative methods of dairying have been adopted with success. The export of beef is the chief aim of extensive cattle breeding on the plains, but hides and other by-products are also marketed. In 1927–8 the value of beef preserved by cold process and exported was £12,700,000. But the industry does not wholly stand on its own feet, being subsidised by the Commonwealth Government as a protection against its rivals in the Argentine and the United States, which have the advantage of being nearer to the British Isles.

Next in importance to the pastoral occupations comes agriculture. Altogether, 19 million acres were under crops in 1927–8, and of these 12½ millions were under wheat, 1 million under oats, and 4 millions under hay and green fodder. No other crop occupied half-a-million acres, though maize, barley, fruit, vines, sugarcanes, and potatoes were of considerable secondary importance. The distribution of wheat is shown on the accompanying map, which makes it clear that the plains of New South

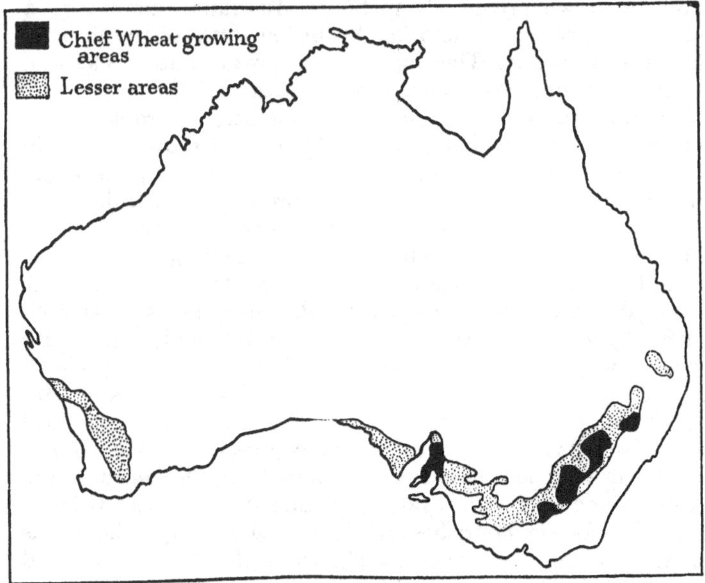

Distribution of wheat crops in Australia

Wales and Victoria, the Adelaide district, and an area in the southwest are the only parts in which the grain flourishes. The total value of the wheat crop in 1927–8 was estimated at £32 million, a figure which placed Australia eighth on the list of the greatest exporters of the grain in those years. The chief buyer is Great Britain, though Italy, Japan, France, and South Africa are also important customers. The average yield per acre is low, but the quality of the grain is good. Dry farming is being increasingly used, and irrigation is spreading in the Murray basin, where the Riverina is almost entirely given up to wheat crops. The Burrinjuck Dam on the Murrumbidgee is the most ambitious work undertaken so far.

Oats and maize are grown chiefly as fodder crops. The former is cultivated mainly in Victoria and Western Australia, while the latter is hardly found outside Queensland and New South Wales. Both crops are used more or less as green forage, particularly in connexion with the dairying industry of the east coast strip. Barley is grown principally in South Australia, but also to a considerable extent in Victoria. About four-fifths of the crop are used in the breweries. The chief potato-growing state is Victoria, Tasmania coming next in order of importance.

Australia produces fruit of the temperate, subtropical, and tropical varieties. Apples, pears, and plums are grown mainly in the southeast, especially in Tasmania and Victoria, the apple crop being the most important in the production of fruit. Subtropical crops flourish especially in the 'Mediterranean' areas of South and Western Australia. The chief kinds grown are the peach, the orange, and the grape, but apricots, cherries, nectarines, and lemons are also cultivated. The vine is cultivated principally in South Australia and Victoria, and from it are prepared wine, raisins, currants, and table grapes. The quantity of wine produced had jumped from one million gallons in 1923 to four million gallons in 1927. Queensland produces tropical fruit, of which the chief kinds are bananas and pineapples. These are difficult to export and are used mainly for home consumption.

Queensland also grows sugarcane and cotton in its tropical parts. The former flourishes in the wet coast strip, where it is cultivated on large plantations. The value of the sugar exported in 1927–8 was £4 million, and naturally great quantities were consumed at home. Alcohol is an important by-product of sugarcane. Cotton plays the same part in Queensland that wheat plays in Victoria and New South Wales, i.e. it is gradually increasing its area at the expense of the herds. At present the quantity produced is not considerable.

Australia has large areas which, though officially reckoned as forest, are yet mere savana land with scattered trees. Real forest is confined to the extreme southwest of Western Australia, the southeast of Victoria, the mountains of Victoria and New South Wales, Tasmania, and the tropical forests of northeastern

Queensland. The timber from these areas is insufficient to supply Australian needs, especially as good softwoods are scarce, and large quantities are imported from New Zealand, the United States, and Japan. Other forest products are eucalyptus oil, which may be distilled from the foliage of all varieties of the eucalyptus tree, and tan barks from various species of eucalyptus, wattle, and mangrove. The last mentioned covers the foreshore in the Gulf of Carpentaria, Arnhem Land, and the east coast of Queensland.

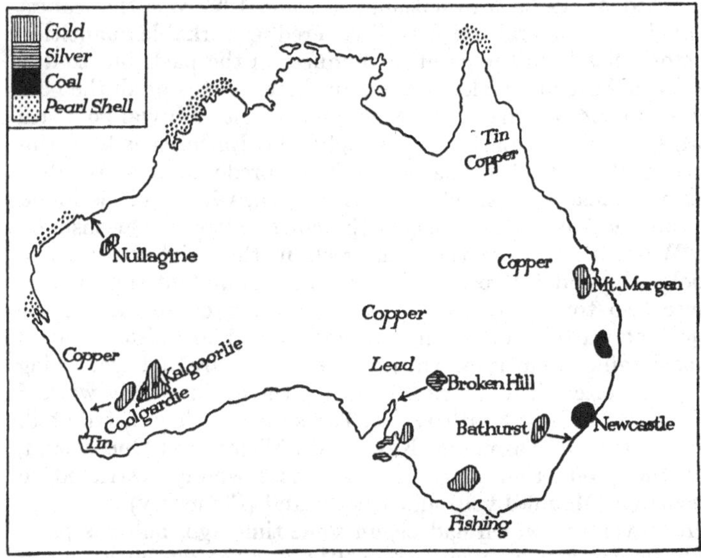

Chief mineral areas in Australia

Only 2¾ million pounds sterling worth of fish (about half of which is imported) was eaten in Australia in 1927, for the Australians are not a fish-eating people. Hence, the fishing industry has turned most of its attention to the pearl-shell, pearl, and trepang catch of the tropical waters. Of these the pearl-shell is the most important, the value of the quantity exported in 1927 being £333,000. Next in importance is trochus shell, nearly all of which is exported to Japan.

Australia has a considerable supply of minerals, the total production of which in 1927 was £23 million. Over half of this represented coal. a mineral which is found chiefly in New South Wales. The port of Newcastle in that state does a large export trade. Next in value is lead, including silver-lead and its cognate ores from which various metals, chiefly silver, lead, and zinc, are obtained. Broken Hill in New South Wales is the chief centre of Australia, but the working of the ores is done in South Australia and exports go through Port Pirie. Gold is now only third in value, though it was the great finds in 1851 which 'precipitated Australia into nationhood', and 1927 was the leanest year since the metal was first discovered in workable quantities. Victoria has been the greatest producer in the past, but to-day Western Australia yields more than three-fourths of all the gold mined in the continent. In New South Wales alluvial gold still forms the greater part of the supply, and Bathurst is the principal centre. In Victoria reef mining predominates, Bendigo being the leading district. Mount Morgan, where gold is found in conjunction with copper, is the chief centre in Queensland. In Western Australia the great reefs in the neighbourhood of Coolgardie and Kalgoorlie give the most abundant supply, and these two towns are just elaborate mining camps set in the midst of waste land. Iron, tin, and copper are also found in considerable quantities, though the mines are not yet being fully exploited. The districts in which the ore is chiefly worked are mostly in New South Wales. Tin is more widely distributed, being got from Tasmania, New South Wales, and Queensland, but the production of copper is almost wholly restricted to Tasmania (Mount Lyell) and Queensland (Cloncurry).

A movement which had begun some time ago, but was painfully slow in development before 1914, was the establishment of manufactures in Australia. The restrictions placed on British exports during the war of 1914–18 gave a stimulus to the movement, and now industrial works are springing up in all the towns in Australia. Government bounties aid the producers of iron and steel goods and encourage other industries. The chief activities either produce food and drink, e.g. breweries and sugar mills; or they are preliminary industries, such as tanneries or saw mills; or else they produce goods like textiles and leatherware. But, though the importance of the industries is rapidly expanding, yet the value of the goods produced is so far relatively small.

Australian overseas trade consists largely of the exchange of foodstuffs (beef, mutton, wheat, and fruit), wool, and minerals for machinery, textiles, and other manufactured goods. The four chief countries with which trade is done are Great Britain, the

Pastoral £125 M	Agriculture £84 M	Poultry etc. £50 M	Mining £23 M	Manufacturing £159 M

Estimated values of production in 1927–8. Total value £453 million

United States, Germany, and France. Australia buys most from Great Britain, but sells most to the United States. Her trade with the former is fostered by reciprocal treaties, such as those arranged at the Ottawa Conference in 1932.

Communications. Motor transport is playing an increasingly important part in the communications of Australia, but for the moment the railways still form the best indication of the main traffic lines. Little foresight was used in the early days of railway construction, nor was there anything like a general plan. Hence,

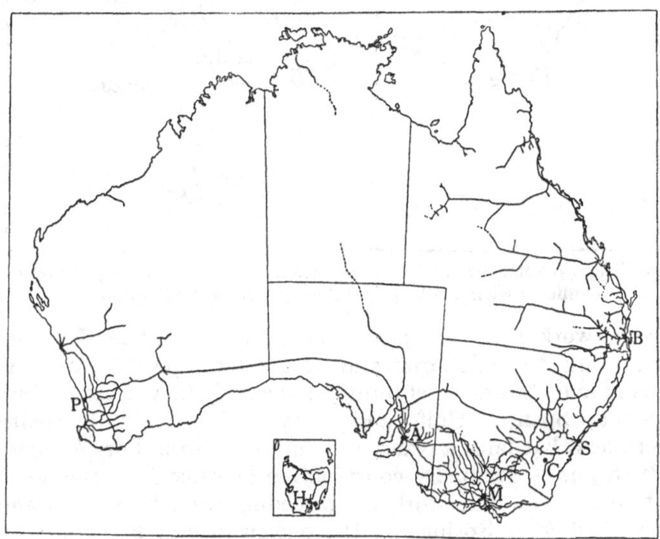

Network of Australian railways

lines were built as occasion demanded and usually took the shortest line between some new centre of production and the nearest harbour. Now that a unified scheme is obviously necessary, the result of this haphazard growth is that there are no fewer than seven different gauges in use, an irregularity which, it is estimated, will cost £21 million to set right. Practically all the railways are now owned by the Commonwealth or the States. The railway map on page 463 shows the closeness

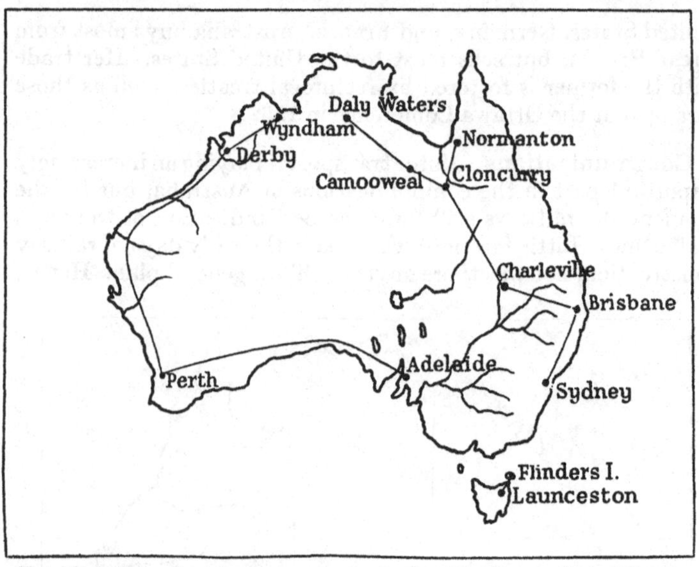

Established regular air services in Australia. The line to Daly Waters connects with the Imperial Airways route to London

of the network in the southeast, the importance of the Kilmore Gate leading to Melbourne and of the focal position of Port Augusta, and the five chief terminal areas: Sydney, Port Phillip, Swanland, Spencer Gulf, and Darwin. The Commonwealth Government has built a transcontinental line from Port Augusta to Perth and is gradually constructing another from the same point to Darwin. The work is proceeding from both ends and has reached Alice Springs on the one hand and Birdum, fifty miles south of Daly Waters, on the other.

The aeroplane has been welcomed as a means of rapid transport in Australia, and the Commonwealth Government subsidises certain airmail lines. The map on page 464 shows those that were in regular operation in 1932. Besides these, there were a number of irregular and occasional services, mostly in the southeast.

Six seaways radiate from Australia, three of which are great ocean routes. The chief is the Suez Canal route to Europe which starts from Fremantle, the port of Perth, and runs through Colombo, the Red and Mediterranean Seas to London. This is the usual way for passengers and light cargo. Heavy cargo travels from Fremantle round the Cape of Good Hope and along the west coast of Africa to England. The third great route leaves from Sydney or Brisbane and, touching at Suva and Honolulu, reaches Vancouver or San Francisco. When this route is continued overland through Canada and then across the Atlantic, it is the quickest, though the costliest, way to the British Isles. Of the three other routes only one leaves the Pacific. This is the original sailing route round Cape Horn, a route still followed by sailing ships laden with grain. The fifth route leaves Melbourne or Sydney and reaches Auckland, while the sixth goes from Brisbane to China and Japan.

Political. Australia forms one of the Dominions of the British Empire. It is governed by a Governor-General appointed by the Imperial Government and by a Cabinet formed on British lines; by a Senate or upper house, and a House of Representatives. The constitution, which came into force on January 1, 1901, laid down matters over which the Commonwealth Government should have jurisdiction, all other things falling to the business of the individual State governments. There are six States: New South Wales, Victoria, Queensland, South Australia, Western Australia, and Tasmania, together with the Northern Territory which is governed by the Commonwealth and for administrative convenience is divided along the 20th parallel of latitude into North and Central Australia. Under the Commonwealth are also the dependencies of Papua and Norfolk Island and the mandated territory of Northeast New Guinea. Since 1927 the seat of the Commonwealth Government has been at Canberra, a town deliberately planned—as was Washington—to be the capital and situated in a small area bought by the Commonwealth from the State of New South Wales.

Each State has a government organised on the same lines as that of the Commonwealth. The several capitals are the focal points to which the States stand in the geographical relation of backlands. Melbourne, for instance, is the natural focus of the Great Valley of Victoria and through the Kilmore Gate is the easiest outlet for the country between the Victorian Highlands and the Murray. Sydney is the equally natural focus for a long coast strip, the Blue Mountain district, and the vast plains between the Murray and the Darling. Brisbane has similar advantages, but lacks the supremacy of Sydney over its neighbours. Hence, it has serious rivals in the more central Rockhampton and Townsville. Adelaide is centrally placed to control the various regions into which South Australia is divided. Perth has, like Adelaide, been built away from the coast for reasons of health, but is fairly centrally placed on the fertile district of Swanland. Hobart suffers from facing away from the mainland and so shares its importance with Launceston.

The aim of each of the Australian settlements, after self-support, was to export wool and precious metals to England, and consequently each settlement expanded to fill the backland of a port which grew with that expansion. The backlands became first separate colonies and afterwards States, the ports being the natural capitals. No sooner had the colonies become conscious of themselves than they began to define boundaries and, since the making of the boundaries outran geographical knowledge, artificial lines were used which effectually prevented the state from having anything like exact coincidence with the backland. In fact, when the minerals of Broken Hill were found, economic necessity tied the working of the mines and the export of the metals to South Australia, although the field actually lay in New South Wales. The backland of Perth-Fremantle is just the southwestern corner of Western Australia. Brisbane is the focus of a relatively small portion of southern Queensland and is therefore seriously rivalled by Rockhampton and Townsville, both of which seem more favourably situated than the official capital.

Friction between the States owing to the imperfection of their boundaries has been common, for there have always been jealousy and other disruptive elements among them. Western Australia was only induced to enter the Commonwealth by the promise of a transcontinental railway which would increase its trade. Now, however, this State feels that it gets little benefit from the union and is agitating for secession. South Australia is also dissatisfied, but has not shown its feelings so openly.

The powerful factors which created the Commonwealth were German aggression in New Guinea and, later, the fear of invasion by Japan. The northern areas of Australia which are tropical have not yet been fully made use of by the European settlers, and it is said that white men cannot live and work as labourers in the Tropics and that Australia can never people and exploit its north. Hence, it has been argued that this area should be handed over to Japan to use for the benefit of the world at large. But the people of Australia, realising the economic, political, and social problems which the presence of a substratum of coloured folk has inflicted on South Africa and the United States, have determined not to allow those difficulties to occur in Australia. They have therefore made it practically impossible for coloured labour legally to enter the country, and to those who accuse them of being 'dogs in the manger' they reply that there is no proof that white men cannot colonise the Tropics. At the same time, they have looked to the British Navy as an important protection, and have been willing to occupy Papua so as to keep a possible enemy from their doors.

NEW ZEALAND

Position and Size. New Zealand lies 1200 miles southeast of Australia. It is at the antipodes of Spain and is thus somewhat closer than the British Isles to the Equator. It comprises two large islands: the North Island (43,131 sq. miles) and the South Island (58,120 sq. miles); and a number of smaller ones distributed round their coasts. Of these the bleak, infertile, and hence unimportant Stewart Island is the largest. The total area of New Zealand is slightly less than that of the British Isles, while the South Island is almost exactly the size of England and Wales. The greatest length of the group is 1000 miles, the greatest breadth 180 miles.

Build and Relief. The main islands of New Zealand have the shape of a riding boot, with a break just above the ankle forming Cook Strait. The North Island is the foot, and the South Island the leg. The long, low Auckland Peninsula is the toe, the underpart of the instep being the Bay of Plenty, and the hollow above the heel Hawke's Bay. On the whole the islands are mountainous. In the extreme southwest there is a worn mass of old fold mountains whose grain runs northwest and southeast. Farther north young fold mountains known as the Southern Alps form a backbone to the South Island. The peaks of this range rise to 10,000 and 12,000 feet above the sea, Mount Cook being the highest (12,500 feet), and their upper slopes are snow-clad and streaked with glaciers. In the tertiary period the mountains were higher than they are now and, owing to this and to the colder climate at that time, vast glaciers covered the slopes. When the climate became milder, the torrents proceeding from the melting snow and ice swept vast quantities of rock waste on to the lower ground to build up the Canterbury Plains and the Otago lowlands, areas which to-day are among the most fertile in New Zealand. These broad lowlands on the east give the South Island an asymmetrical shape, since on the west there is only a narrow coastal strip forming the province of Westland.

In the North Island there is also a line of mountains running in a northeasterly direction, continuing the Kaikoura Mountains of the South Island. The peaks of this range, for which there is no general name, rise to between 5000 and 7000 feet. It is to be noticed that in this island the backbone lies nearer the east coast, where the coastal plain is therefore very narrow. West of these fold mountains there is a large volcanic area, in which the cones of Ruapehu (9175 feet), Mount Egmont (8260 feet), and Ngauruhoe (7515 feet) are the highest peaks in the island. Earthquakes are frequent, and hot springs, geysers and mud volcanoes abound in the district around Lake Taupo, where the hot water is used by the natives for cooking and laundry. The fertility of the soil is largely due to the presence of volcanic rocks. In the south and northwest of the North Island are two coastal plains, the latter of which is drained by the Waikato and Thames Rivers. Beyond it is the long, narrow, and infertile Auckland Peninsula which comprises some of the oldest rocks in New Zealand.

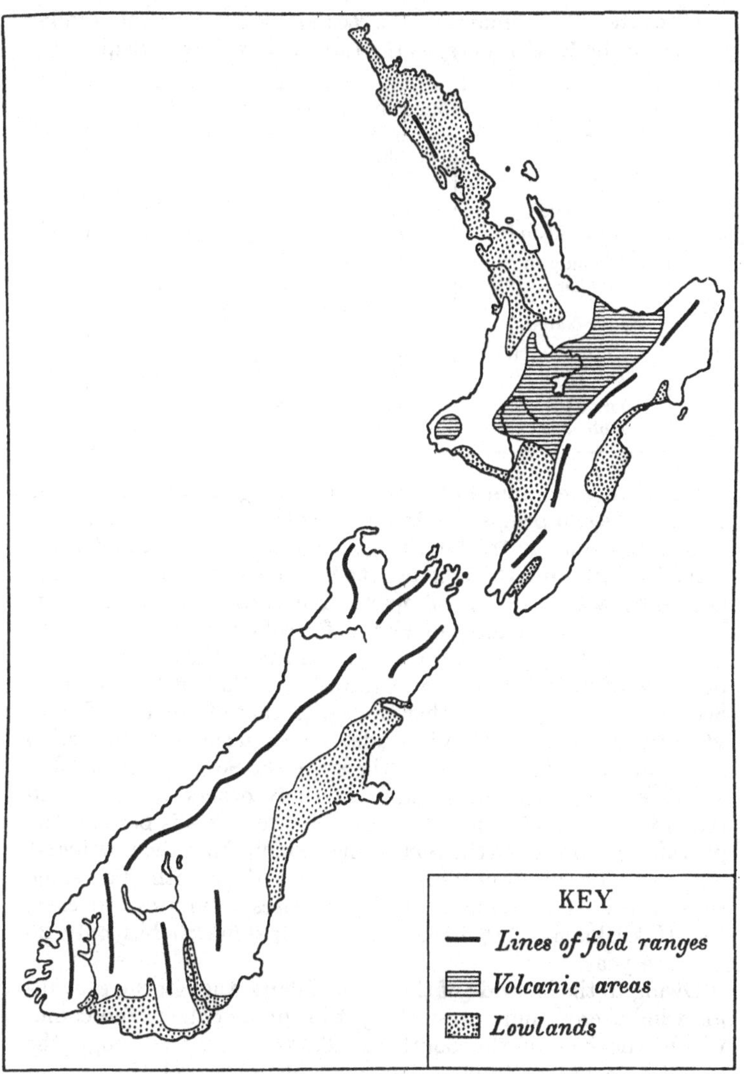

Main structural features of New Zealand

Climate. The climate of New Zealand is on the whole similar to that of the British Isles, as the table below shows adequately:

	Midsummer	Midwinter	Range
	° F.	° F.	° F.
North Island			
Auckland	66½	52	14½
Scilly Islands	60½	45½	15
Wellington	62½	47½	15
London	62½	39	23½
South Island			
Christchurch	61½	42½	19
Dublin	60	42	18
Dunedin	58	42½	15½
Holyhead	58½	42	16½

But there are important differences. The great expanse of the Southern Ocean brings the Antarctic cold nearer to the Equator than is possible for the land-ringed Arctic Sea to do, and there is no Gulf Stream Drift to assist the winds in bringing to the islands the warmth of the Tropics. Hence, the snow line is lower in the South Island and the winters far colder than is the case in the same latitude in western Europe. In fact, the climate of the province of Southland is comparable with that of the north of Scotland or of Norway rather than with that of northern Spain, to which it corresponds in latitude. By an accident in the relief, places like Christchurch under the lee of the Southern Alps have a greater range of temperature than any others, a fact which reminds us of conditions in the British Isles. This is because the prevailing winds over the South Island blow from the northwest and the ridge of mountains lies athwart their path. The same cause affects the rainfall, which decreases from west to east; e.g. Hokitika has 116 inches, while Christchurch has only 26 inches a year.

Owing to the latitude of the North Island, though the prevailing wind there is northwesterly, yet in summer the island comes within the belt of the Southeast Trades. Hence, though the island is in the latitude of the south of Spain, yet the heat of summer is tempered by onshore winds, while at no season of the

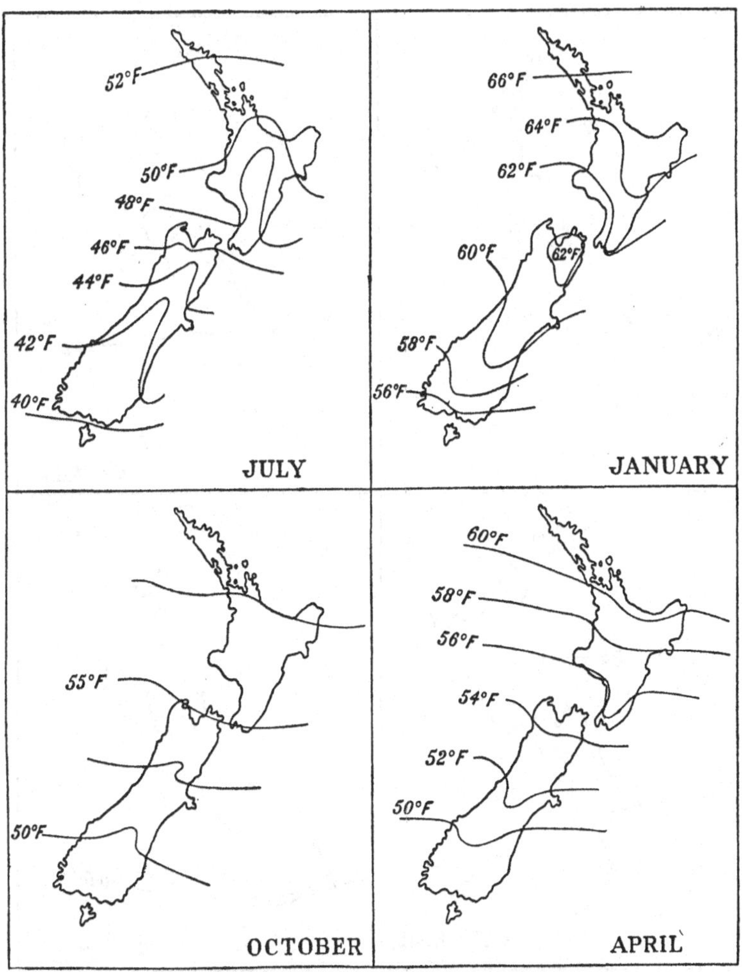

Distribution of temperature in New Zealand

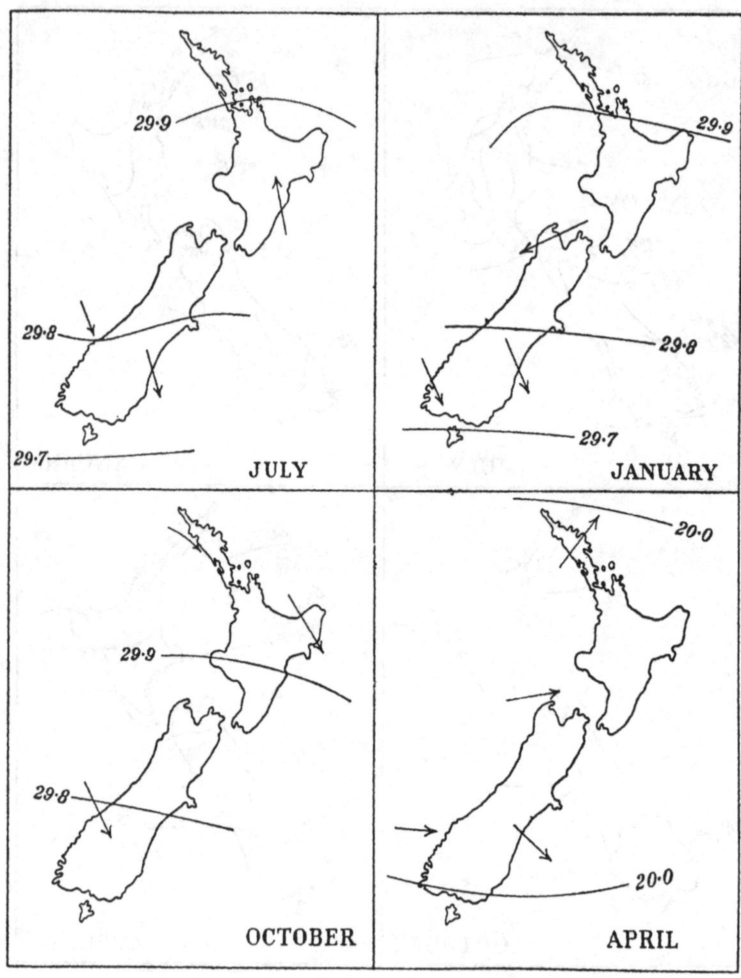

Distribution of pressure and winds in New Zealand

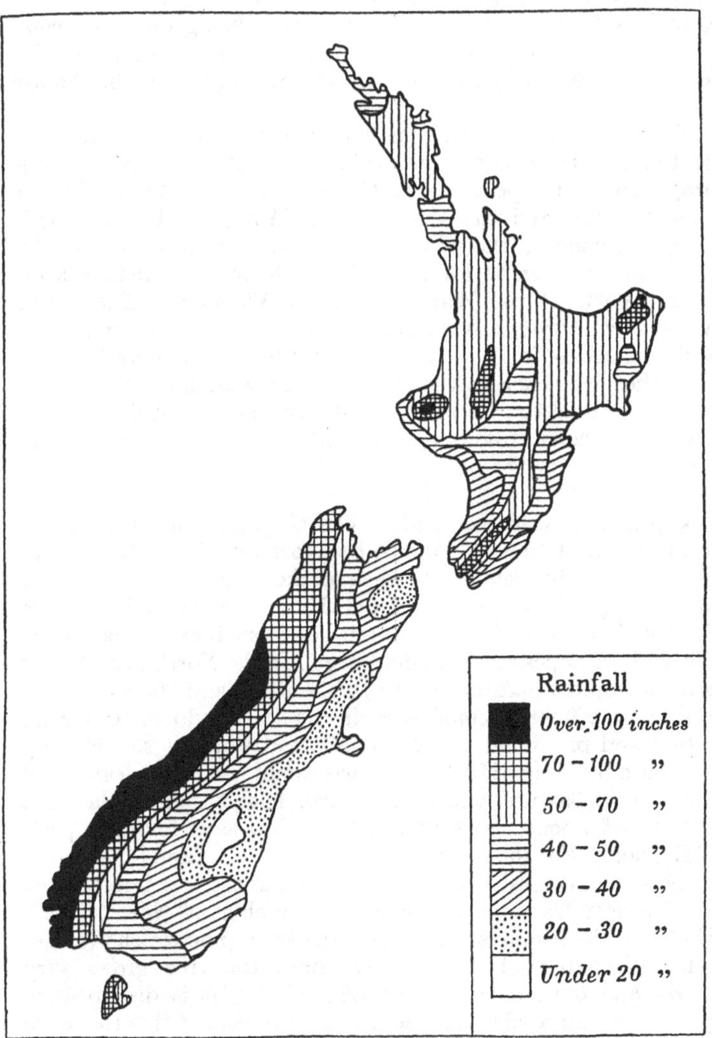

Annual distribution of rainfall in New Zealand

year does the drought of a Mediterranean climate occur. The rainfall is fairly evenly distributed, there being only two considerable areas of markedly less rain than elsewhere. On the whole, the North Island is slightly wetter than the South Island.

The plentiful rainfall throughout the year all over New Zealand gives rise to many rivers which have that equable volume proper to the temperate belt. The largest are the Clutha in the South Island, and the Waikato and Wanganui in the North Island. Owing to the relief of the land, the rivers are on the whole of little value for navigation, though steamers run on certain parts of the Waikato and the Wanganui. The latter, which flows through beautiful scenery and has been compared with the Rhine in its gorge, is much used for holiday boating. A further drawback to the rivers of New Zealand is that either they have no estuaries or their estuaries are blocked by a bar. But the streams provide an excellent water supply for drinking purposes and for the production of hydro-electricity.

Vegetation and Animals. The abundant moisture in the soil has caused New Zealand to be a forest-clad country, except in the driest districts of the east of the South Island; and the forests are not of the gladed parkland type found in western Europe, but form a temperate rain forest such as occurs also on the western slopes of Tierra del Fuego. In the North Island there is a subtropical luxuriance. In the South Island the vegetation is slightly different, conifers replacing leaf-shedders to a great extent and providing a store of softwood which is so necessary for house building. Much of the woodland has been cleared and turned into pasture, and in the North Island there is the same mixture of woods, grassland, and meadow as is found in parts of England. For the most part, the trees are of kinds that correspond to the English oaks and beeches, but the tree fern with its soft feathery fronds is the characteristic plant of New Zealand. Auckland Peninsula is the home of the kauri pine which provides valuable timber. Unfortunately, since the tree grows very slowly and once felled cannot be replaced, it is disappearing from the countryside. In places whence forests of this tree have passed away a fossilised gum is dug from the ground and is used for making polish and lacquer. The quantity is of course limited

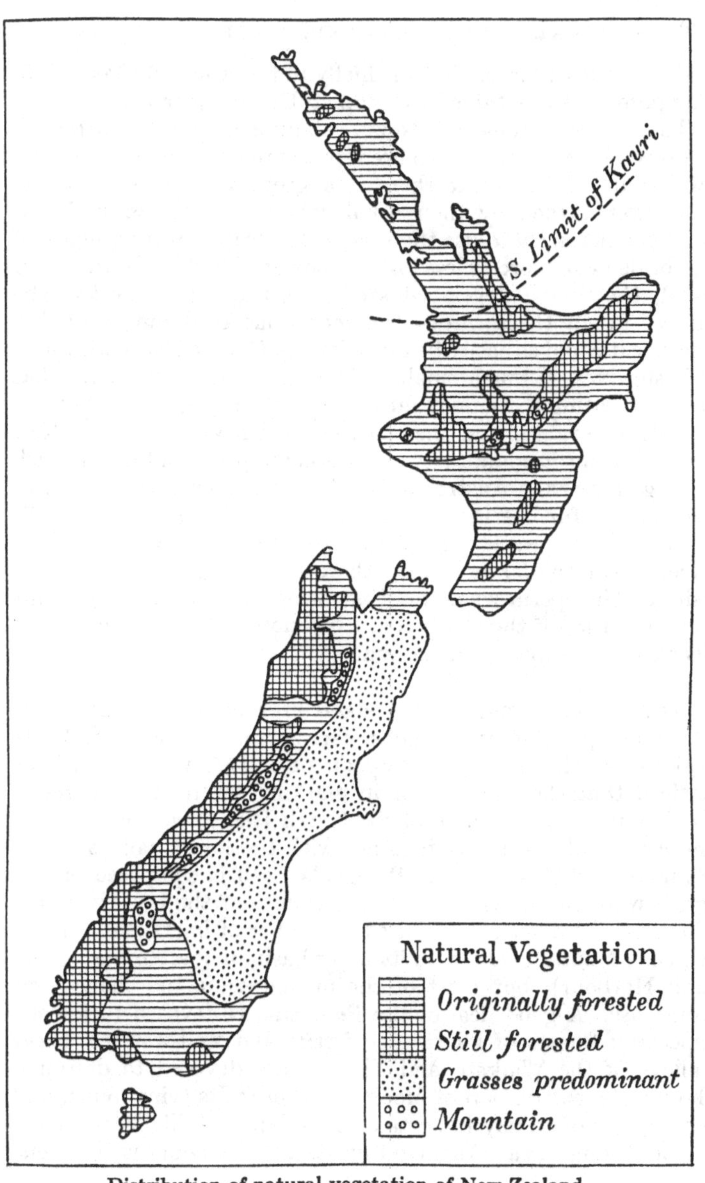

Distribution of natural vegetation of New Zealand

and the industry is carried on chiefly in the Auckland Peninsula. The gum is also obtained by tapping the living trees.

Through long isolation, New Zealand is distinguished by the absence of mammals, which are represented only by the seal and two kinds of bat. But there is a great variety of bird life. Free from attack by their usual enemies on the ground and from the necessity to use their wings to escape capture, many of the birds have more or less lost the power of flight. On the other hand, they have developed strong legs and long necks. The largest of all these birds—the moa—has died out, probably because it could so easily be killed by the Maoris. The kiwi, which still survives, is much smaller. The kea, a kind of parrot, has taken to eating the kidney fat of living sheep and so has become a pest. The rabbit and the sparrow, which were taken to New Zealand from England, have also become pests and do as much damage here as in Australia. The former is now being turned to account as a fur-bearing animal, and its fur is being made into felt.

The shallow waters round the coasts abound with fish, but, except near the larger towns, there is not a great deal of sea-fishing. The sperm whale hunting, which was formerly a profitable business off the south coast, has now decayed owing to the decreasing number of the animals.

Natural Regions. New Zealand may be divided into eight natural or geographical regions, each island having four. These regions are shown on the map on page 477, where it will be noticed that the chief factor in marking off the areas is relief. Beginning with the north of the North Island, we come first to the Auckland Peninsula, together with the lower valleys of the Waikato and Thames. The Peninsula itself is composed of old, much worn rocks, while the lower valleys of the two rivers are alluvial. Subsidence has caused a number of inlets, the chief of which is Hauraki Gulf with its inner basin of Waitemata (Auckland Harbour); but most of the openings are too shallow for much use. A good deal of the Peninsula is devoted to forests, especially forests of kauri. The fertile and well-watered lower valleys of the Waikato and Thames are devoted to dairying, though the cultivation of Mediterranean fruits (vine, olive, and lemon) is moderately important. The Cape Colville Peninsula is a gold-mining area. Auckland on its two harbours is the focus

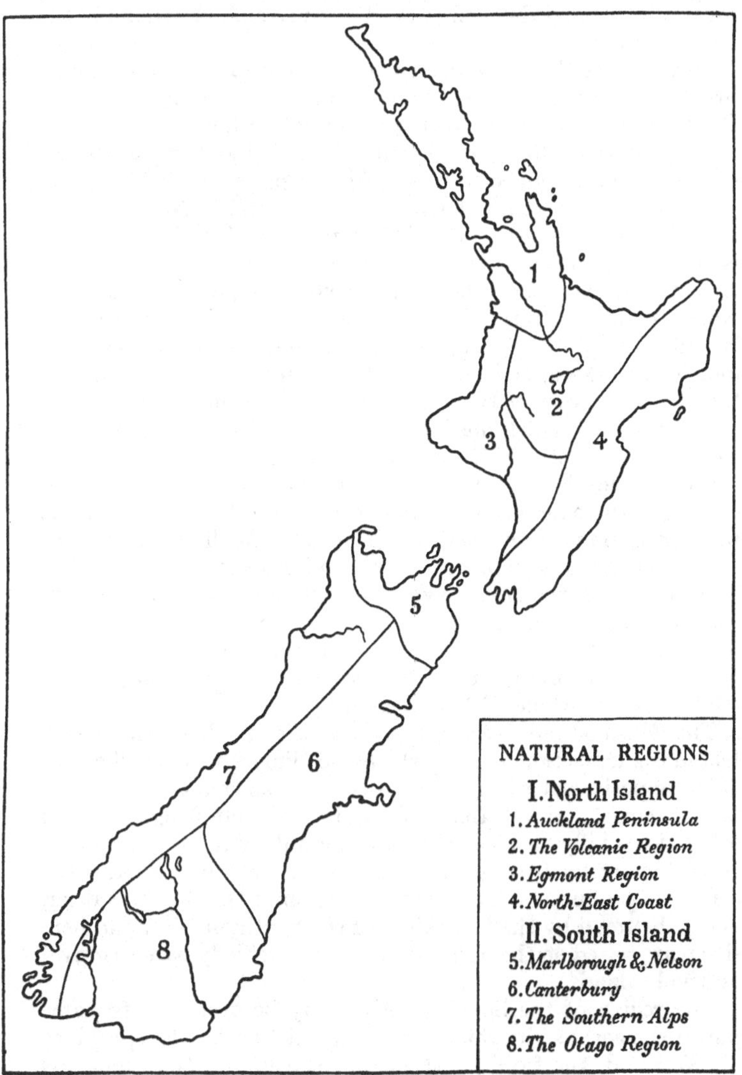

Main natural regions of New Zealand

for the region and the nearest port to Australia, Canada, and the United States.

Next comes the volcanic region centring round Lake Taupo, in which live most of the Maoris. Its only town of any size is Rotorua, a health resort owing to its hot springs, and it is still much forested. More important is the Egmont region which embraces the province of Taranaki and parts of Auckland and Wellington, with some of the most fertile soil in New Zealand. Dairying is the chief occupation at present, though agriculture is increasing on the rich volcanic soils of the lower slopes of Mount Egmont and on the fertile Wellington plain. Palmerston North (pop. 23,000) is the largest town in the region. The scenery is beautiful in the Wanganui valley and wherever the symmetrical, snow-capped peak of Egmont forms a background to the view. The town of Wanganui has a Maori college and a public school run on English lines for Europeans. The last region in the North Island is the northeast coast, which, being sheltered from the Westerlies, is drier than the rest of the island. Hence, sheep-grazing is the chief occupation, except in the Wairarapa valley, where dairying predominates. Fruits of a Mediterranean type are grown to a certain extent around Hawke's Bay. Napier (pop. 19,000) is the chief town, for Wellington is in, but not exclusively of, this region, since it is the outlet for the Egmont region as well. It owes much of its importance to its terminal position on the Main Trunk Railway and to its central position relative to the whole of New Zealand.

The Marlborough-Nelson region in the north of the South Island is a transition area in which the climate is more like that of the opposite parts of the North Island than that of the regions farther south. Small areas of coastal lowland fringe what is otherwise a hilly district and allow the mild climate to be taken advantage of for fruit-growing. The little town of Nelson is the centre of production. The eastern portion of the region is drier, being sheltered by the last ridges of the Southern Alps, and here sheep are raised on the uplands and cereals (chiefly barley) grown on the lowlands.

The region of the Southern Alps may be divided into three parts: the coast strip which forms the chief part of the province of Westland, the fjord coast farther south, and the highlands themselves. The first of these has the most equable, but rainiest climate in all New Zealand. It is, however, too small to be of great importance, though the Greymouth district is the chief mineral area in New Zealand, exporting coal and gold. A railway which crosses the Southern Alps through a tunnel at Arthur's Pass connects the Westland coalfield with Christchurch. The fjords of New Zealand are known locally as 'sounds'. Like their counterparts in Norway, British Columbia, and Chile, they offer beautiful scenery, but lack the backland of fertile lowland which alone could make them of importance. They are backed by the highlands of Southland which pass later into the Southern Alps. The mountains provide sport in the forests of their slopes, on their rivers and glacier lakes, and on their snows; and they also form the gathering ground for the waters of the many rivers which lower down become important sources of hydro-electricity.

The Canterbury region consists of two belts: the rolling country which is formed by the lower spurs of the mountains, and the wide undulating plains formed by glacial action. There is also the little volcanic area of Banks Peninsula. The climate of the region is far more extreme and far drier than that of any other part of New Zealand owing to the protection of the Southern Alps, and as a result the uplands are devoted to sheep-grazing and the lowlands to grain. Christchurch, the third town in New Zealand, with a population of 120,000, is the focus of the region. It is not on the sea, but has rapid communication with its port at Lyttelton.

The last region is that of Otago, which consists of an old, eroded plateau of no great fertility and of some moderately large areas of coastal lowland. The latter are fertile and are being used increasingly for grain, especially oats. The predominance of this crop is due partly to the climate and partly to the Scottish origin of the settlement in the region. The plateau is used for sheep-grazing. The chief towns are Dunedin (pop. 85,000) and Invercargill (pop. 24,000). The southern part of the region, of which Invercargill is the focus, is known as Southland.

Human Geography

Economic. Youth of settlement and poverty in minerals cause New Zealand to depend largely on grazing and agriculture. Five areas stand out as the most productive in the country, namely, the lower valleys of the Waikato and Thames, the east coast strip of the North Island from Featherston to Gisborne, the southwestern plain of the North Island; the Canterbury Plains, and the lowlands of Otago. The absence of drought makes New Zealand an excellent grazing country, and the chief source of wealth still lies in flocks and herds. In addition to the areas of natural grassland, wide stretches of what was once forest have been cleared and turned into pasture. Generally speaking, the wetter pasture lands of the Islands are used for cattle, while the drier east coast pastures are used for sheep.

Sheep farmers are constantly improving their 'runs' by planting soft, luscious English grasses instead of the hard native varieties, and so far they have brought about the change over rather more than half of the country. The product of the farms is either wool or mutton. Up to the economic changes of 1930, the value of the wool exported from New Zealand was far greater than that of the mutton, though the invention of cold storage methods has fostered the export of meat to the British Isles. A successful method of shipping 'chilled' mutton would give even greater stimulus to the trade.

The cattle farmer of New Zealand aims at exporting, not chilled beef, but dairy produce: butter, cheese, and condensed milk. As a rule, the cattle farm is smaller than the sheep farm and is owned by a less wealthy man. There are no areas exclusively for either sheep or cattle, but the latter are preferred in wet districts, and the number of animals is greater in the North Island than in the South Island. The dairying industry is highly organised on co-operative lines, with central factories which receive cream from the farms and turn it into butter, cheese, casein, and dried or condensed milk. Before export, the dairy produce undergoes a searching inspection intended to maintain the high standard of the guarantee implied by the government mark: 'New Zealand Produce'. About four-fifths of the produce of the central factories are exported, mainly to Great Britain. As usual, pigs and poultry form part of the stock on the dairy farm, and sufficient bacon ham, and eggs are produced for home consumption.

Minerals exist, but not in large quantities, in New Zealand, and at present there has been little exploitation of the supplies. There are three chief mineral areas: (1) the Colville Peninsula, (2) the north of the South Island from Hokitika to Nelson, and

Plate XXIX

(*Mondiale*)

SHEEP-SHEARING IN THE RIVERINA, NEW SOUTH WALES

Plate XXX

(*Mondiale*)

CANBERRA, AUSTRALIA

The town is growing up among the streets which have been already laid out

(3) the southeast of the same island. Coal is by far the most important mineral, and its exploitation is increasing steadily. Its quality varies from brown coal to anthracite. It is found in each of the three chief mineral areas, but is worked most near Westport in the province of Westland. In 1931 the value of coal mined was over £2,150,000. Most of the supply is used in the country.

Gold is also found in fair abundance in the same districts, the value of the quantity exported in 1931 being £581,032. In the Otago province the deposits of the metal are alluvial and are reached by 'panning' the soil near the rivers or by dredging the beds of the Clutha and other streams. In the Colville Peninsula the deposits are veins in quartz which must be extracted by pulverising the rock. Small quantities of silver and iron are also obtained, but they are not large enough to be important.

The following table gives the value of the chief exports in 1931:

		£
1. Sheep produce:	Frozen mutton	8,892,555
	Wool	5,515,376
	Skins	805,838
	Tallow	413,080
	Sausage casings	399,418
		16,026,267
2. Cattle produce:	Butter	10,649,527
	Cheese	4,461,293
	Hides	349,047
	Milk	246,483
	Meat	92,054
	Casein	88,720
		15,887,124
3. Minerals:	Gold	581,032
	Coal	83,393
	Kauri gum	128,095
		792,520

Out of a total value of £35 million, no less than £31 million is exported to Great Britain, Australia taking the next greatest quantity to the value of £1 million.

A country of such recent settlement as New Zealand would not be expected to produce large quantities of manufactured goods, since it is easier to export primary products and to import manufactured goods; and this is indeed what takes place. Over half the output of the factories consists of butter, bacon, and other primary products prepared for the market, and manufactured goods, properly speaking, are not turned out in important quantities. Efforts are being made to produce all sorts of things, but, though the country may one day become self-sufficing, most of what is needed is still imported. In 1931 nearly half the imports came from the United Kingdom; the United States, Canada, Australia, and India being also considerable customers. New Zealand is thus still in the early stage of overseas settlement, when a sparse population exchanges foodstuffs and raw materials with the Mother Country for manufactured goods.

New Zealand is fortunately provided with hydro-electric power. The streams are in the youthful stage of the cycle of erosion and are marked by waterfalls, rapids, and other irregularities which give the force needed for turning the turbines. The exploitation of this form of power has been under government supervision and has been part of a system based on six main stations, three being in each island. They are at Mangahoa, Lake Waikaremoana, and the Arapuni Falls of the Waikato in the North Island, and at Lake Coleridge, Waipori Falls, and Lake Monowai in the South Island. Already hydro-electricity is used twice as much as coal for power, and in time this disparity will increase even more.

The Maoris. At the arrival of Captain Cook in 1769, New Zealand was inhabited by the Maoris, a branch of the Polynesian race. These had in the fifteenth century displaced the Morioris, a few survivors of whom live in the Chatham Islands. The Maoris were divided into clans whose social system was communistic. It was largely owing to the failure of the early European settlers to understand that the land was the property of the whole clan, and not of any individual, that the troubles of the middle of the last century were due. The villages were usually built on a bluff near the sea or a river, and they were surrounded by a ditch and stockade of tree-trunks. The houses were one-roomed, and consisted of a wooden frame with a thatched roof and sides. The gables were often elaborately carved.

The dress of the people was mostly of cloth woven from the

fibres of the *phormium*, or New Zealand flax, while the upper classes wore mantles of feathers. The faces and often the bodies of the men were tattooed in elaborate designs, the process of which was exalted into a fine art. Food was chiefly fish, caught either in the sea or in the rivers. Hence, the distribution of population was closely related to these sources. The flesh of dogs and rats introduced by the Maoris was regarded as a delicacy to be reserved for special occasions, while the flesh of man was eaten after victory in war as a religious act. Small plots of land were cultivated near the villages, but the plants used for the crops were of inferior food value. It is probable that the islands were sparsely populated by the Maoris, whose numbers are estimated not to have exceeded 150,000 or 200,000.

Settlement. The earliest European settlers were runaway sailors from trading ships in search of timber, and, somewhat later, missionaries. At first, the islands were under the nominal jurisdiction of the Governor of New South Wales, formal proclamation having been made by Captain Cook of the assumption of British sovereignty; but in 1840 the increase to 5000 of the number of settlers and the trouble that had arisen over the occupation of land led to the Treaty of Waitangi, whereby the native chiefs acknowledged the sovereignty of the Queen of England, and New Zealand was set up as a separate colony. Twelve years later, self-government was granted. Between 1845 and 1848 and again between 1860 and 1870 further trouble over land rights brought war between the Maoris and *pakehas*, or British settlers; but peace was finally established in 1871, and the two races are in process of fusion. Since 1871, the history of New Zealand has been an unbroken record of steady progress with scarcely an outstanding event, save the war of 1914–18.

In 1932 the population was estimated at 1,524,921, of whom 69,893 were Maoris. Just under a million were in the North Island, the South Island counting only something more than half a million. Of the total population, about half a million live in the three largest towns of Auckland, Wellington, and Christchurch, and considerably more than half the whole total dwell in the fourteen biggest towns; i.e. seven out of every thirteen people in New Zealand inhabit a town of over 10,000 persons. Urbanisation is therefore far less than in Australia.

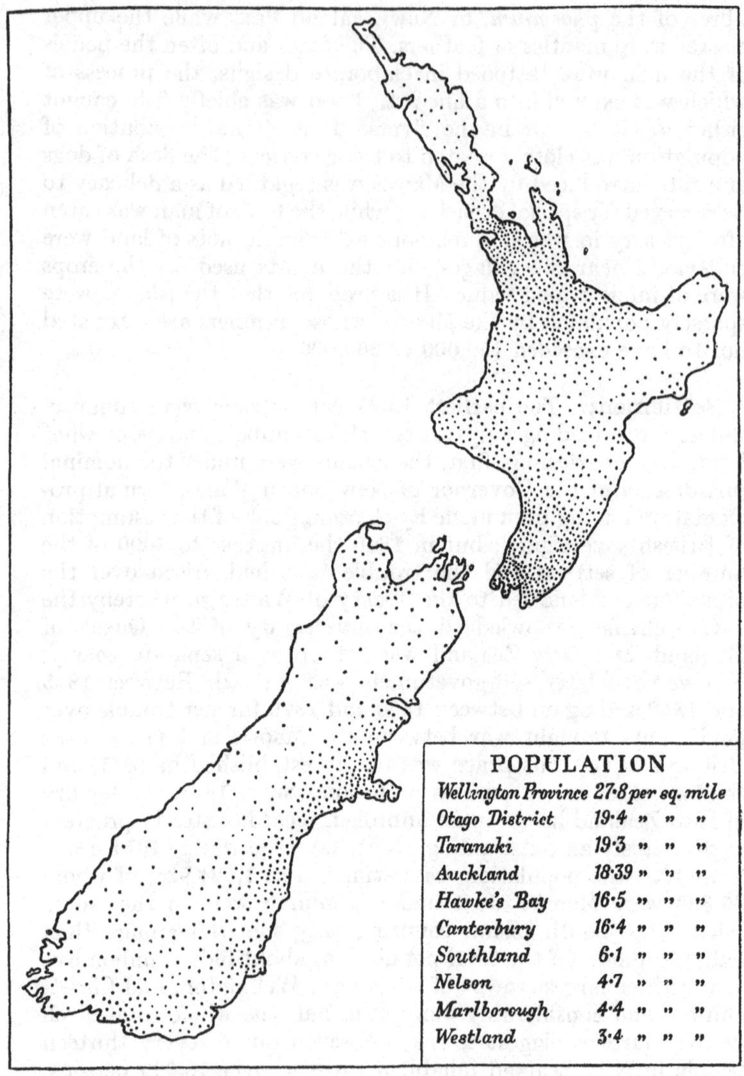

Distribution of population in New Zealand

Communications. Internal communication is by road, rail, and steamer. Regular boats ply daily between Wellington and Lyttelton, and other less important services connect other ports. There are 3000 miles of railway, more than half being in the South Island. In the North Island there are two main lines, both of which start from Wellington: one runs to Napier, the other to Auckland. The latter has an important branch through Taranaki. In the South Island the main line starts from Waiau-ua at the north of the Canterbury Plains and runs southward through Christchurch and Dunedin to Invercargill. Branch lines ascend the various river valleys, the most important crossing the Southern Alps by the Otira tunnel near Arthur's Pass and reaching the sea at Greymouth.

Overseas communication consists of (1) the steamship route through the Panama Canal to the British Isles; (2) alternative lines from Auckland through Suva and Honolulu to Vancouver or from Wellington through Pangopango in Samoa and Honolulu to San Francisco; (3) connexions between Auckland or Wellington and Sydney and between the southern ports of Lyttelton, Dunedin, and Invercargill on the one hand and Melbourne on the other.

Political Divisions and Towns. New Zealand is divided for convenience of administration into nine provinces: Auckland, Wellington, Hawke's Bay, and Taranaki in the North Island; and Nelson, Marlborough, Canterbury, Otago, and Westland in the South Island. Each province is subdivided into counties. Very few of the boundaries, whether provincial or county, are geographical—a fact that might be expected in a newly settled country.

There are no really large towns in New Zealand as yet, only three exceeding 100,000 inhabitants. One other contains more than 50,000 people, and ten more have over 10,000. Most of these fourteen towns are ports, as might be expected in a country settled from over the seas and depending for its trade on its overseas commerce. Auckland (pop. 218,400) is the largest town, having the advantages of (1) a fine harbour, (2) closeness to Australia, and (3) a focal position with relation to the Waikato and Thames valleys. But Wellington (pop. 144,800) is the capital of the Dominion, being in a central position on Cook Strait, having

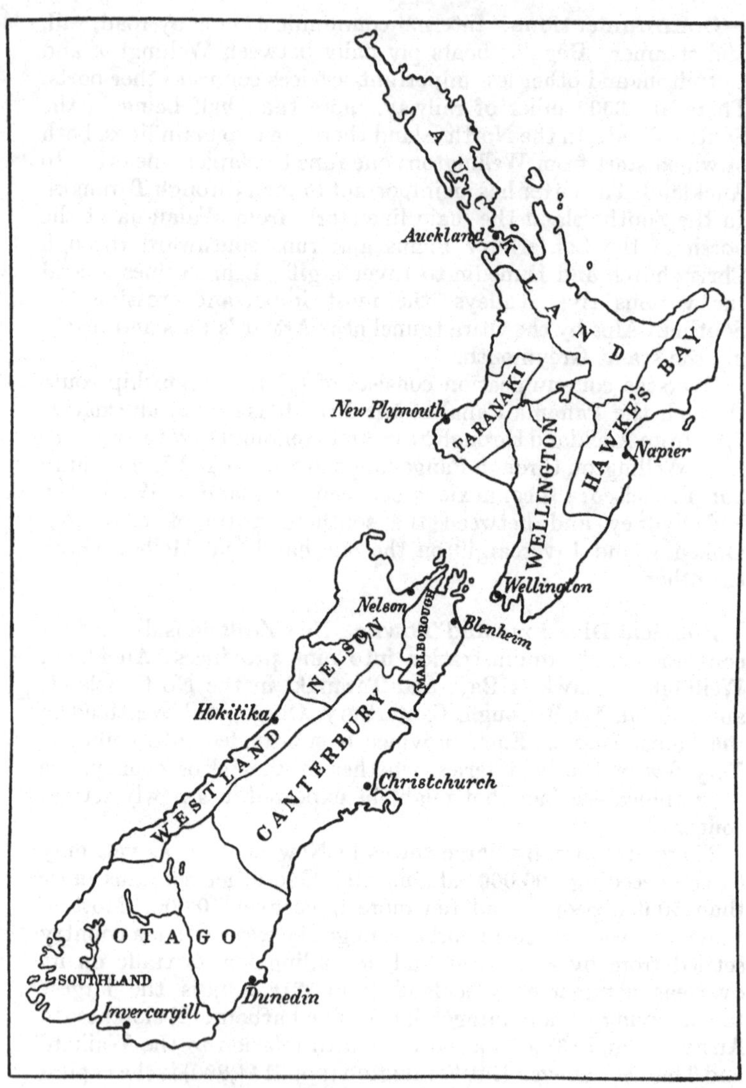

Provinces and towns of New Zealand

a good harbour, and with a large area of fertile lowland behind it. Christchurch is the focus of the Canterbury Plains, but since it is not on the sea, it is forced to have a port at Lyttelton to which it is connected by rail.

The Government of New Zealand is administered by a Governor-General appointed by the British Crown, an upper chamber known as the Legislative Council, and a lower House of Representatives. Executive work is in the hands of a Cabinet. This system, which was established in 1852, is modelled on that of the Mother Country. It is interesting to see a people carry its political organisation to the antipodes.

Outside the limits of New Zealand proper are a number of groups of small islands which are attached to the Dominion politically. The Auckland Islands (uninhabited), the Chatham Islands (pop. 562), and the Kermadec Islands (uninhabited) are groups lying in temperate latitudes and appropriated for political reasons, though they are too bleak and barren to be of any use. Similarly, the sector of Antarctica between 160° E. and 150° W. has been proclaimed to be under the jurisdiction of New Zealand under the name of the Ross Dependency. On the other hand, the Cook Islands and Tokelau have been attached to New Zealand, so that the Dominion may have its own source of tropical products. For the same reason, part of Samoa was taken over on a mandate from the League of Nations in 1920. It is probable that at some future date all the oceanic islands of the Pacific which are in British possession will be handed over to the Dominion.

THE PACIFIC ISLANDS

The Papuan Region

The rest of the area which we are here considering falls into two parts: the Papuan region and the oceanic islands of the Pacific. The former consists of New Guinea and the numerous other islands, large and small, situated on the continental shelf of Australia. The Wallace Line marks the structural separation from the equally extensive continental shelf of southeastern Asia, but this frontier is of no practical significance, since the flora, fauna, and human inhabitants have found it no obstacle and have crossed it at will in both directions. The islands are the highlands of a submerged region whose instability is still marked by earthquakes and the presence of active volcanoes. The fold lines of the Alps-Himalayan system run through the Lesser Sunda Islands and the southern Moluccas to New Guinea, where the latter line reaches the western end of the Australian arcs.

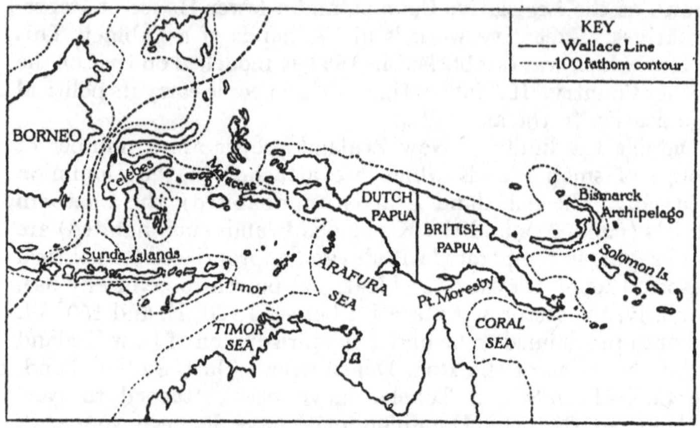

The Australasian islands

The islands are all mountainous, the peaks of New Guinea rising to 16,000 feet, and only the largest of them have any considerable areas of lowland.

The region lies within 12° of the Equator and therefore has an equatorial climate tempered by the sea. The statistics for Port Moresby show an equability of temperature known only in this and similar regions, the annual range being 4° F. Thus:

Port Moresby	Jan. 77	Feb. 76	March 77	April 77	May 77	June 76	Range 4° F.
	July 75	Aug. 74	Sept. 75	Oct. 76	Nov. 76	Dec. 78	

The rainfall is of the equatorial type too, and is mainly influenced by convection; but it is affected in summer by the monsoon winds from the northwest. For some cause that is not yet adequately explained, Timor and the adjacent islands suffer from drought and have a semi-desert vegetation, being as it were an extension of the great Australian desert; but the rainfall increases progressively northwestwards along the Lesser Sunda group. New Guinea and the Moluccas have an abundant pre-

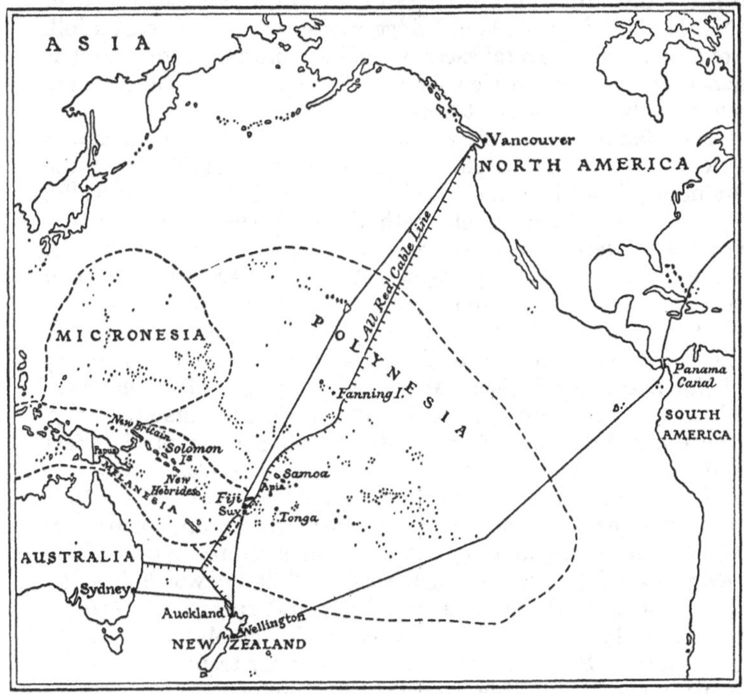

The islands of the Pacific

cipitation. In New Guinea the constant leaching of the heavy rains has removed from the soil much of the chemicals which cause fertility.

The vegetation, dense in most of the region, shows two tendencies. The islands nearer to Australia are characterised by eucalyptus and acacia, but northwestwards these types gradually change to the fig trees and sago palms of the Asiatic flora. This tendency is noticed also in the animals. Secondly, the teak, ebony, and bamboo of the equatorial forest climb high up the mountain slopes, but above 5000 feet they are gradually displaced by oak, laurel, and conifer.

The remoter parts of the forests of New Guinea contain specimens of the various primitive races which spread eastwards over the islands before the land bridge was severed. The coastal folk are a mixture of several races which have crossed from Asia, but generally speaking in New Guinea the population is Melanesian, while farther northeast it becomes Malay.

The islands consist of the two large units of New Guinea and Celébes with their satellites, the Bismarck Archipelago, the Moluccas, Timorlaut, and the Lesser Sunda Islands. Politically, they belong to the Dutch, with the exception of the eastern portion of New Guinea, the Bismarck Archipelago, and part of Timor. The more important settlements are Manado and Kupang in Celébes, Amboina in the Moluccas, and Singaraja in the Lesser Sundas. The eastern portion of New Guinea comprises the Australian territory of Papua and the mandated territory of Northeast New Guinea. Its main settlement is Port Moresby. The Bismarck Archipelago is also mandated Australian territory. The economic value of the Australasian portion of the East Indies is far less than that of the Asiatic section. But the Moluccas are famous for their supplies of spices from which they were once named, and the clove, nutmeg, kardamom, pepper, Kanary nut, cajaput tree, and the cinnamon are indigenous in the group. A general product is sago, while dyewoods (sandalwood and sapan) and teak are fairly widespread. The Dutch have introduced cocoa and sugarcane. In the Moluccas and on the coasts of Papua there is much fishing for trepang and shell, while in the latter place pearl-fishing is important.

The Oceanic Islands

The oceanic islands of the Pacific are to be numbered in tens of thousands and occur in groups or single islands. They are mostly south of the Equator and west of the 150th meridian of West Longitude. Physically, they are of two kinds: (1) 'high' islands, which are the upper portions of submerged highlands, e.g. the Solomons, New Hebrides, New Caledonia, and Fiji; and

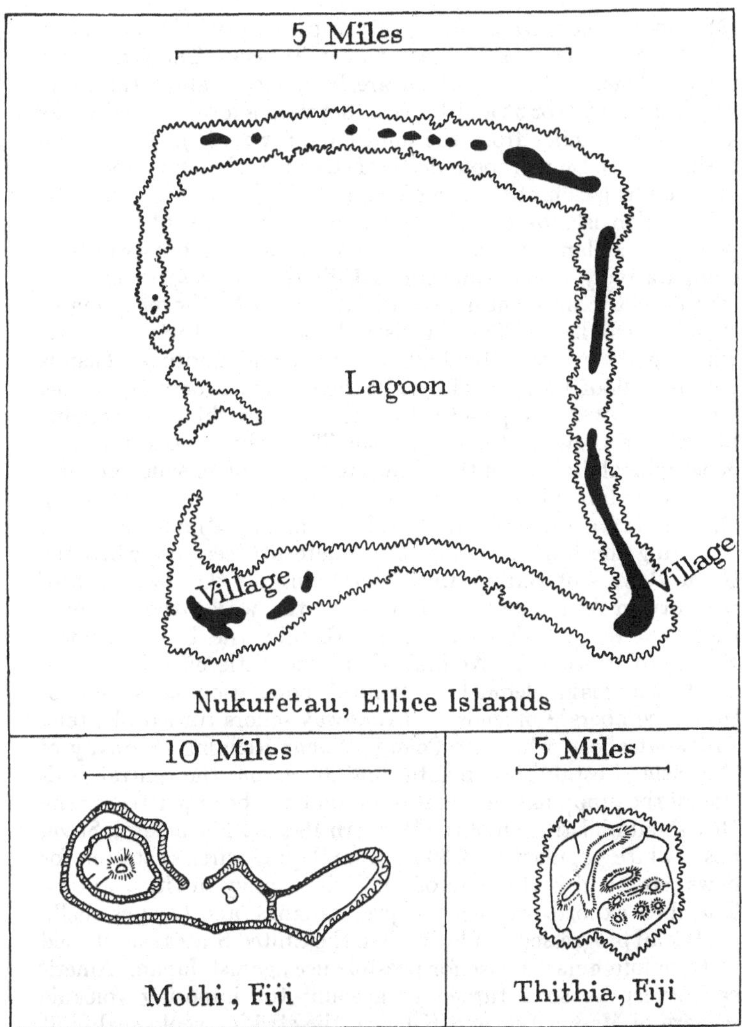

Coral formations in the Pacific. Thithia is a high island surrounded by a fringing reef, while Mothi, which is also a high island, has a fringing reef and is enclosed by a complex barrier reef. Nukufetau is an atoll, or *thakau*, on which the habitable portions are shown in black. The two villages contain about 200 persons each.

(2) 'low' islands, which are atolls or other formations of coral, e.g. Tonga, Gilbert and Ellice Islands, Tokelau. The former are ringed about with reefs which are built either along the shore ('fringing reefs') or at a distance of between a few hundred yards and a dozen miles from the land ('barrier reefs'). Interesting theories have been propounded by Darwin and Murray to account for the formation of these reefs and atolls.

The high islands on the Australasian arcs are inhabited by people of Melanesian race who have spread southeastwards in comparatively recent times. The Fijians still preserve oral accounts of the accidental journey which led to the peopling of their present home. The tiny islands, atolls for the most part, which make up the Marshall, Caroline, and Ladrones Islands contain folk of Malay origin, while the rest of the oceanic islands, when inhabited, are peopled by a tall, yellow-skinned, straight-haired race known as the Polynesian. These three races assist in a geographical division of the islands into categories which are also to some extent physical: Melanesia (='islands of the blacks'), Micronesia (='small islands'), and Polynesia (='many islands').

During the last quarter of the nineteenth century when the Great Powers of Europe were scrambling for the ownership of territory in Africa and other parts of the world, these remote islands were all claimed by Great Britain, the United States, France, or Germany. At first, there was little effective settlement, the main element of 'white' population consisting of 'beach combers',[1] officials, and runaway sailors turned planters. Fiji went ahead when the Colonial Sugar Refining Company of Queensland established plantations there, and the central position of the group has caused it to become the headquarters of the British administration of the Western Pacific. The capital, Suva, has a white population of 3000, is a relaying wireless station for news, etc., and is the base of the British navy in these waters. The chief exports are sugar, copra, bananas, and trochus shell.

Hawaii progressed similarly when the United States established at Honolulu a naval base for possible use against Japan. American enterprise soon turned to account the imposing volcanic scenery of Mauna Loa and Kilauea, the striking geological fault of the Pali, and the wide spread of breaker-lined water off the

[1] The local name for 'down-and-outs' who hang about on the beach near a native village or a settlement and pick up a living by begging or stealing.

Waikiki beach. Hotels were established, tennis courts and golf courses made, and tourists encouraged to visit the islands. A more solid industry was also begun in the planting of pineapples which were tinned and shipped to California.

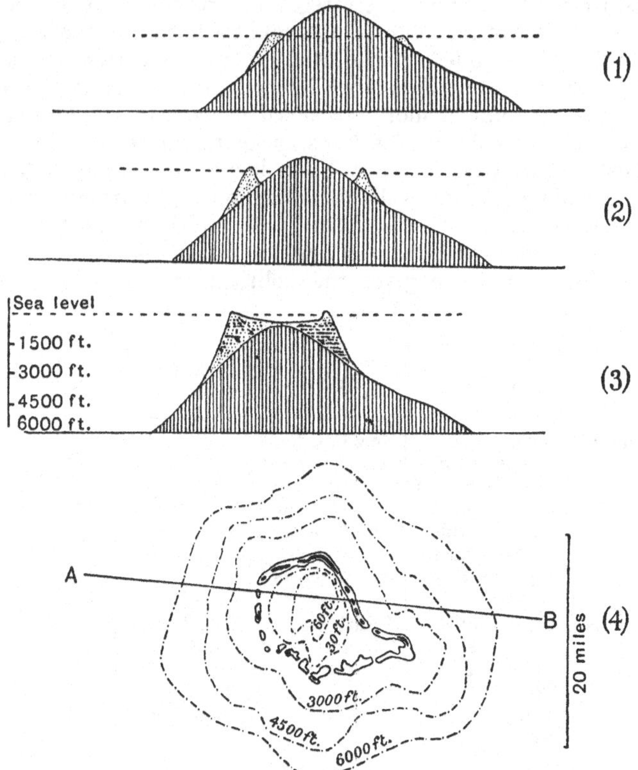

Formation of coral reefs and atolls

Outside these two groups, the settlements are small and few. The main product in Polynesia is copra, which is the dried kernel of the coconut, and in Melanesia copra and sugar. Formerly, the nickel mines in New Caledonia were the richest in the world and were worked by transported convict labour; but this source of labour was cut off in 1898, and the discovery of richer mines in Canada has deprived the island of its importance. It is a French colony, with its seat of government at Numea. Phosphatic guano is obtained in Nauru and Ocean Island.

Apart from the value of the islands for their economic products, there is considerable strategic importance in some of the groups. The use of Honolulu as a naval base has been mentioned, but it should be added that all American ships plying between the United States and Japan or China go out of their way to call here and at Guam, a more advanced outpost in the Marshall Islands. Honolulu is also the first stage on the route from Canada to Australia or New Zealand, Suva being the second, though American ships prefer to substitute Pangopango Harbour for the latter. Papeete, the chief settlement in Tahiti, is a port of call on the way from the Panama Canal to New Zealand.

The following table gives the political distribution of the islands:

Group	British	French	United States	Japan
Melanesia	Solomon Is. New Britain (*cap.* Rabaul) Santa Cruz New Hebrides (with French) Fiji (*cap.* Suva)	New Caledonia (*cap.* Numea) Loyalty Is. New Hebrides (with British) Bougainville (Solomon Is.)		
Polynesia	Gilbert and Ellice Is. Phœnix Is. Tokelau Tonga Samoa (Savaii, Upolu; *cap.* Apia) Cook Is. *Scattered islands*	Tahiti Paumotu Marquesas Tabuai	Tutuila and Manua (Samoa) Hawaii (*cap.* Honolulu)	
Micronesia			Guam (Ladrones)	Bonin Is. Pelew Is. Caroline Is. Marshall Is. Ladrones Is.

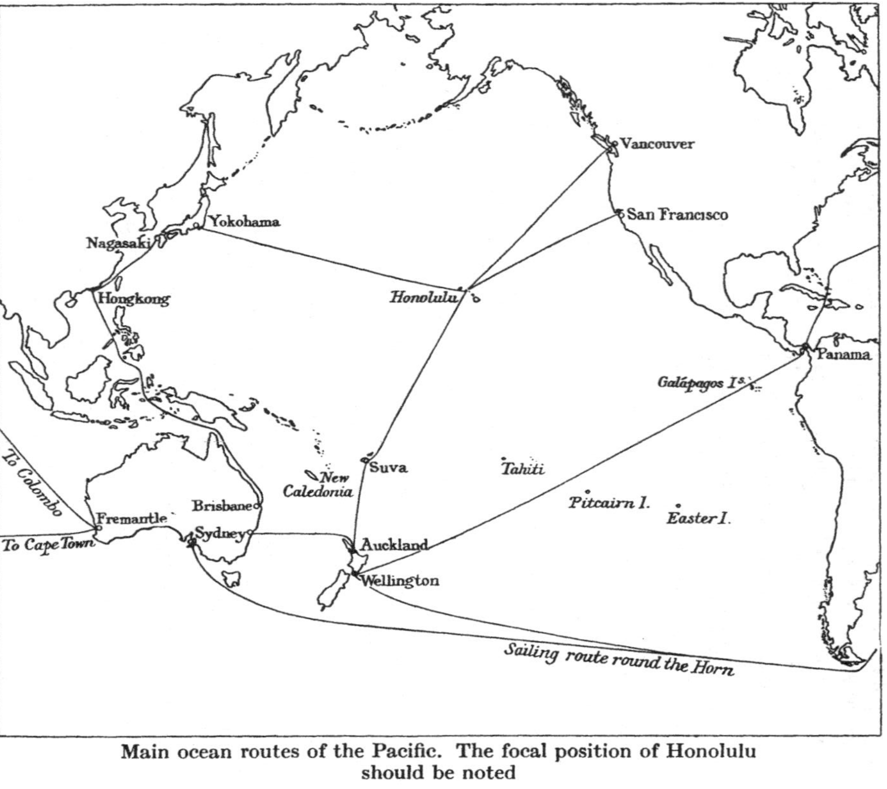

Main ocean routes of the Pacific. The focal position of Honolulu should be noted

Besides what has been said above, there are some interesting facts about particular islands or groups. There is the curious fauna of the Galápagos[1], which get their name from the characteristic animal, the turtle. Then there is Easter Island[2] with its mysterious megalithic statues, Pitcairn Island with its associations with the mutiny of the *Bounty*, and Samoa where R. L. Stevenson spent the last years of his life.

[1] Belonging to Ecuador.
[2] Belonging to Chile.

APPENDIX

AN OUTLINE OF PHYSICAL GEOGRAPHY

(*Intended for the purposes of revision only*)

I. LAND FORMS

A. The Material

The Earth's crust consists of rocks. Rock is composed of mineral substances and possesses three qualities which affect topography: *permeability*, *resistance*, and *solubility*. Some of the commonest rocks are:

Sand	Permeable	Small resistance	Insoluble
Chalk	”	” ”	Soluble
Limestone	”	Resistant	”
Clay	Impermeable	Small resistance	Insoluble
Granite	”	Resistant	”

Rocks may be either *stratified* (i.e. arranged in layers) or *unstratified*. The latter are usually less permeable and more resistant than the former.

The age of rocks is measured in geological periods, as follows:

I. Primary	Archæan sub-period Carboniferous	No fossils Coal	Formation of Hercynian chains
II. Secondary	Sandstone, limestone, chalk beds	Fossils of birds; mammals; modern plants	
III. Tertiary		Man's first appearance	Age of modern folds. Ice ages
IV. Quaternary	Alluvium		

B. The Rough-hewing of Topography

The main features of the relief of the Earth's surface are due to *earth movement*. This may be slow or rapid.

1. **Slow earth movement** gives rise to (*a*) *continent building*, by which vast plains, like the Great European Plain or the

Prairies of North America, are gradually brought above sea-level; and (*b*) *mountain building*. The latter is due to the contraction of the Earth's surface. The resulting horizontal pressure either folds the rock or fractures it. In the former case, parallel ranges of *folded mountains* are formed, consisting of *anticlines*, or up-folds, and *synclines*, or down-folds. All the great mountain ranges of the present day were formed in this way during the Tertiary Period.

When the rock is hard and brittle, a long *fracture*, or crack, occurs, and pressure is relieved by the down-slipping of the rock on one side of the crack. A *fault* is thus formed. A big scale fault causes an *escarpment*, like that on the west side of the Rhone valley. If two faults occur close together, the escarpments may face away from each other, thus forming a *block mountain* or *horst* like the Black Forest. When the escarpments face each other, a *rift valley* is formed, of which the best example is the Great Rift Valley of Africa.

2. **Rapid earth movement** is usually accompanied by volcanic eruptions or earthquakes. By the former volcanic cones are built up; the latter often lead to subsidence or uplift of small areas.

C. The Polish of Topography

The rough-hewn surface of the Earth's crust is modified by the forces of erosion. These are changes of temperature, wind, running water, ice, and waves.

1. **Changes of Temperature.** The expansion of freezing water in cracks in rock causes pieces to be broken off. Under this head is included any rapid change of temperature, e.g. the heating-up of rock in the Sahara often causes it to crack. The cooling at night after the heat of the day has a similar effect in deserts.

2. **Wind.** Wind acts as a transporter of rock-waste, piling up *dunes* on the sea coast and *barkhans* in sand deserts. Sometimes it carries fine particles long distances and covers great areas with fertile soil known as *loess*. Wind also denudes the surface by driving against it blasts of sand. It is typical of wind erosion that the lower parts of exposed rock are more worn than the

upper parts. In deserts, rocks are often worn into fantastic shapes.

3. **Running water.** The action begins with rain which by the impact of its fall tends to break off pieces of rock and with the aid of acids absorbed from the air dissolves the surface on which it falls. Once fallen, rain water either percolates into the ground or runs off on the surface. In the former case, it gathers more acids from the soil and with its help dissolves any soluble rock through which it may pass. Finally, it returns to the surface as a *surface spring*, a *constant spring*, or an *intermittent spring*. The rain water which flows off on the surface sweeps with it loose particles of dust and sand, cutting the ground into a series of temporary runlets. A number of these converge on the line of lowest ground forming a brook or rivulet. A smaller stream flowing into a larger one is a *tributary* or *affluent*. When the flow is sufficiently fast, either because of the steepness of the slope or because of the quantity of water in a narrow channel, the rock-waste swept down by the run-off is carried down towards the sea. Wherever a check in the flow occurs, some of the rock-waste is deposited as *sediment* on the bed of the stream. Thus are formed *alluvial fans*, *flood plains*, *sand banks*, *bars*, and *deltas*.

The stream also *corrades*, or wears out by friction, the ground over which it flows. Pot-holes, gorges, and rapids are the result.

As fast as the stream corrades its bed, the run-off of rain water washes in the sides and back of the basin. Thus, every stream tends to push back the *waterparting* or ridge which divides its basin from those of other streams. Sometimes a stream which flows at right angles to another cuts through the back of its own basin and adds the other stream to itself. This is known as *river capture* and is usually indicated by an *elbow of capture*. The rest of the beheaded stream is too small for its valley, which is then described as *dead*.

Where a river runs through a rainless district, there is no run-off to wash in the banks, and the stream cuts a long gorge with precipitous sides. This is known as a *canyon*.

Thus, the land relief tends to have its higher parts worn away and its depressions filled with rock-waste. Streams tend to wear their basins down to a parabolic curve known as the *graded line*, the lowest point of which is the *base level* of erosion. As soon as the graded line is reached, erosion becomes sluggish or stops, and the topography is said to have reached *old age*. Similarly, topography which is still full of irregularities, and in which the rivers are in the full vigour of erosion, is said to be youthful. The intermediate stage is *middle age*. Earth movement may raise the basin of a graded river and thus cause a *rejuvenation* of erosion. This erosional progression was formulated by W. M. Davis and is known as the *cycle of erosion*.

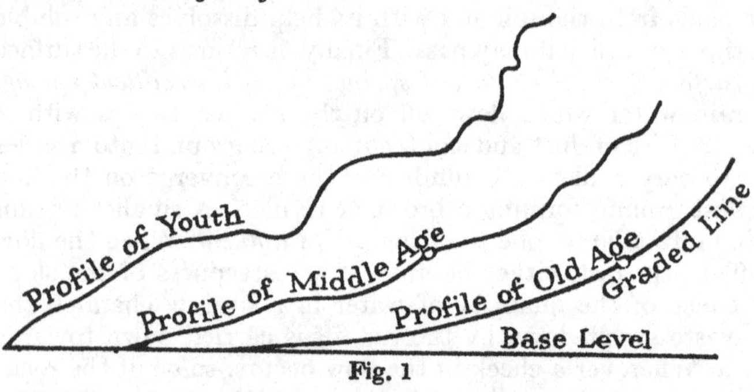

Fig. 1

4. **Ice.** Where the land surface is above the *snow line*, it becomes permanently covered with snow. The lower layers of this are compressed into granular ice. If the area covered is extensive, a *snow cap* is formed, from which the ice moves outwards, shearing off rock-waste and piling it up round the edges of the cap to form a *moraine*. If the edge of the cap reaches the sea, large masses of ice are broken off and float away as *icebergs*. When mountain ranges rise above the snow line, snow collects in hollows, called *cirques* (French) or *corries* (Scotch) or *cwms* (Welsh), between the buttresses of the peaks. Thence the ice moves out in a *glacier tongue* down the valley until it reaches the snow line. Rock-waste falls on the sides of the glacier, forming lines known as *lateral moraines*. These are carried down with the ice and piled up at the end in a *terminal moraine*. Glaciers whose tongues do not reach the foot of the mountains are called *mountain glaciers*. Those which reach the plains and spread out in lobes over the ground are known as *piedmont glaciers*.

Glaciers corrade their beds, but, as there is no run-in from the sides, the banks are steep. Nor are the beds reduced to the graded line, for where an obstacle checks the ice, a deep hole is dug above and sometimes below the obstacle. When a change of climate frees a glacier course from ice, the valley is U-*shaped* (as distinct from the V-*shaped* river valley) in cross-section, is headed by a circular basin often containing a *tarn*, and at intervals may have long, deep lakes. The valleys of tributary streams do not reach the bottom of the main valley and are said to be *hanging*.

5. **Waves.** The waves of the ocean beat against the shore and hurl sand, pebbles, and stones against the land. In time the coast is worn into a cliff. As the resisting power of the rock varies, the progress of erosion is irregular. Rocky islets, known as *stacks* or *skerries*, are cut off from the land. In sheltered places the tides and currents deposit sand and shingle, forming *sand spits* and, when the deposits connect a stack to the mainland, a *tombolo*.

The Influence of Climate. Climate influences erosion in two ways: (*a*) it decides what factor shall perform the work in a given place; and (*b*) by its changes it causes now one agent, now another, to act.

The Influence of Rock. Rock influences erosion by means of its qualities and its stratification. Rock of small resistance, like clay and chalk, gives rounded hills and broad open valleys. In permeable rock, like limestone, rivers tend to disappear underground, and the topography assumes a peculiar appearance known as *karstic*. Erosional escarpments, circular depressions, caves, and a number of minor irregularities of ground form its characteristics. Granite and other very resistant, impermeable, and insoluble rocks tend to give rounded, boulder-strewn hills and marshy, clay-filled valleys under the erosion of running water, and jagged peaks and crags under the action of frost. Sandstone often tends to give a tabular configuration to a region. If it alternates with softer rock, e.g. shale, the sandstone tends to stand out in ridges, whilst the softer rock is eroded. (Where sandstones are laid down horizontally, they often give a rounded appearance to a region, making it very like a granite district.)

II. CLIMATE

The three factors which go to make up the climate of a region are temperature, barometric pressure, and rainfall.

A. Temperature

The degree of heat or cold of a body is called its temperature. In Geography 'temperature' is often used as a short expression for 'temperature of the air'.

The temperature of the Earth's surface, of the water on the surface, and of the atmosphere is regulated by energy received from the Sun's rays. The process by which the energy is received

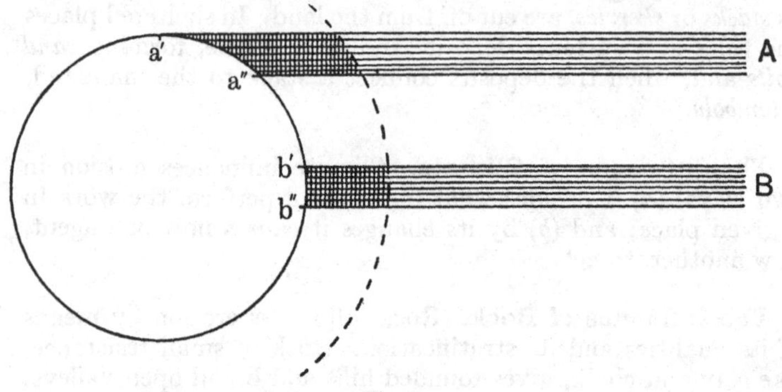

Fig. 2

is known as *insolation*. The amount of insolation is determined by the angle at which the rays reach the ground.

This is shown in Fig. 2. The distance of the Earth from the Sun is so great that the rays may be regarded as parallel on reaching the Earth. Let *A* and *B* represent equal bands of rays. It is clear that *B* gives greater insolation than *A*, since (i) it has less atmosphere to pass through and thus loses less of its energy on its journey, and (ii) on arrival it has a smaller area to distribute its energy over.

Hence it follows that (i) the midday sun gives greater insolation than the morning or evening sun, and (ii) other things being equal, insolation decreases gradually from the Equator to the Poles.

Factors of Temperature in Climate. The climatic temperature of a given area depends on one main and four subordinate factors:

1. *Main Factor.* This, as explained in the last paragraph, is distance from the Equator. But its operation is modified by:

2. *Altitude.* Since the air receives most of its heat from the dark rays radiated from the surface of land and water, temperature decreases with height above sea-level. The decrease is on the average 1° F. for every 325 feet of ascent.

3. *Direction of Prevailing Wind.* Winds which blow from one region to another of different temperature modify the temperature of the region to which they blow by bringing some of the greater heat or cold of the region from which they come.

4. *Nature of Ocean Currents.* Similarly, ocean currents carry the temperature of one part of the ocean to another. Surface currents moderate the temperature of winds passing over them and so may indirectly influence the land.

5. *Differential Heating and Cooling of Land and Water.* Land surfaces cool and warm up more quickly than water. Hence, in times of heat (e.g. summer, midday) water surfaces are cooler than land surfaces, whereas in times relatively cooler (e.g. winter, night) water surfaces are warmer than the land. Winds, therefore, from sea to land moderate the temperature of the latter.

Distribution of Temperature. The last four factors prevent temperature from corresponding exactly to latitude. Distribution of temperature is shown by lines called *isotherms* which join places with equal mean temperature. It will be seen by examining a map of the world showing mean annual temperature that:

(*a*) The temperature decreases roughly from the Equator to the Poles;

(*b*) The decrease is more regular in the southern than in the northern hemisphere;

(*c*) Near the Equator the sea is cooler than the land, while outside the Tropics the reverse is the case;

(*d*) The centres of extreme heat and cold are found on land.

N.B. The influence of altitude is always eliminated from charts showing the distribution of temperature.

Maps showing the distribution of temperature in January and July will make it clear that in January the temperature of the southern hemisphere is greater than that of the northern, and that in the northern hemisphere the sea is warmer than the land. The July map shows that the reverse is true in that season.

The difference between the highest and the lowest mean monthly temperatures at a given place is known as the *range*.

B. Barometric Pressure

That the air has weight may be proved by weighing a toy balloon when inflated and when deflated. The increase in weight of the inflated balloon must be that of the air with which it is filled.

Air Pressure. The atmosphere extends to a height of some 100 miles above sea-level and its weight exerts pressure on all surfaces within it. This pressure acts not downwards only, but in all directions, since the air is fluid. Air pressure is measured by means of an instrument known as a *barometer*, and normal pressure at sea-level is 29·9 inches (or 75 cm. or 1015 mb.). The pressure grows less with altitude, since in higher regions there is less air above to exert pressure, the rate of decrease being 1 inch for every 900 feet of ascent.

Variation in Pressure. The pressure of air varies at a given place owing to:

(*a*) The introduction of water vapour into the air, or the reverse;

(*b*) A variation in air temperature.

Since water vapour is lighter than dry air, air containing water vapour is lighter than dry air. Increase in temperature expands the air, making it lighter bulk for bulk.

Convection Currents. Since air is fluid, any portion which is made lighter at once tends to rise, while cooler (i.e. heavier) air from around tends to flow in from the sides to displace the lighter air. Such a current system when produced by differences in temperature is known as a convection current and will be understood from the diagram (Fig. 3).

Wind. Great air currents are known as wind and are usually caused by convection.

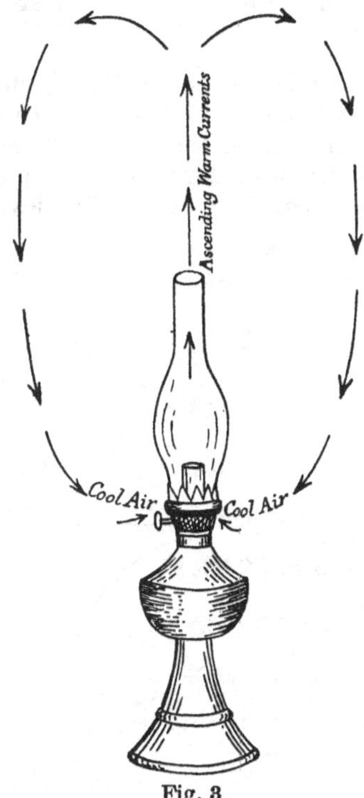

Fig. 3

Planetary Wind Systems. The Equator being on the average the hottest part of the globe, the air nearest it will be hotter and lighter than that farther away. Hence, a vast convection current will be set up, consisting of air rising vertically at the Equator (see Fig. 4, *a*), and cooler air flowing in on the surface north and south to displace the hotter air (see Fig. 4, *b*, *b*). The surface winds are known as *Trade Winds*.

Ferrel's Law. They do not, however, blow due south or north to the Equator, for they are deflected by the rotation of the Earth, according to Ferrel's law that 'every moving body on the Earth's surface is deflected to the right in the northern hemisphere and the reverse in the southern'. Hence, the Trades are northeast winds in the northern hemisphere and southeast winds in the southern.

Planetary Wind Systems (*contd*). The ascending air at the Equator soon cools and can rise no higher. It therefore flows off northwards and southwards as the *Counter Trade Wind*, being deflected in accordance with Ferrel's law. The deflection increases to such an extent that at the Tropics the winds are

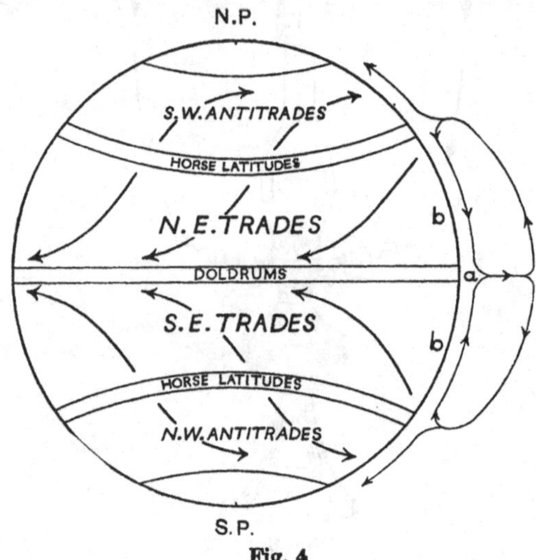

Fig. 4

westerly and can go no nearer the Poles. Hence, a mass of cold air is piled up about 30° N. and S., in what are known as the *Horse Latitudes*. The pressure is very high and so the air tends to flow off on the surface north and south, some joining the Trades going to the Equator and so completing the equatorial cycle, some going towards the Poles. The latter, deflected by the rotation of the Earth, are known as Southwest and Northwest *Antitrades*, which blow in temperate latitudes. The accompanying diagram (Fig. 4) shows the system in theory on a globe of uniform surface and material.

Influence of World Distribution of Temperature. As differences in temperature cause differences in pressure, this theoretical system is modified by the world distribution of temperature. On maps showing distribution of pressure, places of equal mean pressure are joined by lines known as *isobars*. It will be seen that:

(*a*) The theoretical belts of high and low pressure are broken up into 'centres';

(*b*) These 'centres' correspond to areas of low and high temperature;

(*c*) The winds blow from high-pressure 'centres' towards the centres of low pressure.

Seasonal Changes. Examination of charts showing the distribution of pressure in January and July proves that seasonal changes take place to correspond with the changes in temperature. The most important of these gives rise to *Monsoon* winds in Asia, Australia, and North America. In winter those land masses become centres of high pressure from which winds tend to blow outwards, while in summer they become low-pressure centres into which winds tend to blow from the surrounding regions of higher pressure. The rotation of the Earth deflects these winds (see Ferrel's law) and makes them seem to be blowing spirally inwards towards the low-pressure centre and spirally outwards from the high-pressure centre. Hence Buys Ballot's law that 'in the northern hemisphere the centre of low pressure is on the left of a person standing with his back to the wind'.

Swing of the Wind Belts. But a general movement, known as the *swing of the wind belts*, is observed in the pressure centres. In July they all move northwards; in January they are farther south. This swing causes regions at the borderline between the high and low centres to be now in one centre, now in the other—with a consequent change in the prevailing wind. This in its turn leads to special developments in countries lying between 35° and 45° N. and S.

C. Rainfall

Under this head are included not only the actual rain that falls, but also snow, hail, dew, clouds, fog, mist, and other forms of atmospheric moisture.

1. Capacity of Air for holding Water Vapour. Air which has not been artificially dried has as part of its composition a varying amount of water vapour. The amount of water vapour which can be held by a given volume of air depends on the temperature of the air. The accompanying table shows that the higher the temperature of the air the more water vapour it can hold. When air holds as much water vapour as it can contain at a given temperature, it is said to be *saturated*, and the temperature at which the air is saturated is called the *dew point*.

Temperature		Greatest amount of water vapour containable in a cubic metre of air
° C.	° F.	grammes
−10	14	2·28
0	32	4·87
+10	50	9·36
20	68	17·10
30	86	30·08
40	104	50·67

2. Process of Evaporation. The water vapour in the air is derived from (*a*) water surfaces, (*b*) damp ground, (*c*) trees, by evaporation. When water is raised above a certain temperature (normally 212° F.), it changes from liquid into gaseous form. This process, caused by the heat of the Sun, is constantly introducing water vapour into the air.

3. Process of Condensation. When air is cooled down to the dew point, any further cooling will cause the excess moisture to pass from its gaseous state into a liquid. Drops of water form either at or near the surface of the Earth or on dust motes in the atmosphere. In the former case it is called *dew*, or, if frozen, *hoar frost*. In the latter case countless drops form *mist*, *fog*, or *cloud*. The further condensation of the water vapour makes the droplets of a cloud too heavy to remain suspended in the air, and they fall as *rain*. Sometimes the drops are frozen and, when they fall, are known as *snow* or *hail*. A mixture of snow and rain is *sleet*.

4. Natural Causes of Condensation. The fall in temperature which gives rise to condensation may be caused by:

(i) The rise of relatively warm, moist air in a convection current;

(ii) The climbing of air over relatively high ground;

(iii) Contact with colder objects (e.g. other air currents).

5. **Conditions favourable to Rainfall.** Hence rainfall tends to be abundant:

(i) In regions liable to convection currents, e.g. the Doldrums, the paths of temperate cyclonic disturbances, the tropics;

(ii) In regions of high relief, e.g. the west of England, the Himalayas.

6. **Conditions unfavourable to Rainfall.** But rainfall tends to be scanty:

(i) On the lee side of high relief (*Foehn* or *Chinook* effect), where an area of *rain shadow* may be observed (see Fig. 5);

Fig. 5

(ii) In regions protected from the winds by surrounding high relief, e.g. Tibet, Gobi, Colorado;

(iii) At great distances from the ocean, as in Turkistan;

(iv) Wherever the wind seldom blows from the sea on to the land, as in the Western Sahara.

7. **Measurement of Rainfall.** The principle on which rain is measured is that, assuming the water to remain where it falls, the depth is sounded at intervals. A large tank at ground level would prevent the water from running off, but would not stop evaporation. Hence, an instrument called a *rain gauge* is used, which prevents run-off and checks evaporation. The amount or depth of rain that falls is measured in inches or millimetres. A fall of 9 inches of snow is reckoned as equivalent to 1 inch of rain.

8. **Differing Values of a Unit of Rainfall in Different Regions.** As the Sun's heat is the cause of evaporation, and as the effect of evaporation is to remove the rain that falls, regions of great insolation retain less of the rainfall than those of less insolation. Hence, in tropical regions a rainfall of under 50 inches a year is scanty and one of under 25 inches leads to semi-desert conditions, while in temperate regions a fall of 25 inches is plentiful and one of 50 inches is excessive. In temperate regions where the seasons are marked, the same amount of rain which is plentiful in winter may be scanty in summer.

9. **World Distribution of Rainfall.** The amount of rainfall is shown on maps by means of lines known as *isohyets* which join places having equal mean rainfall for a given period. Examination of a map showing the world distribution of rainfall proves the dependence of the amount of precipitation on temperature and wind direction. Thus, generally speaking:

(i) Rainfall is heaviest in equatorial regions and scantiest in the cold belts;

(ii) Within the tropical belt of easterly winds rainfall is heaviest on the east coasts;

(iii) In the temperate belts of westerly winds rainfall is heaviest on the west coasts.

It should be noted that rainfall belts, being dependent on wind belts, swing northwards and southwards with the wind belts (see page 507).

CLIMATIC REGIONS

Owing to differences of temperature, pressure, and rainfall, the Earth's surface may be divided into a number of *natural regions* in each of which the effects of temperature, pressure, and rainfall, i.e. the climate, may be regarded as similar throughout. Differences of temperature give three main divisions: the Hot Belt (within the Tropics), the Temperate Belts (from the Tropics to the Arctic or Antarctic Circles), and the Cold Caps (within the polar circles). Differences of pressure and rainfall subdivide these into:

I. Hot climates:			
	Normal:	1. Equatorial	Amazon Valley
		2. Tropical	Senegal
	Monsoon:	3. Tropical	India
II. Subtropical:		4. Mediterranean	Italy
		5. Steppe	South Russia
		6. Chinese	Eastern China
III. Temperate:		7. Maritime	British Isles
		8. Continental	Poland
		9. Eastern	Newfoundland
IV. Cold climates:		10. Cold maritime	Norway
		11. Cold continental	Siberia
		12. Polar	Baffin Island
V. Desert climates:		13. Hot desert	Sahara
		14. Cold desert	Gobi
VI. Mountain climates:		15. Alpine	Tibet

N.B. The last group is influenced by altitude acting through temperature and group V is caused chiefly by the absence of rainfall.

III. PLANT, ANIMAL, HUMAN

Necessary Conditions of Life. Heat, light, moisture, and air are the four conditions of plant and animal life. Where all four conditions are abundantly fulfilled, as in the equatorial regions, life is most vigorous. But in other parts of the world one or other of the conditions fails more or less. Thus, *heat* is insufficient in polar regions: hence, plant life is reduced to a few mosses and lichens. The summer isotherm for 50° F. is the poleward limit of trees. Animal and human life is restricted to small groups of wanderers. *Light* is insufficient in the polar regions, below 200 fathoms in the ocean, and underground. Hence, subterranean life is restricted to a few primitive creatures, like worms, grubs, beetles, and still fewer animals, like the mole. Deep sea life is not abundant. Absence of light develops blindness. *Moisture* is insufficient in deserts and semi-deserts. *Air* is insufficient at great ocean depths and near the tops of high mountains.

Food is also a prime necessity. Plant life feeds directly on chemical substances in the soil and is abundant in proportion to the amount present in the soil and to the water supply. Rock surfaces, stony ground, and deep water are unfavourable to plant life. Animal and human life gets its food only indirectly from the soil and is dependent in the long run on plant growth.

A. Plants

1. **Vegetation-types.** Various combinations of the primary conditions give rise to the occurrence of groups of plants with marked characteristics. Thus:

(*a*) Equatorial forests with dense, tangled undergrowths, of many kinds of trees, parasites, lianas, and an abundance of animal life.

(*b*) Tropical and Monsoon forests (a mixture of (*a*) and (*c*) due to the occurrence of winter droughts) with less dense undergrowth and wide intervals of grassland.

(*c*) Tropical grasslands, or *savana*, with giant grasses occur in regions of great heat and insufficient rain for forests.

(*d*) Tropical semi-desert, due to lack of moisture, produces xerophilous vegetation, consisting of plants with hard barks, and thorns, no leaves or leaves covered with wax to resist evaporation.

(*e*) Sub-tropical evergreen shrub land with xerophilous plants, whose adaptations are chiefly bulbs, long roots, wax-coated leaves, and thorns.

(*f*) Temperate grassland with short grass which dies away in winter and field flowers like clover, buttercups, etc. (steppe, prairie, pampas).

(*g*) Parkland with forests of broad-leaved trees (oak, beech, etc.) which drop their leaves in autumn and have an undergrowth, but which leave extensive glades of grassland between the forest.

(*h*) Temperate forest with great stretches of *conifers* (pines, firs, larches) without undergrowth.

(*i*) Tundra and Alpine vegetation consisting of plants dwarfed by cold or forced to grow for the short summer only (*annuals*).

2. **Man's Influence.** Man influences natural vegetation by (*a*) clearing away forests (Western Europe), (*b*) restricting growth to the plants he requires (England), (*c*) producing changes in plants by selection (wheat) and grafting (apple pippins), etc., (*d*) transplanting the useful growths of one region to another (rubber from South America to India).

(*Canadian Pacific Railway*)

MAORI GIRLS IN NATIVE DRESS

The costumes are made of feathers and grass. Note the carved gables of the houses

Plate XXXII

PERTH, WESTERN AUSTRALIA

Note the modern buildings in the town and the vertical hanging of the leaves of the eucalyptus trees in the foreground

3. **Economic Plants.** Some plants are of great use to man in general, but can be grown in restricted regions only. The chief of these are:

Plant	Conditions	Areas of chief economic production
I. Hot belt:		
(1) Rubber	Heat and moisture	Amazon, Congo
(2) Rice	Heat and moisture	India, China
(3) Tea	Heat and moisture, drainage	Assam, Ceylon
(4) Coffee	Heat and moisture, drainage, elevation	Brazil, Colombia
(5) Tobacco	Heat and moisture, soil renewal	Brazil, West Indies
(6) Cocoa	Heat and moisture, shelter from wind	West Indies, Gold Coast
(7) Sugar	Heat and moisture	West Indies, Java
II. Subtropical belt:		
(1) Cotton	Heat, moderate rain	United States, India, Egypt, China
(2) Maize	Heat, moderate rain	United States, Hungary, India
(3) Grapes and oranges	Warm summer, mild winter (absence of frost), good drainage	Spain, Turkey, Italy, Florida, South France, Southeast Australia, California
III. Temperate belt:		
(1) Wheat	Hot, dry, sunny summer	United States, Canada, Russia, Western Europe, Hungary, Punjab, Argentine
(2) Oats	As for wheat, but cooler and damper	United States, Canada, British Isles, Germany
(3) Potatoes	Damp	Germany, Russia, Austria, France, United States, Ireland, Chile
(4) Apples	—	East Canada, United States, British Columbia, France
(5) Timber	—	Canada, Norway, North Russia, Southwest Australia

B. Animals

1. **Associations.** Like plants, animals are grouped together in associations. The grouping depends on temperature, the nature of the vegetation, and topography. In the Hot Belt animals are hairless or short haired, in the Cold Caps they are protected by thick coats of fur and layers of blubber under the skin. In equatorial forests the vegetation is too dense to harbour

large animals, but there are swarms of many kinds of insects, troops of monkeys, and other arboreal animals, and large numbers of gaily coloured birds. Grass plains give rise to numerous herds of grazing animals of great speed and powers of endurance, such as the horse, the buffalo, and various kinds of antelopes. In savanas the great abundance of these animals supports numerous carnivorous beasts, such as the lion and the tiger. The elephant and the giraffe live in the more open forests on the border of the savanas. Temperate forests support bears and foxes, and on their fringes wolves.

Just as plains develop swift runners of great endurance, so broken high ground breeds jumpers of the sheep and goat type. Desert conditions have produced the camel and the llama, while marshy areas harbour the alligator, the hippopotamus and wading birds.

2. **Useful and Domestic Animals.** Many of these animals are useful to man. Some provide him with food, others with hides of which he makes articles of clothing, etc.; others he tames and uses for his purposes, e.g. the dog, the horse, the buffalo, the elephant, and the reindeer. Those which are useless or untamable he gradually exterminates. Thus, the wolf has gone from West Europe and even the 'big game' of East Africa is threatened.

3. **Man's Influence.** Man's influence on animal life is fourfold: (*a*) extermination, (*b*) domestication, (*c*) transportation, (*d*) breeding. The first two have been dealt with. Man has carried his domestic animals with him wherever he has gone. The horse was introduced into America, the sheep and the rabbit into Australia and New Zealand, the pig into the South Sea Islands. But he has not been satisfied with Nature's creatures: he has by the process of selection in breeding changed animals like the dog, the horse, the ox, and the ordinary barn-door fowl to suit his purposes.

C. Man

1. **Influence of Environment.** Like the animals, man is influenced by his environment. He cannot live *permanently* (*a*) on the ocean, (*b*) in deserts, (*c*) at high altitudes, or (*d*) in regions of greatest cold. He is dwarfed by life in equatorial forests and shortened in stature by great cold, by mountain life,

or by a severe struggle for existence. His *mode of life* depends on geographical conditions; forest man is a hunter, grassland man a herdsman and nomad; while the dweller in fertile river valleys is an agriculturalist. His *food, costume*, and *dwelling* vary with the region: the herdsman of the grassland feeds on milk and cheese from his herds; the agriculturalist feeds on his vegetable crops; the coast dweller on his catch of fish. The Eskimo wears his garment of fur, the Englishman his cloth of woven wool, the Bengali his loose cotton wrapping. The steppe dweller lives in his skin tent, the forest hunter in his log cabin, the river-valley folk in their houses of mud or brick. Even *character* is formed by environment: the contemplative mind of the Arab was developed by the clear skies of his pathless roaming grounds; the rough hardships of sea life tend to encourage a feeling of personal freedom among coast dwellers like the English and Norwegians. The abundance of food and other necessities makes the tropical negro a sluggard; agriculturalists are peaceful, but steppe dwellers and hunters are warlike.

2. **Man's Conquest of his Environment.** Yet man has learnt to overcome many of the disadvantages of his environment. He resists cold by covering himself with clothes and by producing artificial heat; he turns grassland into large-scale plantations of cereals, and forest into cultivated ground; he irrigates the desert and forces it to support his crops. He transports the food and other products of one region to another and by means of trade and commerce arranges the exchange of specialised products. He counts the seasons and the hours, anticipating favourable and unfavourable periods. He can turn night into day by means of artificial light. By increased facilities of transport he lessens the disadvantages of regions and modifies their influence on him.

3. **Man's Influence on his Environment.** Man has even reacted on his environment. He modifies the topography of his neighbourhood by cutting down forests, by draining swamps, by causing extensive subsidence through his mining, and by piling up veritable hills of slag. He canalises streams, he dams others, and then controls their flow, dredges estuaries, builds causeways and breakwaters. He can even modify climate by the extensive removal of forests, as in Western Europe.

4. **Races of Man.** Man has sprung from a single stock which developed probably in Africa, at least before the quaternary ice ages. As he spread, various factors, of which isolation, climate, and altitude are the chief, produced differences in his appearance and bodily structure. The main differential features are: shape of the head, colouring of the hair and eyes, stature, colour of the skin, and the texture of the hair. Grouping of these features distinguishes three main races:

(*a*) Heads may be long (i.e. with a breadth of less than 75 per cent. of the length) or round (i.e. with a breadth of over 75 per cent. of the length) and may have vertical or prognathous face profiles.

(*b*) Skin colouring varies from almost black through various shades of brown and yellow to a creamy pink.

(*c*) Hair also shades off from jet black through browns to light yellow.

(*d*) Eyes may be black, brown, blue, or grey.

(*e*) Hair may be cylindrical in section, black and lank; very elliptical in section, black and woolly; or moderately elliptical in section, brown or yellow, and wavy.

(*f*) Stature varies from 3 feet 6 inches among pygmies to 5 feet 10 inches among the Lowland Scotch and Polynesians.

Combinations of these features distinguish man into three main *races*:

(i) The White race, whose characteristics are wavy hair, a fair complexion, and well-marked facial features. It includes the three European racial divisions, viz.: the Nordics, with long head, creamy-pink skin, light yellow hair, blue eyes, and tall stature; the Alpine stock, with round head, thick-set figure, and coarse features; and the Mediterranean peoples, with delicate facial features, black hair and eyes, and lithe bodies.

(ii) The Asiatic race: with round head and face, yellow skin, black lank hair, black eyes, and short stature.

(iii) The Negro race: with long head and prognathous face profile, dark brown or black skin, black woolly hair, black eyes, and a variety of statures.

N.B. Purity of race is seldom or never found nowadays. Race must be distinguished from 'people' or 'nation', which is a social group recognising itself as a political unit.

IV. ASTRONOMICAL GEOGRAPHY

1. **The Shape of the Earth.** The spherical shape of the Earth is established by several proofs:

(*a*) The horizon, when unobstructed, is always circular and widens with increasing height of the view point.

(*b*) The higher parts of an approaching ship come into sight before the larger hull.

(*c*) Travel in a constant direction leads ultimately back to the starting point.

(*d*) The shadow of the Earth on the Moon during a lunar eclipse is always circular.

(*e*) The variation in angle of the Pole Star and other stars from different points on the Earth's surface is only satisfied by the assumption of the spherical nature of the Earth.

(*f*) Three poles of equal height placed some distance apart on level ground show that the centre pole rises above the line of sight joining the other two.

The mean curvature of the Earth causes a depression around a given point amounting to 8 inches in 1 mile, 8×2^2 inches in 2 miles, 8×3^2 inches in 3 miles, and so on.

But the Earth is not a perfect sphere. The force of gravity at the Poles is slightly greater than at the Equator, thus proving a greater equatorial radius ($13\frac{1}{2}$ miles).

2. **Size of the Earth.** The Earth was first measured by Eratosthenes (200 B.C.). His method, which is still in use, consists of finding the distance on the Earth's surface subtended by a selected angle at the Earth's centre. Thus, if an angle of 1° is found to subtend 69 miles, the whole circumference must be $69 \times 360 = 24{,}840$ miles. Since the polar diameter is slightly less than the equatorial, the distance subtended at the Poles is slightly greater than at the Equator.

3. Rotation of the Earth. That the Earth rotates on its axis is proved by the apparent daily motion of heavenly bodies as seen from the Earth. Several important facts depend on this rotation:

(*a*) The fact that all places on a given semi-circle stretching from Pole to Pole have midday at the same moment leads to the use of lines known as *meridians of Longitude*, which help to fix the position of places east or west of a chosen origin. (The meridian of Greenwich has been universally adopted as the origin or *prime meridian.*) Other lines drawn at right angles to these and parallel to the Equator fix the position of a place north or south of the Equator. They are called *parallels of latitude*. Thus, the globe is covered in imagination with a network of *co-ordinates* by means of which the exact position of a place can be located.

The principle of finding the longitude of a place is to compare the *local time* with *Greenwich mean time* (G.M.T.). Every four minutes of difference in time gives 1° of difference in longitude. If G.M.T. is earlier, the place is east; if later, the place is west.

The principle of finding latitude is to find the altitude of the midday Sun at one of the Equinoxes and, by subtracting this from 90°, to find the latitude. If the latitude of a place in the northern hemisphere is to be found at any other date than the Equinoxes, then the *declination* for the day in question must be subtracted (between March 21 and September 21) or added (between September 21 and March 21) to the altitude before proceeding further. If the place is in the southern hemisphere, the treatment of declination is reversed.

(*b*) The alternation of day and night is caused directly by rotation. By noting the moment when the Sun crosses the meridian, noon by local time is found. In these days of fast transport, it is usual for a whole country to adopt the local time of a convenient meridian. This is *standard time*.

4. Revolution of the Earth round the Sun. By noting the changes in the positions of the Sun with reference to the fixed stars, it has been found that the Earth revolves round the Sun. This movement takes 365¼ days and gives the period known as a year.

The path, or *orbit*, on which the Earth moves is an ellipse and lies in a plane known as the *ecliptic*. The Sun is in one focus of the ellipse at a mean distance of 92½ million miles from the Earth. The speed with which the Earth moves is greatest in the parts of the orbit nearest the Sun; hence, the amount of energy received is approximately constant. The axis of the Earth is set at an angle of 66½° to the plane of the ecliptic. The effects of revolution combined with the fixed inclination of the axis are:

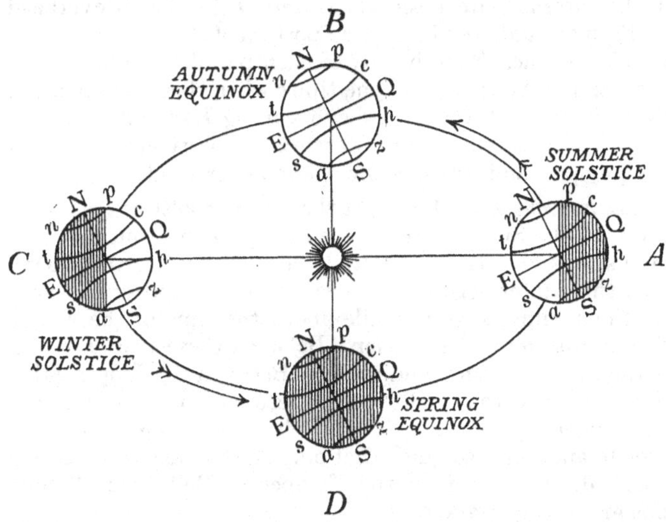

Fig. 6. Causes of the seasons

N = North Pole; *np* = Arctic Circle; *tc* = Tropic of Cancer; *EQ* = Equator; *sh* = Tropic of Capricorn; *az* = Antarctic Circle; *S* = South Pole

(*a*) *The Varying Length of Day and Night.* Fig. 6 shows the Earth in four positions in its orbit. In position *A* more than half of every parallel of latitude in the northern hemisphere is in the illuminated half of the globe. Hence, the length of the period of daylight is longer than that of the night. North of 66½° N. daylight lasts for a complete rotation (24 hours). This parallel is known as the *Arctic Circle*, while the parallel (23½° N.) which receives the Sun's rays vertically is called the *Tropic of Cancer*. This position is reached on June 21, a date known as the

Summer Solstice. In the southern hemisphere conditions are reversed: less than half of every parallel is in the illuminated half of the globe, and hence night is longer than day, while south of $66\frac{1}{2}°$ S. (the *Antarctic Circle*) the night will be 24 hours long.

At position *C*, which is reached at the *Winter Solstice* (December 21), all the conditions of position *A* are reversed. The Sun is overhead at the *Tropic of Capricorn*, and day is longer than night in the southern hemisphere.

At the intermediate positions *B* and *D* the Sun is overhead at the Equator and the illuminated portion of the globe stretches from Pole to Pole. Exactly half of each parallel is illuminated, and day and night are equal. The times when these positions are reached are called the *Autumn* and *Spring Equinoxes* and are reached on September 21 and March 21 respectively. Between these equinoxes and solstices the change is gradual.

(*b*) *The Seasons of the Year.* At the summer solstice the longer exposure to the Sun and the more nearly vertical incidence of the Sun's rays give the northern hemisphere greater insolation than at the winter solstice. Hence, at this period of the year occurs the summer season, while about the time of the winter solstice occurs the winter season. Between these two extremes occur the milder autumn and spring. Between the Tropics (Hot Belt) the change is too small to be noticed, and one long summer occurs. In the polar caps (Cold Belt) the change from one extreme to another is so quick that only the two extreme seasons are marked. It is only in the Temperate Belt that all four seasons are clearly marked.

INDEX OF PLACE-NAMES

Black type denotes that the pages mentioned contain the principal description of the place; italics denote a reference to the Appendix.

Where it has been considered useful, the pronunciation of names has been added, first according to the system of the P.C.G.N. and then by using letters in the ordinary English way. An accent (′) is written to show the syllable on which the stress of the voice falls; e.g. A rébel rebéls against the government.

GENERAL INDEX FOR HELP IN REVISION

Physical Geography

Structure and Earth Movement

Relief and Topography

Climate

Geographical Conditions and Human Life